The Art of Cryogenics

The Art of Cryogenics
Low-Temperature Experimental Techniques

Guglielmo Ventura
and
Lara Risegari

ELSEVIER

Amsterdam • Boston • Heidelberg • London • New York • Oxford
Paris • San Diego • San Francisco • Singapore • Sydney • Tokyo

Elsevier
Linacre House, Jordan Hill, Oxford OX2 8DP, UK
30 Corporate Drive, Suite 400, Burlington, MA 01803, USA

First edition 2008

British Library Cataloguing in Publication Data
A catalogue record for this book is available from the British Library

Library of Congress Cataloging-in-Publication Data
A catalog record for this book is available from the Library of Congress

ISBN: 978-0-08-044479-6

Printed and bound in Great Britain
07 08 09 10 10 9 8 7 6 5 4 3 2 1

Contents

Preface

The aim of this book is to provide a guide for physicists, chemists and engineers who wish to approach the field of low-temperature physics.

The book is made of six parts: the first part (Chapter 1) briefly deals with the basic vacuum techniques which are used in cryogenic equipments. In the second part (Chapters 2–4), the low-temperature properties of liquids and solids are reported together with many experimental data. The third part (Chapters 5–7) is devoted to the refrigeration techniques and to the physics on which they are based. The fourth part (Chapters 8–10) deals with the definition and measurement of temperature. The heart of the book is the fifth part (Chapters 11–13) where a detailed description of eight low-temperature measurements is presented. To our knowledge, no book exists up to now which covers this area that has got a renewed interest from the production of new technological materials. The sixth and last part (Chapters 14–16) of the book reports about the applications of cryogenics to different areas and describes in detail two large cryogenic experiments.

The content of the book is based on the 6-h-per-week lecture course which one of us (G. Ventura) has given at the University of Florence between 2001 and 2006. In authors' intents, the book should stand alone as a practical guide to cryogenics. For this reason, a brief chapter about vacuum techniques precedes the cryogenic items.

In order to keep the book to a manageable size, we did not insert an analogous chapter dealing with electronics. Nevertheless, a chapter about the electronic instrumentation was included in the fourth part.

Most of the measurements described in the fifth section have been carried out using a low-power dilution refrigerator. This apparent constraint obliged the authors of the experiments to look for original configurations which are fully discussed in the text. We avoided to make a list of the suppliers of cryogenic equipments as happens in most books, since this type of information is at present easily obtained through internet.

The book does not pretend to be a 'Handbook of Cryogenics': several items are simply cited and the reader is addressed to 'classic' textbooks.

Subsections devoted to tutorial problems have been avoided, but some examples are scattered along the text.

We are deeply indebted to our colleagues and friends who, in different ways, have contributed to the realization of this book. Among them, we wish to mention especially Marco Barucci, Girgl Eska, Giorgio Frossati, Andrea Giuliani, Andrea Peruzzi and Kurt Uhlig.

We are particularly grateful to Ilaria Peroni for a lot of useful suggestions and for the final revision of the manuscript.

G. Ventura and L. Risegari

PART I

1 Vacuum Techniques

Contents

1.1 Introduction

Vacuum technology is necessary to cryogenics. The vacuum techniques are needed, for example, to eliminate gas convection in cryostats (Chapters 4 and 5), to open thermal switches (Chapter 4), to decrease the pressure above the surface of a liquid to lower its temperature (Chapter 5), to circulate helium in dilution refrigerators (Chapter 6), for the purpose of leak detection (see Section 1.9.1). The ultimate performance of any refrigerator depends on the right choice of pumps and connecting tubes.

A cryogenist must be sufficiently experienced in vacuum physics to be able of projecting leak tight vacuum equipments, of avoiding gas leaks, or if they occur, of locating and getting rid of them. Hence, the cryogenist must know how to solder and glue. As a matter of fact, cryogenic equipments undergo very high mechanical stress due to great thermal gradients and due to different expansion coefficients of the various materials used.

A peculiar characteristic of vacuum technologies as applied to cryogenics is that vacuum equipments are built and tested at room temperature, then evacuated and cooled. Defects that may develop cannot be fixed at low temperature. Moreover, some problems that occur at low temperatures disappear when the equipment has been warmed up again: hence, they cannot be easily located.

Let us remember also that cold surfaces adsorb gases: if a small leak to atmosphere exists, air will condense on cooled surfaces. If the amount of gas adsorbed is large, during warm up, the pressure in the 'vacuum' space may become very high. A release valve must be therefore present in the system.

We will hereafter describe only techniques and equipments of interest for cryogenics. More information can be found in ref. [1–4].

A container is said to be under vacuum when the inside pressure is lower than the outer one, usually the atmospheric pressure. If no container exists, vacuum is a space region at a pressure lower than the atmospheric pressure.

Vacuum may be 'natural' or artificially produced. Natural vacuum, for example, occurs on the lunar surface or in the interstellar space where one should however speak of numerical density of particles ($\sim$1 particle/cm^3) instead of pressure. In the intergalactic space, the density is around 1 particle/m^3. Natural vacuum laboratories in use are, for example, the Space Shuttle or Space Stations, but we will deal only with artificial vacuum, produced by pumps inside a container.

The measuring unit of pressure in SI system is called pascal (Pa) = newton/m^2. One Pa corresponds to a very small pressure. Also for the latter reason, several other units are commonly used in vacuum practice and instrumentation. In Table 1.1, the conversion among the most frequently used pressure units is reported.

Vacuum can be divided into three regions depending on the pressure of the gas:

- low or rough vacuum (from the atmospheric pressure down to about 1 Pa);
- high vacuum (1 Pa–10^{-4} Pa);
- ultrahigh vacuum (below 10^{-4} Pa).

Table 1.1
Conversion table of some pressure units

	pascal	torr	atm	mbar	psi	kg/cm^2
pascal (newton/m^2)	1	7.5×10^{-3}	9.87×10^{-6}	10^{-2}	1.45×10^{-4}	1.02×10^{-5}
torr (mm of mercury)	133	1	1.32×10^{-3}	1.333	1.93×10^{-2}	1.36×10^{-3}
atm (atmosphere)	1.013×10^5	760	1	1013	14.7	1.033
Mbar (millibar)	100	0.75	9.87×10^{-4}	1	1.45×10^{-2}	1.02×10^{-3}
psi (lb/in^2)	6.89×10^3	51.71	6.85×10^{-2}	68.9	1	7.03×10^{-2}
kg/cm^2	9.81×10^4	735.6	0.968	981	14.2	1

Table 1.2
Composition of dry air at 1 atm (760 torr) expressed by partial pressure of components [5]

Constituent	Partial pressure	
	torr	Pa
Nitrogen	593.43	79 104.22
Oxygen	159.19	21 220.03
Carbon dioxide	0.25	8.33
Argon	7.10	946.43
Neon	1.38×10^{-2}	1.84
Helium	3.98×10^{-3}	5.3×10^{-1}
Krypton	8.66×10^{-4}	1.15×10^{-1}
Xenon	6.61×10^{-5}	8.81×10^{-3}
Hydrogen	3.80×10^{-4}	5.06×10^{-2}
Methane	1.52×10^{-3}	2.03×10^{-1}
Nitrous oxide	3.80×10^{-4}	5.06×10^{-2}

Such more or less arbitrary classification reflects three different physical situations. In the low vacuum range, the number of molecules in the volume is much larger than that adsorbed on the internal surface of the container. In the high-vacuum range, the mean free path (see Section 1.3) of the molecules is of the order or larger than the dimensions of the container, and most of molecules are on the container inner walls. In the ultrahigh-vacuum range, the flux of molecules onto an initially clean surface is so low that there is enough time to carry out experiments before it is covered by a monomolecular layer of gas.

The simplest vacuum system consists of a vacuum chamber, an interconnecting tube (vacuum line or pipe) and a vacuum pump which produces a pressure gradient along the tube (see Fig. 1.5).

To obtain a high or ultrahigh vacuum, both the chamber and the tube must be clean, since impurities (like water, with its high vapour pressure, and its unavoidable presence in the air) slacken the reaching of the final vacuum.

Let us also anticipate that, if a vacuum chamber initially contains air at the atmospheric pressure (with a typical composition like that reported in Table 1.2), during the pumping process the composition remains approximately the same in the low-vacuum range. Then, the composition changes, becoming usually richer in light molecules.

1.2 Vapour pressure

Any liquid or even solid material always produces a gaseous phase in equilibrium with the denser phase. The pressure of the gaseous phase is called ‘vapour pressure’. The final vacuum in an evacuated chamber is often controlled by the vapour pressure of the most volatile material present in the system.

The dependence on temperature of the vapour pressure p can be approximately expressed as:

$$p \propto \exp\left(-L(T)/RT\right) \tag{1.1}$$

where $L(T)$ is the molar latent heat of evaporation.

Table 1.3
Some vapour pressure data at 20°C

Liquid	Vapour pressure (torr at 20°C)
Benzene	74.6
Ethyl alcohol	43.9
Methyl alcohol	96.0
Acetone	184.8
Turpentine	4.4
Water	17.5
Carbon tetrachloride	91.0
Mercury (Hg)	1.2×10^{-3}
High-vacuum pump oil	10^{-7}

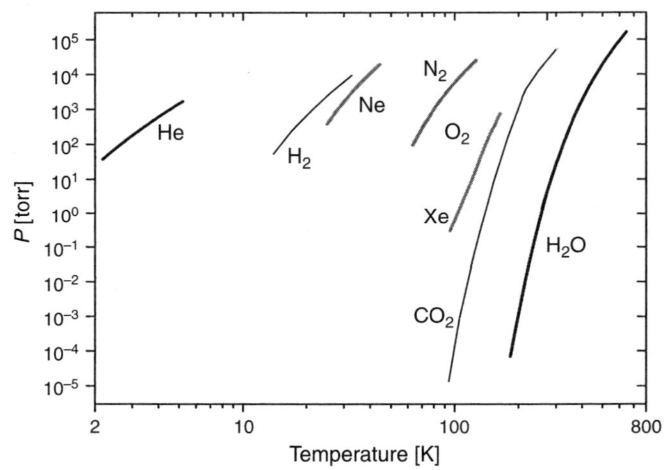

Fig. 1.1. Vapour pressure of some substances as a function of temperature [5].

Some vapour pressure data at 20°C are reported in Table 1.3. In Fig. 1.1, the temperature dependence of vapour pressure of some substances is reported. A special attention is to be devoted to water vapour pressure which is very high.

1.3 Mean free path and viscosity

In the kinetic theory, the gas molecules are represented by hard spheres colliding elastically with each other and with the container walls. Details of this theory are given, for example, in ref. [1]. An important parameter that can be calculated by this model is λ, the mean free path of a molecule between collisions. The mean free path λ of molecules is:

$$\lambda = \frac{1}{\sqrt{2} \cdot \pi n d^2} \tag{1.2}$$

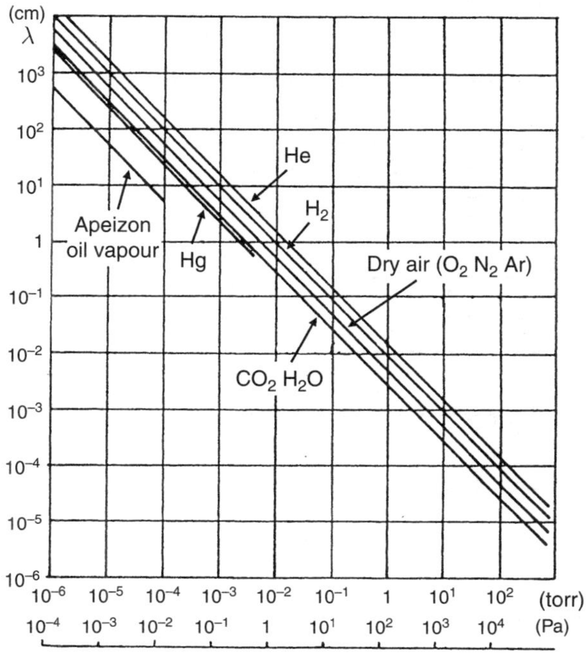

Fig. 1.2. Mean free path of some gases as a function of pressure.

where d is the equivalent hard-sphere diameter of the molecule ($d = 0.22$ nm for ^{4}He [6]) and n is the number of molecule for unit volume.

In the case of an ideal gas:

$$\lambda = \frac{k_B T}{\sqrt{2}\pi d^2 p} \tag{1.3}$$

where k_B is the Boltzmann constant and p the gas pressure. Figure 1.2 shows the mean free path λ at 293K for some gases of interest in vacuum technologies.

For the viscosity, the hard-sphere model for a gas gives [6]:

$$\eta = \frac{2}{3}\sqrt{\frac{m k_B T}{\pi^3}}\frac{1}{d^2} \tag{1.4}$$

where m is the mass of one gas molecule. The SI unit of viscosity is Pa s.

1.4 Gas flow

When a gas is removed from a container through a tube, the type of gas flow depends on pressure. In the low vacuum range, the gas flow is controlled by the collisions among molecules (viscous flow). If some molecules are removed from a region (by a pump), other molecules will refill the 'empty' region. In this situation, the diameter of the pumping line

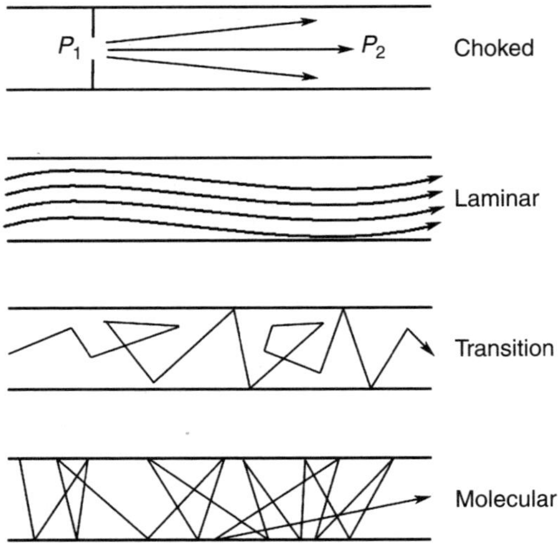

Fig. 1.3. Schematic picture of the gas flow.

is not crucial. In the high- and ultrahigh-vacuum range (molecular range), the mean free path is larger than the dimensions of the container, and interactions among molecules are rare. The pumping tube diameter must be large to increase the probability that molecules reach the pump. For a tube of radius r, the flow is viscous if $\lambda/r < 10^{-2}$ and is molecular if $\lambda/r > 1$.

The different types of gas flow are depicted in Fig. 1.3.

A choked flow takes place when the pressure ratio p_2/p_1 of gas through an orifice is very small. In the choked flow, the gas speed beyond the orifice is close to the sound velocity, and a change of p_2 does not modify the flow rate. This regime finds application in the production of molecular beams.

In the laminar regime, the flow is very slow and shallow, and the viscous forces dominate: the gas velocity is constant along the tube with no orthogonal component. When pressure decrease, the viscous forces vanish, and the flow is controlled by the collisions with the walls. In the molecular flow, the mean free path is greater than the distance between the surfaces of the tube. In this regime, all the surfaces, even if macroscopically levigated, are in reality covered with projections that are of atomic or molecular dimensions and irregularly distributed over the surface. Consequently, a gas molecule, on striking the surface, is repelled in a direction that is totally independent of the direction of incidence.

Whatever the particular regime, the gas flow in a tube is usually described in terms of the 'volumetric flow' or 'throughput':

$$Q = p\frac{\mathrm{d}V}{\mathrm{d}t} \tag{1.5}$$

where $\mathrm{d}V/\mathrm{d}t$ is the volume of gas crossing an isothermal plane perpendicular to the flow per unit time and p is the pressure at which the flow rate is measured. Often the throughput is given as [torr l s]. Dimensionally, the throughput is a power (1 W = 7.50 torr l s) and represents the power which must be delivered to the pump to produce Q (neglecting losses).

The conductance C of a pipe is defined by:

$$C = \frac{Q}{p_1 - p_2} \tag{1.6}$$

where p_1 is the pressure at the upstream end of the tube and p_2 is the pressure at the downstream end.

The parallel connection of two pipes of conductance C_1 and C_2 gives a conductance:

$$C = C_1 + C_2 \tag{1.7}$$

For series connection:

$$\frac{1}{C} = \frac{1}{C_1} + \frac{1}{C_2} \tag{1.8}$$

1.4.1 Conductance of an orifice and of a pipe for molecular flow

Let us consider the stationary flow between two chambers at different pressures and connected by an orifice or a pipe. The conductance for the orifice of area A is [1]:

$$C = 3.638\left(\frac{T}{M}\right)^{\frac{1}{2}} \cdot A \text{ l/s} \tag{1.9}$$

where A is in cm^2 and M is the molecular mass in g/mol.

As example, for nitrogen at 293 K, $M = 28$ and the conductance C of an orifice of area A is:

$$C = 11.77 \cdot A \text{ l/s} \tag{1.10}$$

An orifice is a useful means to create a known conductance for the measurement of pumping speeds or calibration of pressure gages.

The conductance of a pipe of length L and diameter $D(L >> D)$ is:

$$C = 3.81\left(\frac{T}{M}\right)^{\frac{1}{2}} \cdot \frac{D^3}{L} \text{ l/s} \tag{1.11}$$

where D and L are expressed in cm. for Nitrogen at 293 K:

$$C = 12.32\frac{D^3}{L} \text{ l/s} \tag{1.12}$$

If the tube is short, the conductance of the entrance aperture cannot be neglected: formulas (1.9) and (1.11) give overestimated values of C [1].

It is worth noting that $C \propto T^{1/2}$, and this accounts for the long time needed when pumping helium gas used in the 4.2 K thermalization of cryogenic apparatus.

1.4.2 Conductance of an orifice and of a pipe for viscous flow

The analysis of the conductance of an orifice in the viscous flow range is quite complicated and is reported in ref. [7,8]. The conductance of a long pipe for the laminar flow of an incompressible gas is given by the Poiseuille equation:

$$C = \frac{\pi}{128\eta}\frac{D^4}{L}\cdot\overline{p}\ \mathrm{m^3/s} \tag{1.13}$$

where $\overline{p}$ is the average pressure in the pipe.

For N_2 at 293 K, with L and D in cm, and $\overline{p}$ in torr:

$$C = 188\frac{D^4}{L}\cdot\overline{p}\ \mathrm{l/s} \tag{1.14}$$

The case of short tubes is reported, for example, in ref. [9].

Often the flow in a pumping line is neither molecular nor viscous.

Formulas for the conductance in this transition regime (see Fig. 1.3) are quite complicated and can be found in ref. [9]. It is worth noting that the flow in the transition region is usually greater than that calculated assuming a molecular flow.

Example 1.1. We wish to evaluate the series conductance C_S of the four elements shown in Fig. 1.4 for nitrogen at 293K and for a pressure $p = 10^{-4}$ torr.

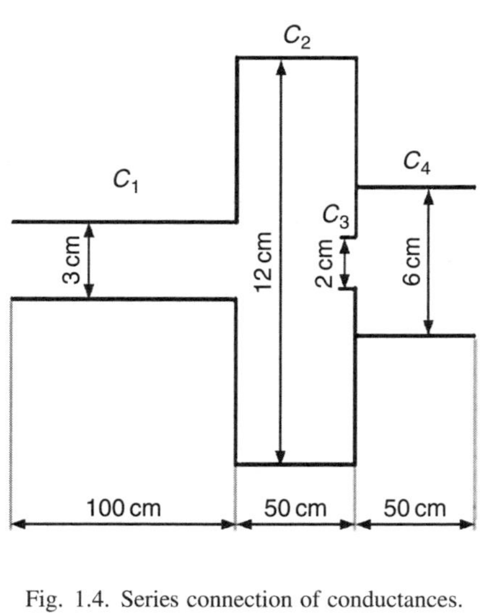

Fig. 1.4. Series connection of conductances.

From Fig. 1.2, we see that at $p = 10^{-4}$ torr, the mean free path λ is about 1 m, longer than the diameter of the largest tube (12 cm). Hence the flow is molecular. The conductances C_1, C_2 and C_4 can be calculated from eq. (1.12):

$$C_1 = 12.32\frac{D_1^3}{L_1} = 3.32\ \text{l/s}$$

$$C_2 = 12.32\frac{D_2^3}{L_2} = 425.78\ \text{l/s}$$

$$C_4 = 12.32\frac{D_4^3}{L_4} = 53.22\ \text{l/s}$$

From eq. (1.10):

$$C_3 = 11.77\,A = 36.98\ \text{l/s}$$

From eq. (1.8), the series conductance C_S, slightly lower than the smallest conductivity C_1, is:

$$C_S = 2.86\ \text{l/s}$$

1.5 Evacuation of a lumped volume

We will now describe the evacuation of a vacuum container in terms of lumped constants. Referring to Fig. 1.5, this means that the volume V of gas is that inside the container, while the gas volume in the pipe and in the pump is disregarded. Also pressure gradients in the volume V will be neglected, and leaks and degassing are supposed to occur only in the vacuum container.

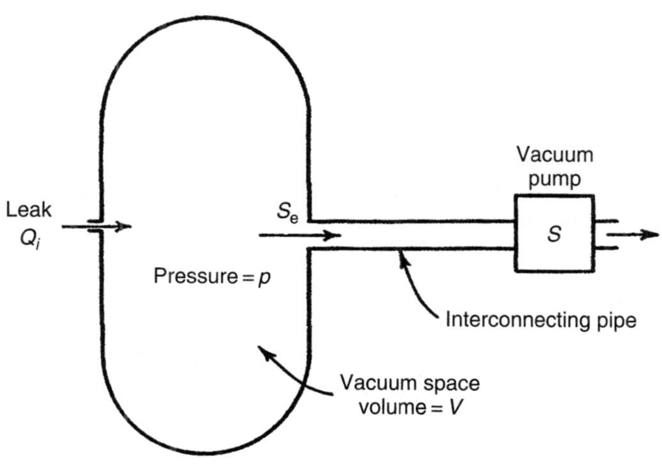

Fig. 1.5. Schematic representation of the evacuation of a container of volume V.

The evacuation of a distributed volume can be described only through much more complex formulas (see for example ref. [9], p. 84), and this is beyond the scope of this short introduction to the vacuum technology.

The pump action is usually defined in terms of a pumping speed S:

$$S = \frac{Q}{p_i} \tag{1.15}$$

S is the volume of gas removed from the container each second at the existing pressure p_i. The pumping speed of many pumps is fairly constant over a few decades of pressure (see Fig. 1.7).

The pumping speed of a series connection of a pump and a pipe is:

$$S_e = \left(\frac{1}{S} + \frac{1}{C}\right)^{-1} \tag{1.16}$$

C is usually larger than S.

At low pressure, two main sources of gas are present: Q_W gas desorbing from surfaces and Q_L gas leaking into the container from outside (imperfect tightness of gaskets, permeation, back-streaming from pumps, and so on). If Q_i represents the sum of all these inputs, the change of pressure p with the time t is described by:

$$\frac{dp}{dt} = \frac{Q_i}{V} - \frac{S_e p}{V} \tag{1.17}$$

Equation (1.17) is independent on the flow regime and type of vacuum system.

Since the flow is viscous during the initial pump down, C is not constant in the beginning (see eq. (1.13)) and the integration of eq. (1.17) is impractical.

A simple mathematical solution of eq. (1.17) occurs if both S_e and Q_i are constant. The former hypothesis is realistic over a wide range of pressure; the latter is usually only a rough approximation.

From eq. (1.17), with $dp/dt = 0$, we obtain the final pressure p_f reached by the system:

$$p_f = \frac{Q_i}{S_e} \tag{1.18}$$

The time t_p needed to reach a pressure p_2 starting from a pressure p_1 is:

$$t_p = \frac{V}{S_e} \ln\left(\frac{p_1 - p_f}{p_2 - p_f}\right) \tag{1.19}$$

In high-vacuum systems, the limit pressure p_f is usually controlled by the degassing rate (Q_i) that decreases with time.

1.6 Vacuum pumps

The main components of a vacuum system are the pumps. The types of pumps most commonly used in low-temperature experiments are:

1. Roughing pumps: pressure range 10^5–1 Pa
   - Rotary vane pumps;
   - Booster (Roots) pumps;

- Scroll pumps;
- Sorption pumps (cryopumps).

2. High-vacuum pumps: pressure range 100–10^{-6} Pa

- Oil diffusion pumps;
- Turbomolecular pumps;
- Molecular drag pumps.

To achieve very low pressures (high vacuum), at least two pumps (one roughing pump and a high-vacuum pump) are connected is series. To get high pumping speed, a parallel of pumps both in rough and in high vacuum is sometimes used. Nowadays, the vacuum industry offers integrated vacuum systems containing several vacuum pumps in order to cover an extended range of operating pressure with a compact apparatus.

In the following sections, the functioning principles of the single pumps used in cryogenics will be described.

1.6.1 Rotary vane oil-sealed mechanical pump

A volume of gas is enclosed in the space bounded by the rotor, the stator and the two vanes (see Fig. 1.6). The pump removes the gas by compressing it to a pressure slightly higher than the atmospheric pressure. This overpressure opens the spring-loaded outlet valve, and the gas escapes to the atmosphere. A thin film of oil makes the final seal; therefore the ultimate pressure depends also on the oil vapour pressure. With one stage, the lowest attainable pressure is about 10^{-2} torr and with two stages in series 10^{-3} torr.

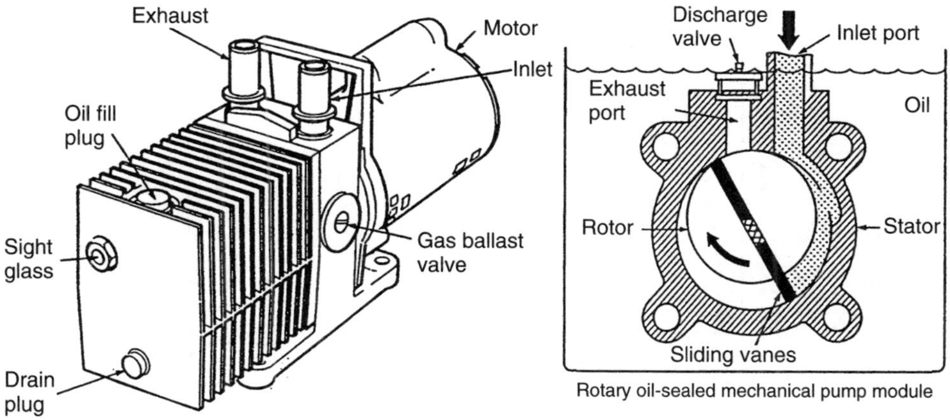

Fig. 1.6. Scheme of a rotary vane oil-sealed mechanical pump.

Due to the friction of the sliding vanes, the biggest pump size available (water cooled) is about 400 l/s. Smaller pumps (0.5–5 l/s) are air cooled.

Condensation of vapours must be avoided since it causes a deterioration of the oil: water vapour condense at 234 torr at 70°C, which is the typical working temperature of the pump. To avoid condensation, most pumps have a 'ballast' valve in the high-pressure stage. After the gas has been closed off from the inlet, some gas is bled in through the ballast valve. This causes the discharge valve to open sooner, with a reduction of the compression ratio. Rotary piston pumps have similar characteristics.

Most mechanical pumps exhibit vibration that may represent a crucial drawback for very low-temperature refrigerators.

Another drawback of oil-sealed pumps is the back-streaming of oil vapour into the roughing line, which may occur at low pressure. Contamination by back-streaming oil can be drastically reduced by using proper traps like molecular sieve traps with zeolite (see Section 1.6.4).

The speed of rotary vane pumps is nearly constant from 1 atm down to 10^{-2} torr (see Fig. 1.7). To obtain higher speeds in the 10^{-2}–10^{-3} torr range, booster pumps are used in series with rotary vane pumps.

1.6.2 Booster pumps

The booster (blower or Roots) pumps are high-throughput pumps with a low compression ratio. A booster pump is schematically shown in Fig. 1.8. The gap between the two counter-rotating rotors is of the order of 0.1 mm. The rotation (about 50 rps) of the rotors,

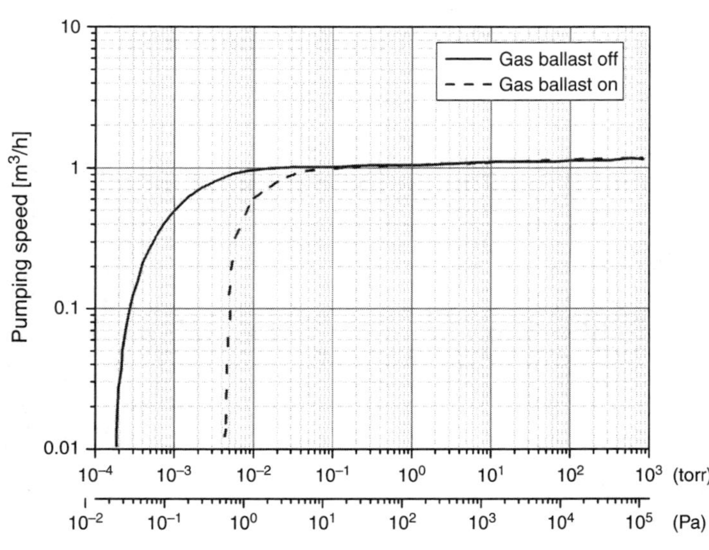

Fig. 1.7. Typical pumping speed vs. inlet pressure of a two-stage rotary vane pump.

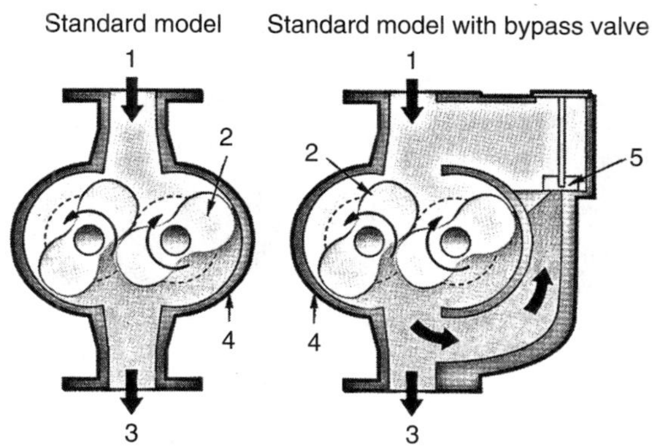

Fig. 1.8. Scheme of a Roots pump. Main components: 1, inlet; 2, two-lobe piston; 3, exhaust; 4, pump casing; 5, bypass valve.

mounted on parallel shafts, is synchronized by gears to avoid contact. No oil is used to seal this gap. The starting pressure of booster pumps is about 10 torr. Heating of the pump becomes excessive when it is operated continuously at high pressure. Therefore, a bypass is used for high-pressure roughing. Booster pumps with throughput up to 600 l/s are available.

To overcome the problem of the low compression ratio, multistage systems are produced in which several booster units are cascaded. The system presents both an overall high compression ratio and a high throughput. Such multistage system is capable of reaching 10^{-2} torr and works from the atmospheric pressure without the need of rouging pump as in the case of a single booster pump. The throughput can be up to about 300 l/s.

A system of cascaded booster pumps has the characteristics of a medium-size rotary vane pump but does not present the drawback of the back-streaming. For this reason, it finds application in cryogenics, for example, for the circulation of the He mixture in dilution refrigerators.

1.6.3 Scroll pumps

A scroll vacuum pump uses two interleaved Archimedean spiral-shaped scrolls to pump or compress gases (see Fig. 1.9). One of the scrolls is fixed, while the other orbits eccentrically without rotating, thereby trapping and compressing gases between the scrolls and moving it towards the outlet.

The device was first patented in 1905 but did not become efficient for 50 years.

Also for these pumps, multistage systems (e.g. Varian TriScroll Pump) are used which allows for final pressure around 10^{-2} torr and pumping speed of about $30\,m^3/h$ (see Fig. 1.10). The typical application of scroll pumps systems is the backing for turbomolecular pumps (see Section 1.6.6).

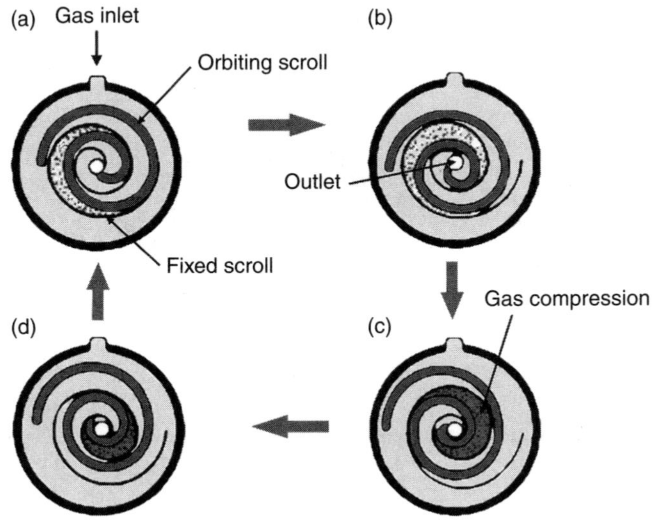

Fig. 1.9. Scheme of a scroll mechanism.

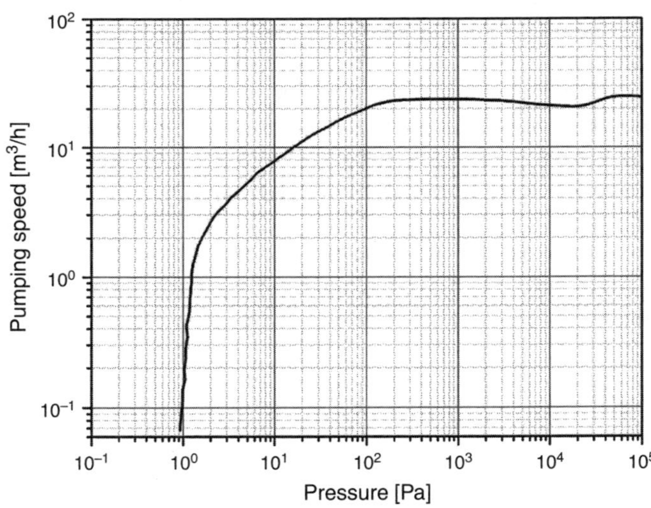

Fig. 1.10. Pumping speed vs. pressure for a three-stage scroll pump (Varian TriScroll 600).

There are two main drawbacks in scroll pumps: the production of Polytetrafluoroethylene (PTFE) powder dust due to the friction between the fixed and the moving part; moreover, since in the present commercial versions the motor is not in vacuum, the tightness is not the best.

1.6.4 Sorption pumps

In a sorption pump, the gas is trapped within the adsorbing material (zeolites or active charcoal) called molecular sieve. Zeolites are porous aluminium silicates which adsorb large amount of gas when cooled to low temperature (usually 77K). The pump is filled with zeolite and put in a bucket containing liquid nitrogen (see Fig. 1.11).

Gases that are condensable at 77K are trapped by 'cryocondensation'. ('Cryosorption' is instead the trapping of gas with a lower melting temperature inside the pores of the molecular sieve that has a huge surface/volume ratio, typically about $700\,m^2/cm^3$.)

If gases like H_2, He and Ne are to be trapped, the pump must be cooled to 4.2 K. Detailed information on cryopumping can be found in ref. [10]. Ultimate pressure of the order of 10^{-5} torr can be achieved.

When cooled with liquid He, this type of pump can reach pressure below 10^{-8} torr.

The sorption pumps are 'clean' but are 'one shot', that is, two pumps in parallel and connected by valves alternatively are needed for a continuous pumping. When the first pump is saturated, the second pump is started, while the first is regenerated: removing the liquid nitrogen, the trapped gas is expelled through the blow-off valve. The pump (with zeolite) is heated to 200–300°C to remove water vapour. Charcoal pumps are heated to about 100°C.

Another advantage of this kind of pumps is the very low level of vibration.

1.6.5 Oil diffusion pumps

In a diffusion pump, the dense oil vapour produced by the boiler (see Fig. 1.12) is ejected into the vacuum at high (or even supersonic) speed through the nozzles.

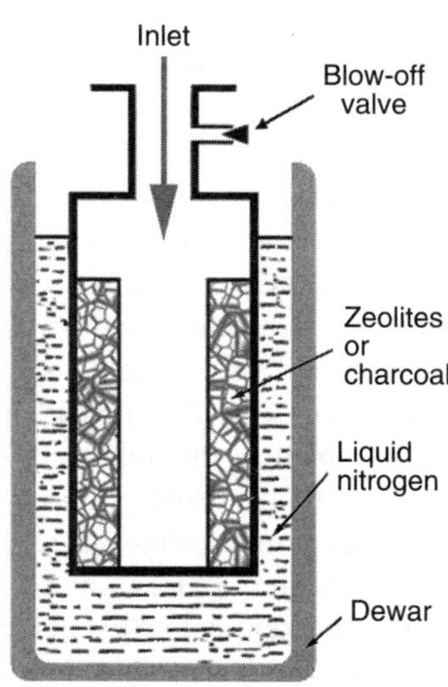

Fig. 1.11. Scheme of a sorption pump with liquid nitrogen refrigeration.

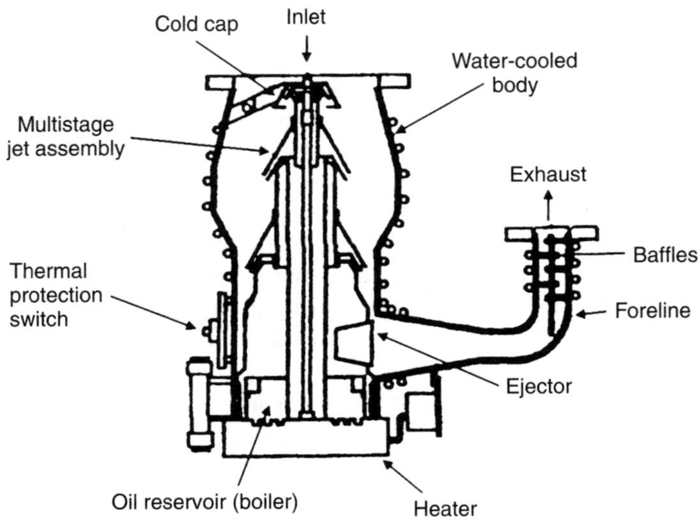

Fig. 1.12. Scheme of an oil diffusion pump.

It collides with the gas communicating a large downward momentum to the gas molecules. The top jet operates at the lowest pressure. The body of the pump is typically cooled by water or air. When the vapour reaches the cooled walls, it condenses back into a fluid which drops into the pump boiler for reboiling.

A cold cap is usually mounted on the top of the pump assembly to prevent vapour from reaching the vacuum chamber. A thermal protection switch is often used. The maximum working pressure of a diffusion pump is about 10^{-3} torr. The ultimate pressure of a diffusion pump can be around 10^{-9} torr and heavily depends on the oil vapour pressure ($p < 10^{-8}$ torr at room temperature for very good oils). Pumps with very large pumping speed (up to 10^4 l/s) are commercially available.

A drawback of oil diffusion pumps is the so-called back-steaming. It is the flow of a small quantity of oil vapour towards the inlet of the pump. A water-cooled baffle like that shown in Fig. 1.13 can be put above the inlet. Baffles are made up of arrays of optically dense fins cooled by a continuous water flow. A baffle always reduces the pumping speed.

To reduce to a minimum the amount of fluid reaching the vacuum chamber, the use of a liquid N_2-cooled baffle is the best choice (see Fig. 1.14).

A diffusion pump needs a forepump (usually a rotary pump). Oil can also migrate from the forepump into the vacuum chamber through the diffusion pump, when the rotary pump works in the molecular regime ($\sim 10^{-2}$ torr).

If an operating diffusion pump is exposed to atmosphere, even for a short time, a strong oxidation of the fluid takes place. In extreme cases, combustion or explosion is possible.

The pressure in the foreline must be kept below a value called critical forepressure. If this pressure is exceeded, oil vapour will be injected into the vacuum chamber in great amounts. A 'dumping' of the pump can occur. For this reason, the size of the rotary pump must be properly chosen.

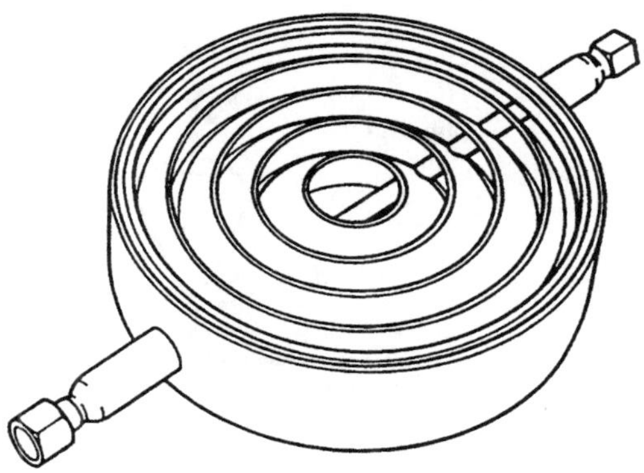

Fig. 1.13. Water-cooled baffle.

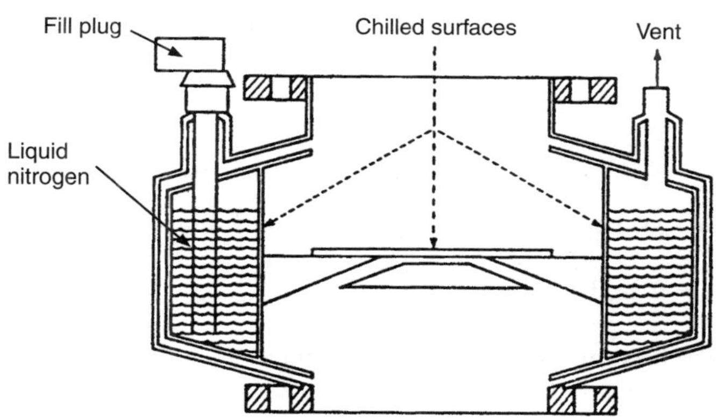

Fig. 1.14. Nitrogen-cooled baffle.

Note that a decrease of the pumping speed, due to a reduction of the heater power to the boiler, decreases the critical forepressure.

Example 1.2. A diffusion pump with constant speed $S_1 = 30\,\mathrm{l/s}$ below 10^{-3} torr has a critical forepressure $p_K = 0.12$ torr. It is connected to a rotary pump by a tube of length $L = 150\,\mathrm{cm}$ (see Fig. 1.15).

We wish to choose conservative values for the tube diameter D and the pumping speed S_3 of the rotary pump.
The throughput is maximum for $p_1 = 10^{-3}$ torr:

$$Q_{\max} = p_1 \cdot S_1 = 3 \times 10^{-2}\,\mathrm{l\,torr/s} \tag{1.20}$$

The pressure p_2 at the exhaust end of the diffusion pump must be less than p_K.
Let us choose $p_2 = 0.08$ torr as a conservative value.

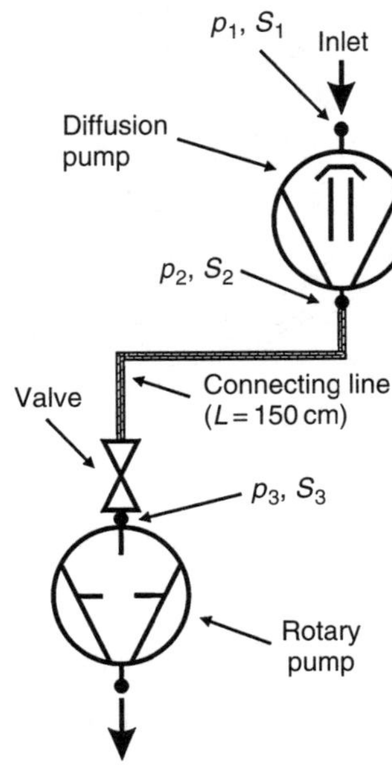

Fig. 1.15. Scheme of the pumping system.

The pumping speed S_2 at the exhaust port of the diffusion pump must be at least:

$$S_2 = \frac{Q_{\max}}{p_2} = 0.375\,\mathrm{l/s} \tag{1.21}$$

Of course, the pressure p_3 at the inlet of the rotary pump must be less than 0.08 torr, say 0.04 torr. Hence the speed S_3 at the inlet of the rotary pump must be at least:

$$S_{3\,\min} = \frac{Q_{\max}}{p_3} = 0.75\,\mathrm{l/s} \tag{1.22}$$

To obtain a speed $S_2 = 0.375\,\mathrm{l/s}$, the conductance of the tube must be:

$$C = \frac{S_2 S_3}{S_3 - S_2} = 2\,\mathrm{l/s} \tag{1.23}$$

The flow regime in the tube is in the transition region. From eq. (1.14) for the viscous flow $D = 1.5\,\mathrm{cm}$. In the molecular regime (from eq. (1.12)) $D = 2.2\,\mathrm{cm}$.

The latter value is a conservative choice.

1.6.6 Turbomolecular pumps

Turbomolecular (turbo) pumps are very clean (especially magnetically levitated version) mechanical pumps, with pumping speed up to more than 7000 l/s.

The pumping action is due to a high-speed rotating surface that transfers momentum to the gas molecules (see Fig. 1.16).

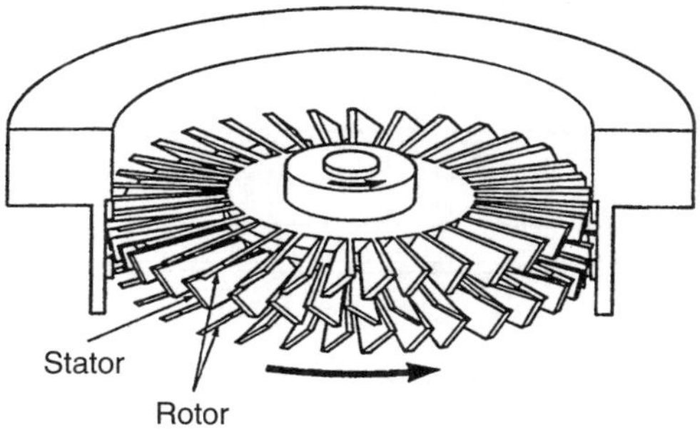

Fig. 1.16. Scheme of a turbomolecular pump.

Fig. 1.17. Internal view of a turbomolecular pump (courtesy of Alcatel).

With metal gaskets and moderate bakeout, turbo pumps can reach pressure below 10^{-9} torr without traps. They can be started at a pressure up to 1 torr. The time required to reach full pumping speed ($\sim$1 min) is much shorter than for diffusion pumps. Also these pumps must be backed by a primary pump.

The turbo pumps are made up of 10–40 rotating (rotor) and fixed (stator) disks (see Fig. 1.16 and Fig. 1.17) alternatively arranged. Each disk has 20–60 blades with proper tilt.

The motor (usually working in vacuum) is moved by a special current power supply. The rotor turns at 10^4–10^5 rpm, usually a multiple of the line frequency. The pumping speed of a turbo pump unit depends on its rotational speed. High-speed turbo pumps need more frequent maintenance interventions. In some turbo pumps, a 'low-speed mode' allows operation up to 10^{-1} torr. However, full rotational speed is achieved at pressures

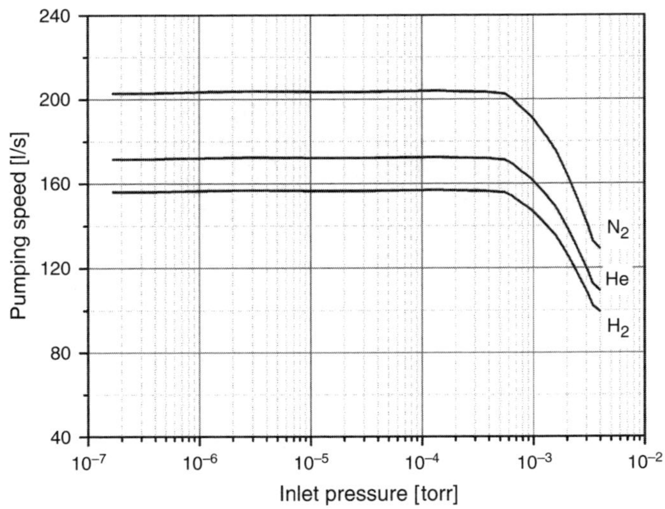

Fig. 1.18. Typical pumping speed of a turbo pump for nitrogen, helium and hydrogen.

below 10^{-3} torr. Low-level vibrations in kHz range are always produced due to residual unbalance of the rotor.

The pumping speed of a turbo pump is almost constant over a wide range of pressure and depends on the gas species as qualitatively shown in Fig. 1.18.

Light gases with higher thermal velocity are pumped less than heavier ones. This is why turbo pumps produce an (almost) oil-free vacuum. The lubrication of turbo pumps is made with a special vacuum grease. Pumps with magnetically levitated rotor are available, but they are more expensive.

1.6.7 Molecular drag pumps

Molecular drag pumps vary from turbo pumps in that the momentum is transferred to the gas molecules not by blades but by a rapidly rotating solid surface. The stator is, in this case, a fixed surface very close to the rotor (see Fig. 1.19).

The molecules being pumped are dragged along the surface, hence the term molecular drag pump. These surfaces can take the form of a rotating spiralled drum (Holweck) or slotted rotor discs (Gaede).

An important difference between molecular drag pumps and turbo pumps is that a drag pump will start pumping at a higher inlet pressure (up to 30 torr) and does not require as low a backing pressure for full operation. In general, commercially available molecular drag pumps reach an ultimate pressure of 10^{-5} torr or slightly below, due to the practical impossibility of providing closer clearances between the moving and fixed surfaces. Moreover, molecular drag pumps have a much lower pumping speed (roughly 1/8–1/3) than a turbomolecular pump with equivalent inlet dimensions.

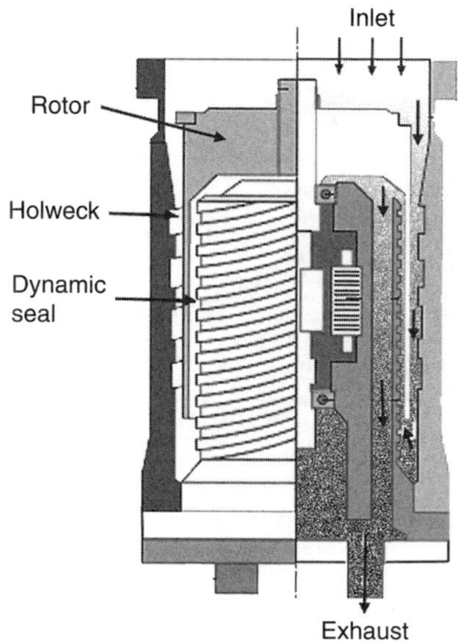

Fig. 1.19. Cross-section of a molecular drag pump (courtesy of Alcatel).

1.7 Other vacuum components

Even the simplest vacuum system like that of Fig. 1.5 consists of several separate components (vacuum vessel, pump, connecting pipe, detachable joints, valves, etc.) which are usually commercially available.

Detachable joints are needed to assemble the various parts. They are made up of flanges pressed together with a gasket between. Among the commercial components, the three mostly used flanges for joints in vacuum technology are:

- ISO-KF
- ISO-K
- CF

The choice of the type of joints to be used depends on the dimensions and on the requested vacuum level.

In the first type (ISO-KF), the connections are made with conic profile flanges joined together by clamps, with an elastomer gasket (see Fig. 1.20).

The standard dimensions for this type of flanges are from 10 to 50 mm of internal diameter (DN10–DN50).

In the second type (ISO-K), the flanges have a grooved profile, and the connections are realized by claw clamps (see Fig. 1.21). The number of clamps depends on the dimensions of the flange. The standard dimensions for this type of flange are from 63 to 500 mm of internal diameter (DN63–DN500).

The third type (CF, see Fig. 1.22) is used in ultrahigh-vacuum application ($p < 10^{-7}$ torr) where heating is needed to speed up the degassing process of the inner walls of the

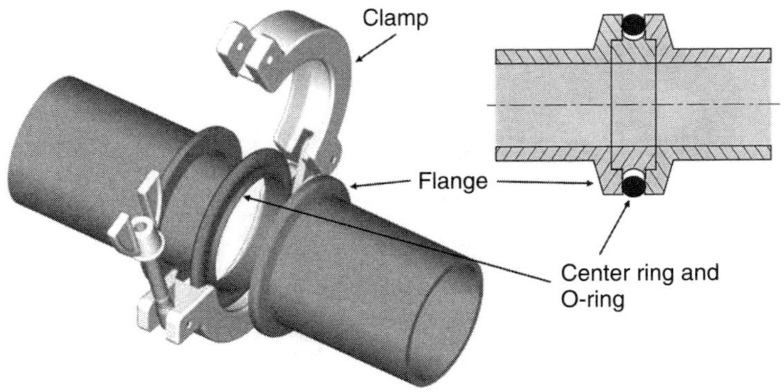

Fig. 1.20. Example of an ISO-KF joint between two pipes.

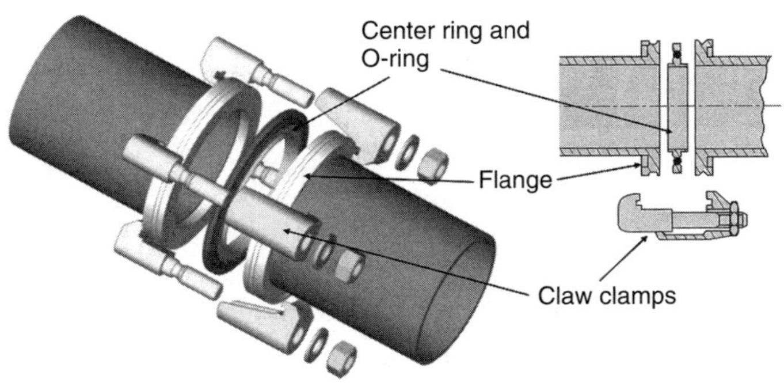

Fig. 1.21. Example of an ISO-K joint between two pipes.

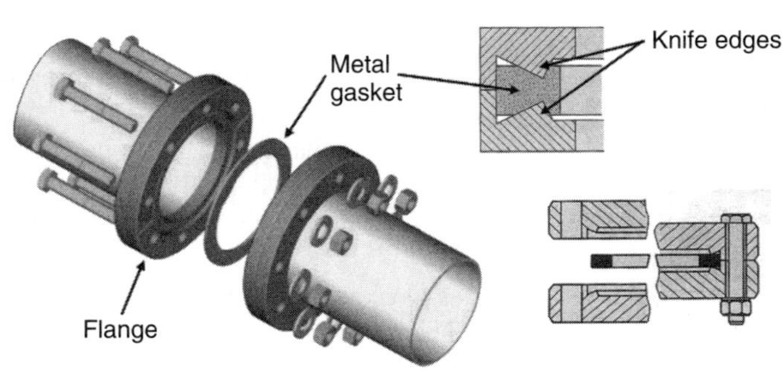

Fig. 1.22. Example of a CF joint between two pipes.

system. In this case, all the components of the system are metallic. The maximum baking temperature is ~450°C.

As regards the vacuum valves, the three mostly used types are shown in Fig. 1.23. The gate valve gives the maximum of conductance for a fixed diameter but cannot be operated with a pressure difference more than about 30 mbar.

Other commercially available useful components are (a) cross, (b) linear or (e) right-angle tubes, (c) flexible hoses, (d) adapters, etc. (see Fig. 1.24).

We have seen that several kinds of gaskets are used with the various types of flanges. For low and high vacuum at room temperature, elastomers O-rings (rubber, ethylenpropylen, silicon etc.) are used. Small pressures of flanges are needed. The maximum bakeout

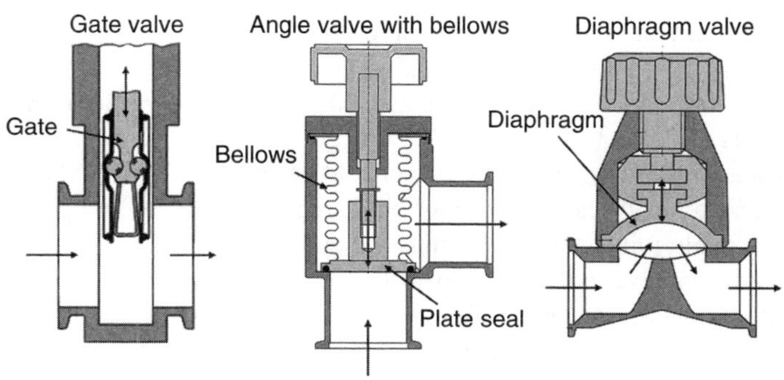

Fig. 1.23. Scheme of three different types of vacuum valves.

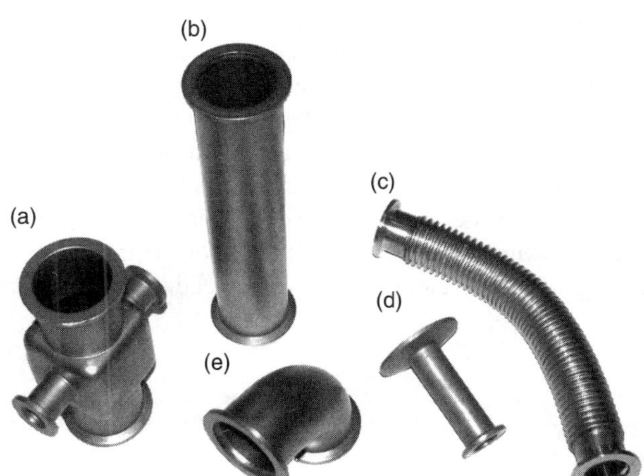

Fig. 1.24. Image of some standard vacuum components.

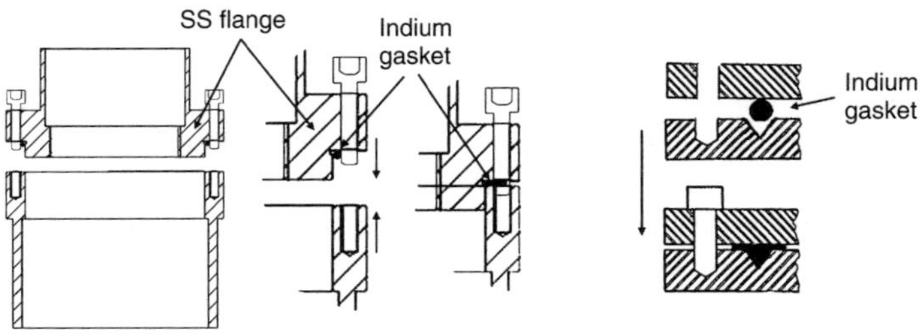

Fig. 1.25. Mounting of an indium gasket for cryogenic temperatures.

temperature is between 90°C and 250°C. Elastomer O-rings are used down to –60°C (butil). Some of them show a strong permeation to He.

In ultralow ($p < 10^{-7}$ torr) vacuum, metallic gaskets are used.

At cryogenic temperatures, ductile gaskets (indium, Kapton, vacuum grease) are to be preferred (see Fig. 1.25).

It is important to point out that any detachable component (valve, flange, and so on) always introduces a small leak that cannot be disregarded. To reach a good vacuum, the number of detachable joints and valves must be as low as possible.

For example, a valve usually has two leaks, one between the two vacuum sides, the second towards the ambient, which often is even more important.

Typical value of leak for a good-quality valve of 40 mm diameter is $<6 \times 10^{-10}$ torr l/s for the body and $<10^{-9}$ torr l/s between the two vacuum sides.

1.8 Pressure gages

The precise measurement of the total or partial pressure of gas is an extremely arduous task. Pumps, pressure gages and vacuum chamber walls deeply influence the total pressure and composition of gas under measure; selective pumping, chemical reactions and physical processes continuously change the gas parameters, and in most cases only an approximate knowledge of the gas pressure is possible. Fortunately, the order of magnitude of pressure is often an adequate information in most problems. An accuracy of $\pm 10\%$ is usually considered quite good. A further difficulty occurs when the pressure of a gas at low temperature is to be known: the gage is usually at room temperature, and gradients of temperature and pressure occur along the tube connecting the measuring head to the low-temperature gas.

1.8.1 Total-pressure gages

We shall describe only a few of the most commonly used total-pressure gages. For more details on pressure measurements see ref. [1,11,12]. All gages described, except the McLeod and the diaphragm gage, measure density rather than pressure.

1.8.2 McLeod gage

The McLeod gage is shown in Fig. 1.26. By rising the mercury reservoir, the gas in the volume V at the pressure p to be measured is trapped in the bulb B.

A further rising of the reservoir causes a compression of the gas in the capillary C (closed). Capillary D is open and connected to the vacuum system. The difference Δh between the two mercury heights corresponds to a pressure difference $\Delta p = \rho g \cdot \Delta h$ (Δh in mm gives numerically Δp in torr); ρ is the density of mercury. If the compression of the gas in B and C is isothermal, we can write:

$$pV = (p + \rho g \Delta h) \cdot V_C \tag{1.24}$$

where V_C is the volume of the capillary C. Hence:

$$p = \frac{V_C g \rho \cdot \Delta h}{(V - V_C)} \tag{1.25}$$

The McLeod gage is a primary vacuometer; moreover its readings are independent of the type of gas, except condensable vapours. It covers a wide range of pressures with a good accuracy ($\pm 10^{-4}$ torr around 0.1 torr, $\pm 2 \times 10^{-7}$ torr around 10^{-6} torr) [13].

The main drawback is represented by the fact that a continuous monitoring of the pressure is impossible. Bakeable McLeod gages have been built [14].

The correct operation of the gage for accurate measurements is discussed in ref. [15].

The main application of this vacuometers is for calibration of other gages.

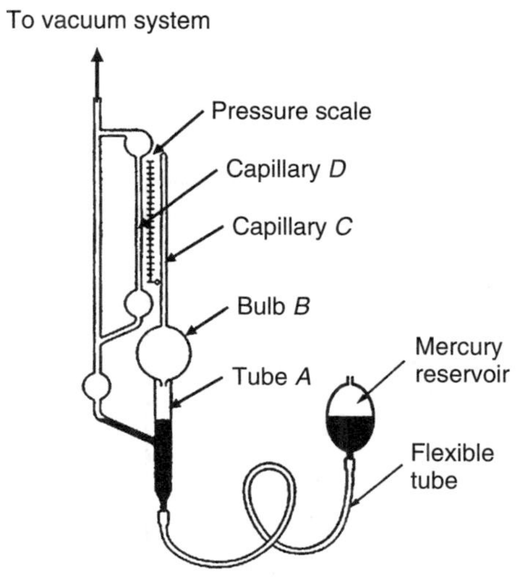

Fig. 1.26. Scheme of a McLeod pressure gage.

1.8.3 Bourdon gage

It is a simple and reliable gage covering six decades of pressure (not with one gage). Below 1 torr, its sensitivity is very low. The gage is usually made up of a bended flexible metallic tube (see Fig. 1.27) of elliptic cross-section. When pressure inside the tube is different for the outside (atmospheric) pressure, the tube bends. A gear and lever system moves the needle.

1.8.4 Diaphragm gage

It measures differential pressures. A thin diaphragm separates two chambers (see Fig. 1.28). One of the two chambers is kept at a constant pressure. A pressure difference causes the bending of the diaphragm which forms, together with a fixed electrode, a capacitor. The change in capacitance is measured by a very sensitive capacitance meter (measurement of $\Delta C \approx 10^{-19}$ Farad are possible).

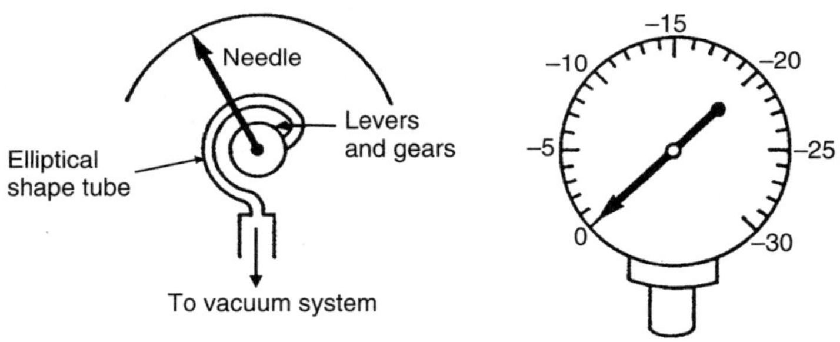

Fig. 1.27. Scheme of a Bourdon-type pressure gage.

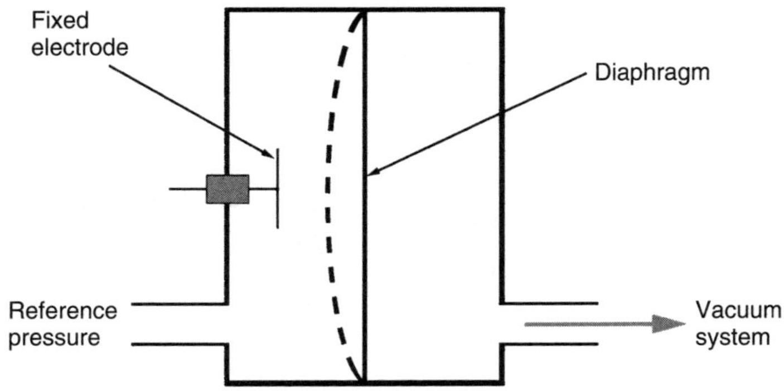

Fig. 1.28. Scheme of a diaphragm capacitance pressure gage.

Pressure down to 10^{-4} torr can be measured. The accuracy of the diaphragm gage is limited by the dependence of the capacitance on temperature (typical 1% of full scale). The time response can be around 10^{-3} s. Bakeable units are commercially available.

1.8.5 Thermal conductivity gages

In these gages, a wire inside the gas whose pressure is to be measured, is electrically heated by a constant power (see Fig. 1.29). As the gas density decreases, the heat loss from the filament to the envelope walls decreases and hence the filament temperature increases (not linearly). The temperature (200–300°C) is read by a thermocouple in thermal contact with the wire.

Pirani gage is similar: here the change in temperature of the wire is measured by the variation of the wire electrical resistance. The lowest measurable pressure is around 10^{-3} torr. The calibration (usually for dry air) is a function of the heat conductivity of the gas. Figure 1.30 shows a typical correction to be introduced for different gases.

1.8.6 Hot cathode ionization gage

This is a high-vacuum gage. Figure 1.31 shows the 'triode' and inverted configuration gage with indicative biasing. As shown in Fig. 1.31(a), electrons emitted by the filament are accelerated towards the grid; they collide with the gas molecules and ionize them. The ions are attracted by the collector. The ion current is a measure of the gas density. The ratio of the ion current to the pressure is called sensitivity of the gage (ampere/torr). It depends primarily on the geometry of the gage, biasing and ionization cross-sections of the gas. The ionization efficiency of electrons for some gases as a function of energy is shown in Fig. 1.32.

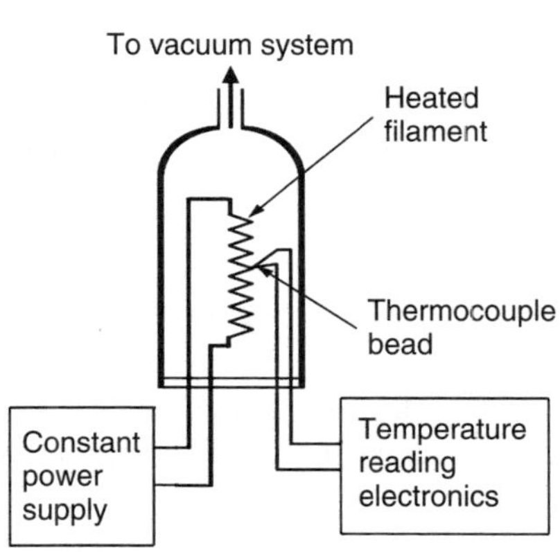

Fig. 1.29. Scheme of a thermal conductivity gage.

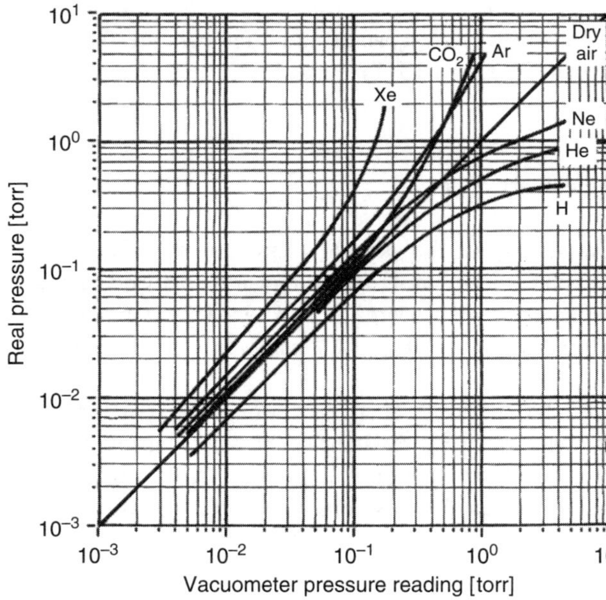

Fig. 1.30. Typical pressure correction for a Pirani gage for different gas species (courtesy of Alcatel).

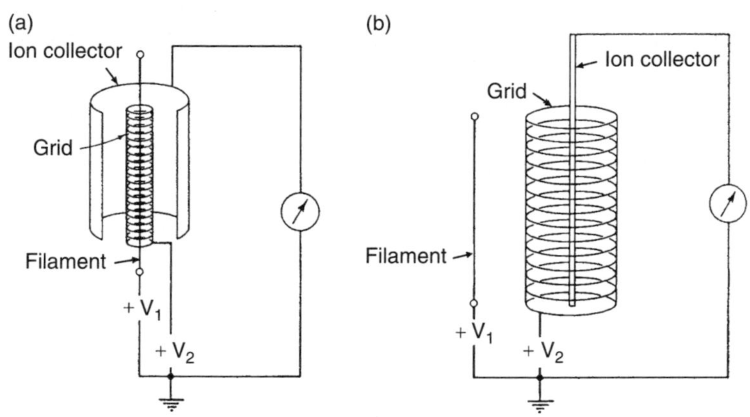

Fig. 1.31. Scheme of ion gages: (a) conventional ion gage; (b) inverted ion gage (Bayard-Allpert gage).

When the grid is bombarded by electrons, soft X-rays are emitted that, in turn, produce photoelectrons at the collector. This phenomenon limits the minimum pressure that can be measured to about 10^{-8} torr.

The inverted configuration of Fig. 1.31 (b) reduces the contribution of the photoelectron current, extending the range to about 10^{-10} torr.

Due to the presence of a high-temperature filament, frequent recalibrations are necessary.

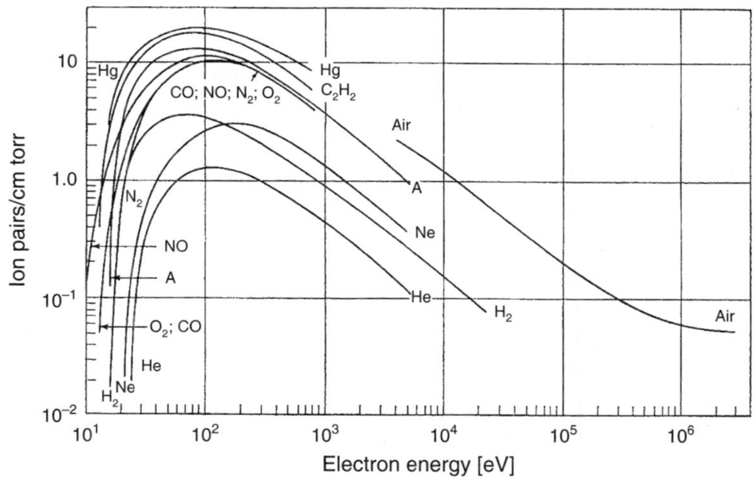

Fig. 1.32. Ionization efficiencies of electrons for some gases as a function of electron energy [16].

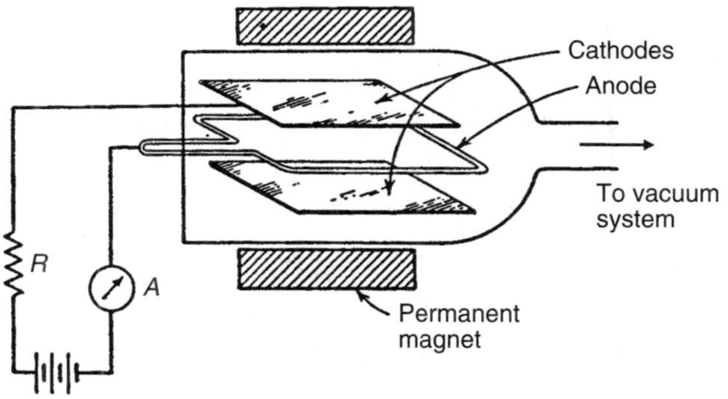

Fig. 1.33. Cold cathode of Philips vacuum gauge. The microammeter *A* gives an indication of the pressure.

1.8.7 Cold cathode gage

It is a high-vacuum gage made up of two cathodes and one anode placed in a magnetic field produced by a permanent magnet (see Fig. 1.33). Electrons due to natural radioactivity or to field emission start a discharge in the gas. The presence of the magnetic field produces paths about 100 times longer than the distance between the electrodes. Positive ions are collected by the cathodes.

The main advantages of this gage are its simplicity, ruggedness and high sensitivity. Its pressure range is limited between 10^{-2} and 10^{-8} torr.

Drawbacks are the occurrence of instabilities in the discharge [17] and the presence of a magnetic field.

1.9 Measurement of partial pressures

A mass spectrometer is often indispensable for a complete analysis of low-pressure gases, but a description of the various types of spectrometers is beyond the purpose of this book, but see, for example, ref. [18]. We simply remind that a mass spectrometer consists of three parts: an ion source where the neutral gas is ionized (usually by electron bombardment); an analyser where ions are selected according to their mass to charge ratio; and a collector with an amplifier to measure the weak ion current.

1.9.1 Leak detectors (LDs)

A cryogenist does not usually need a general purpose mass spectrometer, but the cryogenist cannot work without an LD which is made up of a small vacuum system (rotary pump or diaphragm pump in series with a turbo pump) and a mass spectrometer for the detection of light gases (H_2, ^{3}He and ^{4}He).

Usually ^{4}He is blown through a capillary towards the zone of the system under test where a leak is suspected. The system or vacuum component is pumped by the LDs and helium penetrates. LD and the mass detector reveal the presence of He. The art of leak detection is described in ref. [19,20].

A typical detection limit is 10^{-11} l torr/s.

The slits in LD (see the schematic representation of Fig. 1.34) are quite large, to increase the sensitivity. Consequently, the resolution is not very high, and overlapping of two adjacent mass peaks always occurs.

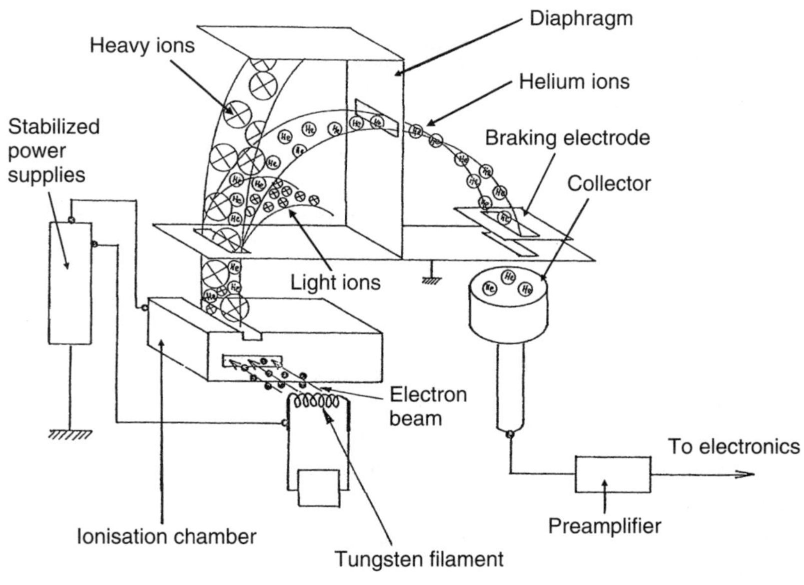

Fig. 1.34. Scheme of a leak detector analysis cell (courtesy of Alcatel).

When searching for leaks of ^{3}He and ^{4}He mixture, a great care must be used in interpreting the reading of an LD: even if the leak of only one of the two gasses is in reality present (e.g. ^{3}He), due to the overlapping of peaks, the LD apparently reveals the presence of the other one (^{4}He). The apparent flux of ^{4}He is usually about 10% of the real flux of ^{3}He.

We wish also to remind that leak detection in an ambient where the atmosphere is rich in the test gas (He) is usually a waste of time. A control of the background flux level of the test gas before blowing it to the leak point is wise.

LDs usually give also an approximate reading of the total pressure (ion current collected by the diaphragm electrodes of Fig. 1.34).

References

[1] S. Dushman: *Scientific foundations of vacuum techniques*, Wiley, New York (1962)
[2] L. Holland, W. Steckelmacher, J. Yarwood: *Vacuum manual*, E. & F.N. Spon, London (1974)
[3] A. Roth: *Vacuum technology*, 2nd ed., North-Holland, Amsterdam (1982)
[4] L.N. Rozanov: *Vacuum technique*, Taylor & Francis, New York (2002)
[5] R.C. Weast (ed.): *Handbook of chemistry and physics*, 53rd ed., CRC Press, Cleveland (1972–73)
[6] E.H. Kennard: *Kinetic theory of gases*, McGraw Hill, New York (1938)
[7] A. Guthrie, R.K. Wakerling: *Vacuum equipment an techniques*, McGraw Hill, New York (1949)
[8] H.W. Liepmann: J. Fluid Mech. **10**, 65 (1961)
[9] G. Lewin: *Fundamentals of vacuum science and technology*, McGraw Hill, New York (1965)
[10] R.A. Haefer: *Cryopumping, theory and practice*, Clarendon Press, Oxford (1989)
[11] J.H. Leck: *Pressure measurements in vacuum systems*, Chapman & Hall, London (1964)
[12] G. Grigorov, V. Kanev: *Le Vide Pussé*, Masson & Cie (1970)
[13] J.C. Simons Jr.: 1963 *10th AVS Vacuum Symp. Trans.*, p. 246, Macmillan, New York (1964)
[14] G.M. Monk, W.W. Stickey III, A.T. Callo: *1961 Vacuum Symp. Transp.*, p. 531 (1962)
[15] W.B. Nottingham, F.L. Torney Jr.: *Vacuum Symp. Trans.*, p. 117 (1961)
[16] R. Jaeckel: *Encyclopedia of Physics*, Vol. **12**, p. 535, Springer Verlag OGH, Berlin (1958)
[17] J.M. Lafferty: J. Vac. Sci & Tech. **9**, 101 (1971)
[18] J.H. Gross: *Mass spectrometry*, Springer Verlag, Heidelberg (2004)
[19] W.L. Briggs, A.C. Jones, J.A. Roberts: *1958 Vacuum Symp. Trans.*, p. 1081 (1959)
[20] John F. O'Hanlon: *A user's guide to vacuum technology*, 3rd ed., John Wiley & Sons, Canada (2003)

PART II

2 Cryoliquids

Contents

2.1 Cryogenics: Introduction and history

Cryogenics, 'production of cold', is a branch of physics which studies the methods to obtain temperatures well below room temperature and the methods to measure such temperatures. This technology is nowadays fundamental in different areas, from the familiar conservation of food by freezing to the most advanced scientific application such as quantum computing. When dealing with cryogenics, an immediate observation is evident: this is a technology in which man has largely surpassed nature. Figure 2.1 shows the range of temperature existing in nature or obtainable in laboratory: note that the range covers 15 decades!

In nature, the lowest temperature existing in the universe is 2.7 K. This background temperature depends on the presence of 'fossil photons' from the 'big bang'. On the other hand, in laboratory, it is possible to freeze samples of materials down to about 10^{-6} K. Moreover, in condensed matter laboratories, it is nowadays possible to cool a small number of atoms or molecules ($\sim 10^6$) down to $\sim$500 pK.

Figure 2.2 reports the historical development of refrigeration temperatures of matter.

The first noticeable experiments in cryogenics are probably those carried out around 1825 by Michael Faraday who used freezing mixtures to liquefy some gases such as CO_2 and NH_3. The Faraday's method (high pressure and moderately low temperatures), however, did not work with gases such as oxygen, nitrogen or hydrogen. Hence the term permanent gases was coined for these three gases.

It was only in the 1860s that Thomas Andrews showed the conditions needed for the liquefaction of a gas by establishing the existence of a critical point in the isotherms of CO_2 (304 K). However, at that time, none knew the critical temperatures of N_2, O_2 and H_2. The year 1877 saw two very different events both important for the future of cryogenics: the

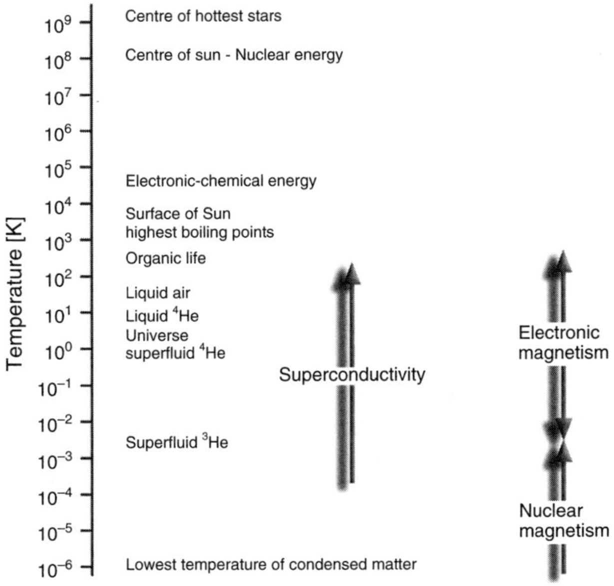

Fig. 2.1. Range of temperature existing in nature or obtainable in the laboratory.

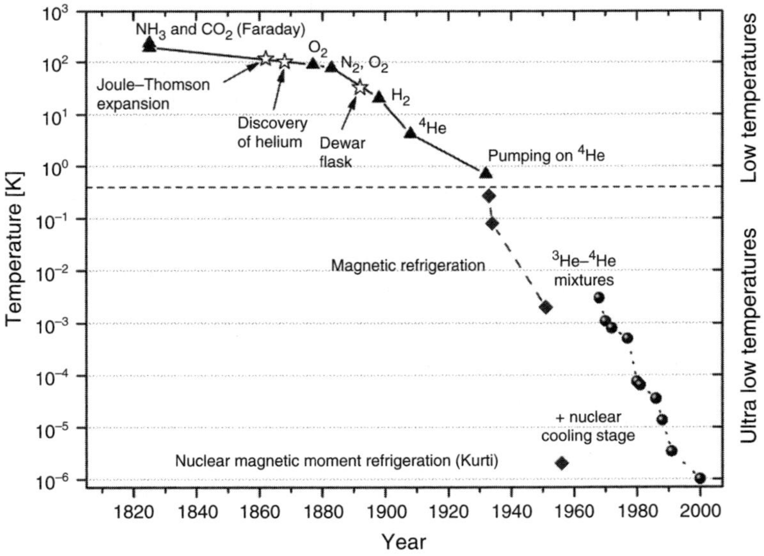

Fig. 2.2. Historical development of refrigeration.

first was the successful transport from Argentine to France of meat refrigerated by means of a NH_4 refrigerator. The second was the liquefaction of a permanent gas (O_2) obtained independently by Pictet (Geneve) and Cailletet (Paris). Both experiments (by Pictet and Cailletet) were extremely difficult because of the lack of a good thermal insulation. It was only few years later that a definite advance in this direction was achieved. A well-visible bath of liquid oxygen and liquid nitrogen was produced for the first time in 1883 in Cracow by S. von Wroblewski and K. Olszewski. (The method was the same as that used for the liquefaction of O_2 and CO_2.) In 1892, James Dewar introduced his double-walled vacuum flask which now brings his name. Thanks to the improvement in thermal isolation, in 1898 James Dewar was able to liquefy the hydrogen. It must be however noticed that this historic step in liquefaction depended not only on the invention of the dewar, but also on the use of the process of expansion discovered by Joule and Thomson in 1862. The beginning of the twentieth century sees the birth of the first industries for the distillation of air. In 1902, Claude and Delorme founded the society ‘Air Liquide’; in 1907, Linde installed the first liquefaction station of air in USA.

In 1908, Kamerling-Onnes got the liquefaction of helium (discovered by Janssen e Lockyer during the solar eclipse of 18 August 1868). Kamerlingh-Onnes obtained in Leiden 60 cc of liquid helium extracted from several tons of monazite sable imported from India. Kamerlingh-Onnes himself discovered the λ-transition and the superfluidity in ^{4}He and in 1911 the superconductivity of Hg, a particularly pure substance at that time. In the race towards lower and lower temperatures, Kamerling-Onnes, pumping on liquid ^{4}He, obtained 0.7K in 1926.

In 1927, Debye [1] and Giauque [2] independently suggested the use of disordered magnetic dipoles for refrigeration purposes. By the process of isothermal magnetization and adiabatic demagnetization, temperatures well below 1K were obtained by means of the adiabatic demagnetization of paramagnetic salts (0.27 K in 1933). The demagnetization of nuclear magnetic moments, based on the same principle, led to the range of the microkelvin [3].

Among the important achievements in the history of cryogenics, it is worth pointing out the invention of the dilution refrigerator (see Chapter 6) proposed by London in 1951 [4] and realized in the 1970s. This is still the only refrigerator based on a continuous cycle (i.e. not ‘one shot’), which can produce temperatures below 1 K. In fact, this apparatus is capable of maintaining temperatures of a few millikelvin for a very long time. Until now, the coldest dilution refrigerator reached 1.7 mK [5]. A refrigerator similar to the dilution refrigerator (but ‘one shot’) which can work in the absence of gravity has been proposed by Benoit, Caussignac e Pujol [6] in 1994 and will be used in Plank satellite experiment to be flown in 2007 [7] (see Section 6.8).

The combination of a dilution refrigerator and a nuclear cooling stage using a superconducting magnet [8] paved the way for lower temperatures. Refrigerators of this type have achieved temperatures in metals around 1 μK and nuclear temperatures below 250 pK [9,10].

In 1994, a type of ‘electric refrigeration’, still in its early phase of development, has been proposed (see Section 7.5 and ref. [11]).

For the detailed history of cryogenics until 1992 in the various countries, see ref. [12].

2.2 Cryoliquids

Nowadays, temperatures around 2 K can be easily obtained by means of mechanical refrigerators (e.g. pulse tubes, see Section 5.8). Cooling power more than 1W can be supplied by one pulse tube at 4.2 K. Moreover, to achieve a higher cooling power, such refrigerators can be put in parallel. Prototypes of three-stage pulse tube refrigerators using ^3He have been built [13], and temperatures close to 1 K have been reached. The only drawback of these mechanical refrigerators appears for the moment to be the high level of vibration induced by the high-pressure pulses. 'Dry dilution refrigerators' born from the coupling of pulse tubes and Joule–Thomson dilution refrigerators (see Section 6.7) start to be commercially available. It is likely that in the next future, mechanical refrigerators will replace N_2 and ^4He thermal baths.

Nevertheless, the simplest way to produce low temperature is still the use of cryoliquids (e.g. nitrogen, helium). It must be considered that most low-temperature equipments existing in a laboratory are designed for the use with cryoliquids, and the change to the new technologies is definitely expensive. Also for this reason, we shall briefly describe the properties and the use of cryoliquids used in low-temperature experiments and in particular helium (liquid or gas as used in pulse tubes) which practically intervenes in all refrigeration processes below 10 K.

The reader interested in the liquefaction technologies can see, for example, ref. [14,15]. We will only remind that in most cases, the gas cooling is obtained by the Joule–Thomson process: an isothermal compression of the gas is followed by an expansion. This procedure leads to a cooling only if the starting temperatures are lower than the inversion temperature $T_i = 6.75\,T_{cr}$ (for a Van der Waals gas), where T_{cr} is the critical temperature.

In Table 2.1, some properties of cryoliquids are compared with the thermodynamic properties of water. The boiling temperature T_{bp} and the latent heat L are of particular importance for the refrigeration processes.

2.2.1 Liquid oxygen and hydrogen

Liquid oxygen and hydrogen are rarely used in refrigeration: the former because of its high chemical reactivity, the latter again for its flammability. There is a second reason that limits the use of liquid hydrogen: the boiling temperature of hydrogen (20 K) can be easily obtained with most mechanical cryocoolers.

Liquid oxygen and hydrogen find instead precious application in triple-point reference cells for metrological use (see Chapters 8 and 9).

Both cryoliquids find also important applications in the space activity and in the industry (see Chapter 14).

Hydrogen, however, deserves a special attention because of the problems that it poses when working at very low temperatures. In a H_2 molecule, the two nucleus (two protons with spin ½) can be coupled in two ways forming molecules that can have total spin $I = 0$ or $I = 1$. Hence, two types of hydrogen exist:

1. Para-H_2: $I = 0$ with an antisymmetric nuclear state;
2. ortho-H_2: $I = 1$ with a symmetric nuclear state.

Table 2.1
Thermodynamic properties of some liquids [16,17]

Substance	T_{bp} [K]	T_m [K]	T_{tr} [K]	p_{tr} [10^5Pa]	T_{cr} [K]	p_{cr} [MPa]	T_i [K]	p_i [10^5Pa]	Latent heat L [kJ/l]	Volume % in air
H_2O	373.15	273.15	273.16	0.00610	647.096	22.064	—	—	2252	—
NH_3	293.8	195	195.40	0.0607	405.5	11.35	—	—	—	—
SO_2	263	200.75	197.68	0.00167	430.8	7.88	—	—	—	—
CO_2	194.6	216	216.5	5.173	304.1	7.38	—	—	—	—
C_2H_4	169.5	104.1	104.0	0.00120	283.2	5.03	—	—	—	—
Xe	165.1	161.3	161.4	0.82	289.77	5.841	—	—	303	10^{-5}
Kr	119.9	115.8	114.9	0.73	209.41	5.50	—	—	279	1.1×10^{-4}
CH_4	111.8	90.8	90.67	0.117	190.6	4.6	>500	533	—	—
O_2	90.2	54.4	54.36	0.015	154.6	5.04	742	570	245	20.9
Ar	87.3	83.8	83.81	0.67	150.87	4.898	—	—	224	0.93
N_2	77.4	63.3	63.15	0.13	126.2	3.39	621	380	160	78.1
Ne	27.1	24.5	24.56	0.43	44.4	2.76	260	—	110	1.8×10^{-3}
n-D_2	23.7	18.7	18.69	0.17	38.2	1.650	—	—	50	—
n-H_2	20.3	14.0	13.80	0.07	33.19	1.315	200	164	31.8	0.5×10^{-4}
^{4}He	4.21	—	—	—	5.195	0.227	43	39	2.56	5.2×10^{-4}
^{3}He	3.19	—	—	—	3.32	0.115	—	—	0.48	—

T_{bp}: boiling point temperature a $p = 1$ bar, T_m: melting temperature at $p = 1$ bar, $T_{tr}(p_{tr})$: triple point temperature (pressure), $T_{cr}(p_{cr})$: critical point temperature (pressure), $T_i(p_i)$: inversion temperature (pressure) L: latent heat of vaporization at T_b.

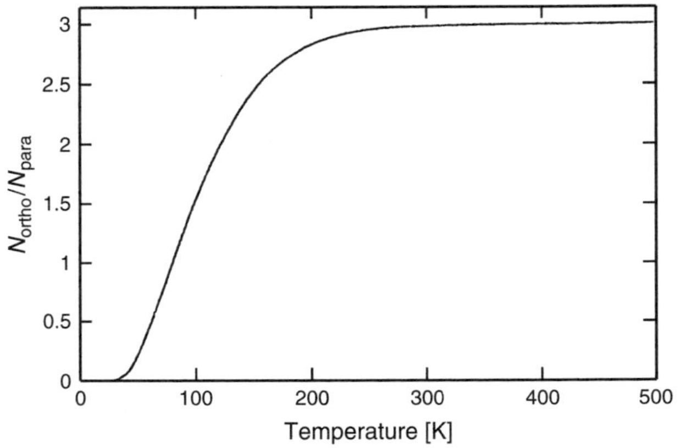

Fig. 2.3. Ratio of ortho to para hydrogen as a function of temperature [18–20].

When the equilibrium is reached, the ratio of the two forms of H_2 is 1:1 at about 77 K. At lower temperature, the form para prevails over the form ortho as shown in Fig. 2.3.

The ortho to para conversion of hydrogen [18,19] is exothermic: the 'heat release' is 1.06 kJ/mole starting from an ortho concentration of 75% and 1.42 kJ/mole for 100%. Such conversion can be very slow, for example, 1.9%/h for solid H_2 at the melting pressure [20].

There are at least three situations in which the thermal release from the ortho to para conversion can be important:

1. A liquid with a high ortho hydrogen concentration evaporates even without any external input of power.
2. Some metals such as Pd and Nb can dissolve hydrogen in atomic form in their lattice. For other metals as Cu, Ag, Au, Pt, Rh, this phenomenon is usually absent. If these latter metals contain traces of hydrogen (10–100 ppm, due to the production process), there is the formation of small gas bubbles with a typical diameter around 10^{-4} mm [21]. The pressure of hydrogen, which is in the molecular form, inside the bubbles, is very high, and hydrogen becomes liquid or solid when the metal is cooled. Hence also in this case, a heat release due to the ortho to para conversion takes place [22,23]. The thermal release is of the order of 1 nW/g; nevertheless it may be important in experiments at extremely low temperatures.
3. In the practical realization of the H_2 triple point: the cell containing H_2, for example, must be kept about 24 h at a temperature $T > T_{triple}$ to speed the conversion. Otherwise, the thermal release may spoil the reproducibility of the triple point.

We wish also to mention that solid hydrogen is usually responsible for the block of capillaries in low-temperature apparatus, for example, dilution refrigerators.

2.2.2 *Liquid nitrogen*

The nitrogen boiling temperature under 1 bar pressure is 77.4 K.

Liquid nitrogen is obtained from air in large liquefaction and separation plants.

Because of its low price and no toxicity, it finds a lot of applications (see Chapter 14). In particular, because of its high latent heat of evaporation, it is commonly used in the precooling of cryogenic equipments (see Chapter 5).

A drawback in the use of liquid nitrogen is displacement of oxygen in an enclosed volume.

2.2.3 Liquid helium

Nowadays, helium is no longer extracted from minerals, but from natural gas sources.

Liquid helium is inert, colourless, odourless, non-corrosive and non-flammable. Helium will not react with other elements or compounds under ordinary conditions. Helium is the most interesting element and the most used in cryogenics.

There exist two stable isotopes of helium, both of strong cryogenic interest: ^{4}He, which is a boson (nuclear spin $I = 0$) and ^{3}He, which is a fermion (nuclear spin $I = 1/2$).

^{3}He is present in natural gases with a concentration of $\sim 10^{-7}$ of that of ^{4}He and $\sim 10^{-6}$ of the helium in the atmosphere. The separation is very expensive. Hence ^{3}He is instead obtained as by-product of tritium production in nuclear reactors. Tritium in fact produces, by beta decay (the half life is 12.26 years), ^{3}He; the separation of ^{3}He is obtained through a diffusion process.

In Table 2.2, some properties of the two isotopes are summarized.

Table 2.2 clearly shows the strong differences between the two 'quantum liquids'. It is worth noting that both isotopes have very low boiling and critical temperatures and a low density (the molar volume is more than the double than that corresponding to a classic liquid). Figure 2.4 shows the p–T phase diagrams; besides the presence of a superfluid phases it is to be noted for both isotopes the missing of a triple point.

Table 2.2
Properties of liquid helium [24]

Properties	^{3}He	^{4}He
Density ρ [g/cm^3] ($T = 0$ K)	0.0823	0.146
Boiling point T_b [K]	3.191	4.215
Critical temperature T_c [K]	3.324	5.20
Critical pressure p_c [bar]	1.165	2.289
Critical density ρ_c [g/cm^3]	0.0413	0.0693
Melting pressure p_m [bar] ($T = 0$ K)	34.39	25.36
Minimum melting pressure p_m [bar]	29.31	25.32
Maximum superfluid transition temperature T_c [K]	0.0025	2.177
Gas-to-liquid volume ratio (liquid at 1 K, gas at 300 K and 1 bar)	662	866
Classical molar volume V_m [cm^3/mol] (at saturated vapour pressure and $T = 0$ K)	12	12
Molar volume V_m [cm^3/mol] (at saturated vapour pressure and $T = 0$K)	36.84	27.58

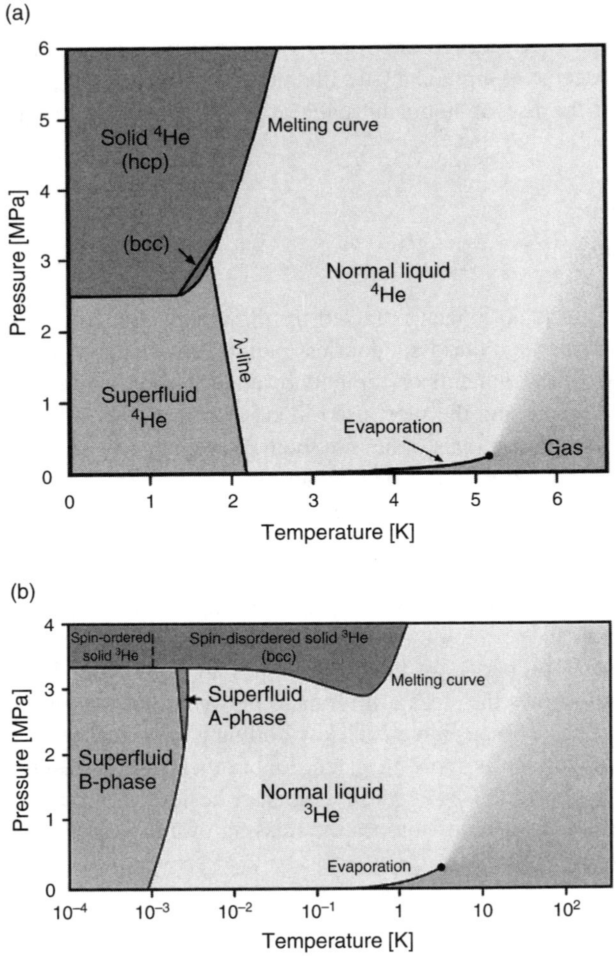

Fig. 2.4. Phase diagram of ^{4}He (a) and ^{3}He (b). Note the different temperature scales.

The minimum pressure necessary to get solid ^{3}He and ^{4}He is 34.4 and 25.4 bar respectively. Note also in Fig. 2.4(b) a zone in the melting curve with changing slope which presents a minimum at ~0.32 K (see Chapters 7–8). These properties are due to two peculiar helium characteristics:

1. The Van der Waals force between the atoms are very weak;
2. Helium has a very large quantum mechanical zero-point energy owing to its small atomic mass ($E_0 \simeq \hbar^2/2mV^{2/3}$ where V is the atomic volume and m atomic mass).

Since the zero-point energy is positive, that is of repulsive type and quite large, it is capable of practically balancing the effect of the attractive forces. Hence, solidification is possible only under elevated pressures.

2.2.4 Helium physics properties

We shall now describe the main physical properties of helium:

- vapour pressure and latent heat of evaporation;
- specific heat;
- thermal conductivity and viscosity.

2.2.4.1 Helium vapour pressure and latent heat of evaporation

The latent heat of evaporation L and the vapour pressure p_{vap} are fundamental parameters when using these two cryoliquids in the refrigeration process. Figure 2.5 shows L of ^{3}He and ^{4}He as a function of temperature. Note that L ($\sim$20.9 J/g for ^{4}He) is very small in comparison, for example, with that of hydrogen (445 J/g) or of nitrogen (200 J/g). Note also the minimum at $\sim$2.2 K in the graph for ^{4}He, in correspondence with the superfluid transition.

A low value of L means a small cooling power, a serious drawback when cooling by evaporation.

The vapour pressure can be approximately be calculated from the Clausius-Clapeyron equation:

$$\left(\frac{\mathrm{d}p}{\mathrm{d}T}\right)_{\mathrm{vap}} = \frac{(S_{\mathrm{gas}} - S_{\mathrm{liq}})}{V_{\mathrm{m(gas)}} - V_{\mathrm{m(liq)}}} \tag{2.1}$$

where S is the entropy and V_m the molar volume.

The difference between the entropies is L/T and the molar volume of the gas is much larger than that of the liquid. Since approximately $V_{gas} = RT/p$, from eq. (2.1) we get:

$$\frac{\mathrm{d}p}{\mathrm{d}T} = \frac{L(T)}{RT^2} \tag{2.2}$$

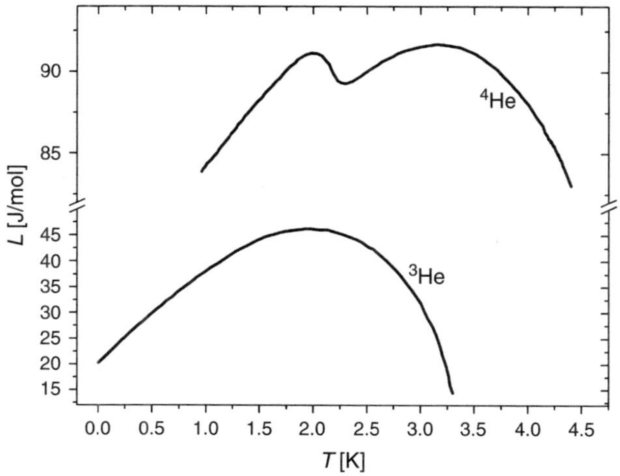

Fig. 2.5. Latent heat of vaporization of ^{3}He and ^{4}He. Data from [24]. Note the change in the vertical scale.

If we suppose L is approximately constant:

$$p = k \cdot \mathrm{e}^{-(L/RT)} \tag{2.3}$$

i.e. the vapour pressure increases exponentially with T.

If we pump over a liquid (^{4}He, for example), molecules of the liquid will evaporate producing the cooling of the remaining liquid. The cooling power will be:

$$\dot{Q} = \dot{n} \cdot L \tag{2.4}$$

where $\dot{n}$ number of molecules evaporated in the time unit.

If a volumetric pump is used:

$$\dot{n} = k \cdot p_{\mathrm{vap}}(T) \tag{2.5}$$

From eq. (2.5), (2.3) and (2.4):

$$\dot{Q} = k \cdot L p_{\mathrm{vap}} = k \cdot \mathrm{e}^{-(L/RT)} \tag{2.6}$$

Equation (2.6) shows that the cooling power decreases exponentially when temperature decreases, because p_{vap} and the pump throughput vanish.

A limit temperature is finally reached at which the cooling power supplied by the evaporating molecules is balanced by the external power input. In practice, this limit temperature is about 1.3 K for ^{4}He and 0.3 K for ^{3}He (see Fig. 2.6 and Chapter 5).

We shall see in Chapter 8 that the vapour pressure of ^{4}He and ^{3}He is used in thermometry (see Chapter 9) and in the realization of ITS-90 (Chapter 8).

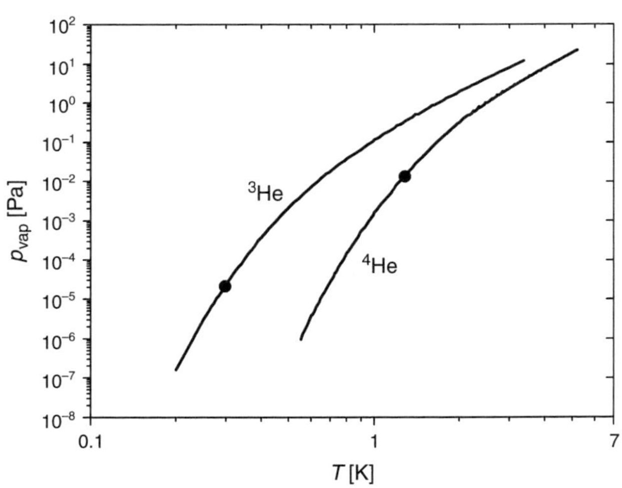

Fig. 2.6. Vapour pressure of liquid ^{3}He and liquid ^{4}He. (•) indicate the lowest temperature that can be practically obtained reducing the vapour pressure on these liquids.

2.2.4.2 Helium specific heat

Figure 2.7 shows the specific heat of He and Cu for $T < 1$ K. Down to 0.1 K, the specific heat of both isotopes is larger than that of Cu. For example at 1.5 K (Tables 2.3 and 2.4), the specific heat of both isotopes is about 1.5 J/gK, whereas that of Cu is about 10^{-5} J/gK. This observation is very important in practical cryogenics: for example, it means that the thermal time constant of an apparatus depends on the quantity of helium contained in it.

Moreover at low temperatures, the latent heat of vaporization L (20 J/g at 1.5 K for ^{4}He) is still very large compared to the specific heat of most materials. Hence, it is possible to cool these materials (metals) using L.

The properties of the two helium isotopes in the liquid state are strongly influenced by quantum effects. In Fig. 2.8, the specific heat of ^{3}He, calculated from the ideal gas Fermi model ($T_F = 4.9$ K) with the liquid ^{3}He density, is compared with the experimental data. The inadequacy of this model is evident. A better fit, especially at the lower temperatures, is obtained by the Landau theory [25].

An 'anomalous' behaviour is found also for ^{4}He, as can be observed from Fig. 2.9: the ^{4}He specific heat shows a sharp maximum around ~2.2 K, corresponding to the transition to the superfluid state (He II). The characteristic shape of the C–T curve has baptized this transition with the name of 'λ-transition'.

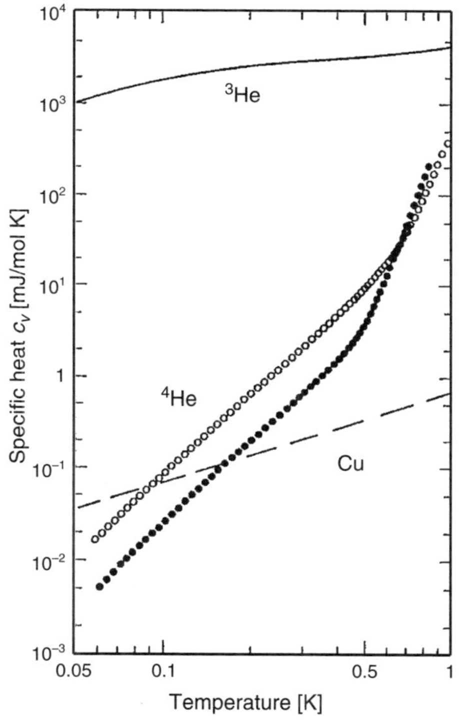

Fig. 2.7. Specific heat of liquid ^{4}He at vapour pressure (27.58 cm^3/mol, ○) and at about 22 bar (23.55 cm^3/mol, •) [26,27] compared to the specific heat of liquid ^{3}He at vapour pressure [28,29] and of Cu.

Table 2.3
Thermodynamic properties of liquid ^{4}He at saturated vapour pressure [30]

	Temperature [K]	Pressure [atm]	Density [g/cm^3]	Entropy [J/g K]	c_{sat} [J/g K]
	1.4	0.002835	0.1450	0.132	0.780
	1.6	0.007487	0.1451	0.284	1.572
	1.8	0.01641	0.1453	0.535	2.810
	2.0	0.03128	0.1456	0.940	5.18
T_λ	2.172	0.04969	0.1462	—	—
	2.2	0.05256	0.1461	1.671	3.16
	2.4	0.08228	0.1453	1.898	2.25
	2.6	0.1219	0.1442	2.068	2.10
	2.8	0.1730	0.1428	2.225	2.22
	3.0	0.2371	0.1411	2.382	2.42
	3.2	0.3156	0.1393	2.542	2.67
	3.4	0.4100	0.1372	2.706	2.95
	3.6	0.5220	0.1348	2.875	3.28
	3.8	0.6528	0.1321	3.050	3.68
	4.0	0.8040	0.1290	3.234	4.19
	4.2	0.9772	0.1254	3.429	4.88
T_{Nbp}	4.224	1.000	0.1250	3.454	4.98
	4.4	1.174	0.1213	3.640	5.86
	4.6	1.397	0.1163	3.873	7.44
	4.8	1.648	0.1101	4.144	10.5
	5.0	1.929	0.1011	4.495	19.1
T_c	5.201	2.245	0.0696	5.589	—

Note: 1 atm = 760 torr = 1.013 bar = 1.013×10^5 Pa.

Table 2.4
Thermodynamic properties of liquid ^{3}He at saturated vapour pressure [28,29,31,32]

Temperature [K]	Pressure [atm]	Density [g/cm^3]	Entropy [J/g K]	c_{sat} [J/g K]
0.100	—	0.08192	—	0.558
0.120	—	0.08193	—	0.658
0.150	—	0.08195	0.829	0.774
0.200	1.6×10^{-8}	0.08198	1.069	0.893
0.250	3.23×10^{-7}	0.08201	1.275	0.954
0.300	1.98×10^{-6}	0.08204	1.454	0.996
0.350	9.46×10^{-6}	0.08206	1.609	1.028
0.400	3.71×10^{-5}	0.08207	1.748	1.056
0.450	8.39×10^{-5}	0.08208	1.875	1.084
0.500	1.872×10^{-4}	0.08209	1.990	1.106
0.550	3.682×10^{-4}	0.08209	2.097	1.126
0.600	7.181×10^{-4}	0.08208	2.195	1.148
0.650	0.001091	0.08207	2.288	1.179
0.700	0.001704	0.08205	2.376	1.210
0.750	0.002536	0.08203	2.461	1.242
0.8	0.003819	0.08199	2.575	1.275
0.85	0.005005	0.08195	2.656	1.309

Table 2.4
Continued

Temperature [K]	Pressure [atm]	Density [g/cm^3]	Entropy [J/g K]	c_{sat} [J/g K]
0.9	0.006721	0.08191	2.745	1.342
0.95	0.008808	0.08185	2.815	1.392
1	0.01167	0.08178	2.884	1.437
1.1	0.01767	0.08163	3.028	1.507
1.2	0.02662	0.08143	3.125	1.605
1.3	0.03685	0.08119	3.265	1.702
1.4	0.05084	0.0809	3.390	1.800
1.5	0.06640	0.08056	3.516	1.911
1.6	0.08642	0.08017	3.642	2.023
1.7	0.1083	0.07972	3.767	2.149
1.8	0.1353	0.07921	3.893	2.274
1.9	0.1646	0.07863	4.018	2.400
2.0	0.1995	0.07797	4.144	2.539
2.2	0.2807	0.07641	4.409	2.860
2.4	0.3810	0.07444	4.674	—
2.5	0.4392	0.07327	4.813	—

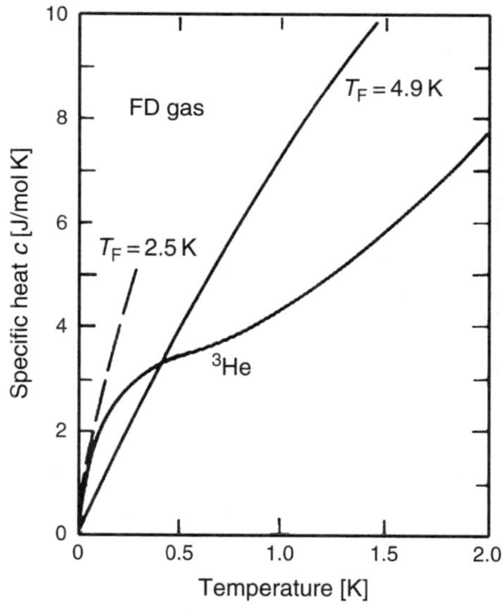

Fig. 2.8. Specific heat of liquid ^{3}He compared with curves for the ideal Fermi gas [30].

The transition temperature is $T_\lambda = 2.1768\,\mathrm{K}$ at the saturated vapour pressure and decreases, when pressure increases, up to $T_\lambda = 1.7673\,\mathrm{K}$ at fusion line (see Fig. 2.4 (a)).

Between 1 and 2 K, the ^{4}He specific heat shows a steep dependence on temperature due to the so-called rotonic excitation. Below 0.6 K, ^{4}He behaves like a Debye insulator with a specific heat proportional to T^3.

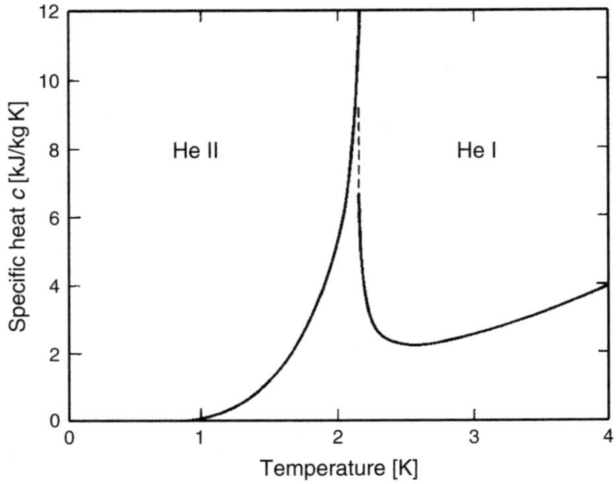

Fig. 2.9. Specific heat of liquid ^{4}He at temperature closed to its normal superfluid phase transition. Data from [33].

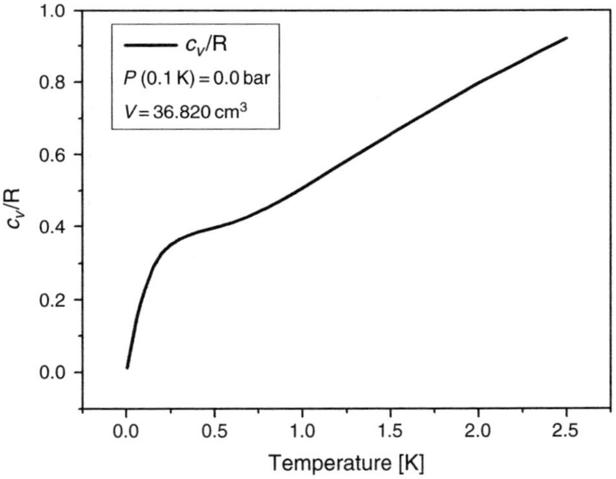

Fig. 2.10. Smoothed value for the ^{3}He specific heat (in units of the gas constant R) measured at molar volume V corresponding to nominal sample pressure of 0.0 bar. Data from [28].

Note that instead for ^{3}He, the specific heat below 0.1 K is linear with T (Fig. 2.11) down to the temperature of the superfluid transition $T_c = 0.929$ mK (for 0.0 bar [28]) (Fig. 2.10).

The idea that new phenomena could be present in ^{3}He at very low temperatures arose from thermal measurements. The first observation was the anomaly in the specific heat at the normal superfluid transition which reminded the behaviour of specific heat at the superconductive transition in metals (Fig. 2.11) [34–36].

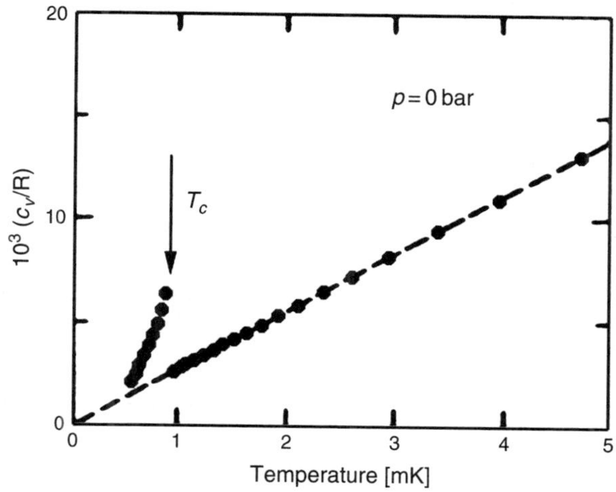

Fig. 2.11. ^{3}He specific heat (in units of the gas constant R) plotted versus the temperature at pressure of 0 bar. The transition temperature T_c is 0.929 mK [29].

Afterwards, measurements of the specific heat have been carried out in a wider pressure range [29,37–39]. It was thus evident that there is a discontinuity at the transition temperature which, at the melting pressure, gave a $\Delta C/C \sim 2$, (well above the 1.43 value expected from BCS theory) which decreased as the pressure diminished.

Such transition is called 'phase A transition' ($T = 2.505$ mK at the melting pressure); this is a second-order transition, with a specific heat discontinuity more similar to that of normal superconductors than to the ^{4}He λ-transition (Fig. 2.11).

When temperature decreases, a second transition to a distinct superfluid phase B is observed. The latter is first-order transition at $T = 1.948$ mK at the melting pressure.

2.2.4.3 Transport properties of liquid ^{4}He*: thermal conductivity and viscosity*

Liquid ^{4}He, above ~2.2 K (He I), because of its low density, shows transport properties similar to those of a classic gas (Fig. 2.12). The same happens for ^{3}He above 0.1 K. Above the λ-point, ^{4}He has a low thermal conductivity (a factor of 10^{-4} compared to Cu and a factor of 10^{-1} compared to stainless steel) and boils with strong bubbling.

When pumping on ^{4}He I (and also on ^{3}He), bubbles of vapour form inside the liquid which is warmer than the surface cooled by the evaporation process.

In many experiments, strong thermal gradients are present in the liquid which cannot be disregarded (see Section 9.3). On the other hand, below ~2.2 K, that is in the superfluid state, the thermal conductivity is extremely high (theoretically infinite, see Fig. 2.13), and bubbles do not form in the liquid during the evaporation process, since the temperature gradient is very close to zero.

The thermal conductivity of superfluid ^{4}He below 0.4 K and 2 bar is given by:

$$k = 20 \cdot d \cdot T^3 \propto C_{ph} \quad [\mathrm{W/cm\,K}] \tag{2.7}$$

where d is the diameter (cm) of the ^{4}He column (cm) [40].

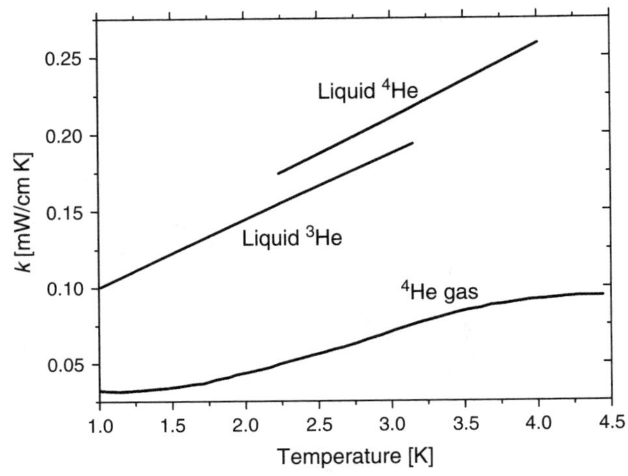

Fig. 2.12. Thermal conductivity of liquid helium and helium gas. Data from [41].

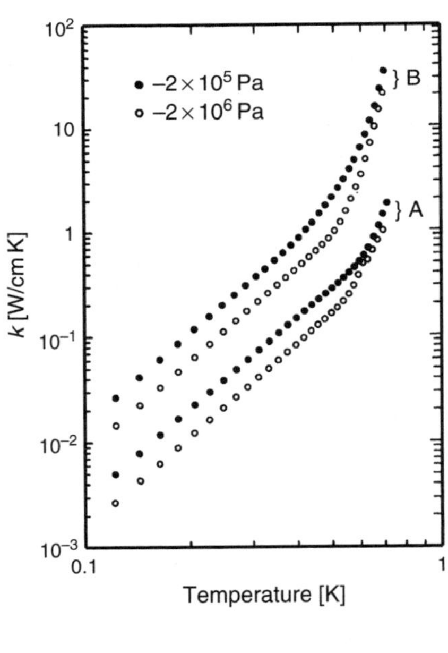

Fig. 2.13. Thermal conductivity of liquid ^{4}He at two different pressures in tubes of 1.38 mm (A) and 7.97 mm (B) in diameter. Data from [40].

It is easy to check that sometimes k can exceed the Cu conductivity. The very high thermal conductivity of ^{4}He II allows for the propagation of temperature waves (second sound) [42–48].

At very low temperatures, the thermal excitations in liquid ^{4}He (the long-wavelength phonons) are scattered only from the walls of the container with a ballistic movement.

Hence the heat transport, in this case, depends on the dimension and shape of the liquid container. As we can see in Fig. 2.13, the thermal conductivity (and the specific heat) of liquid ^{4}He decreases when pressure increases and scales with the tube diameter. At temperatures below 0.4 K, the data of thermal conductivity (eq. 2.7) follow the temperature dependence of the Debye specific heat. At higher temperatures, the thermal conductivity increases more steeply because of the viscous flow of the phonons and because of the contribution of the rotons.

In the superfluid state, the ^{4}He viscosity is very low: this property let the liquid pass through microscopic holes (superleaks), whereas the gas (viscous) cannot pass.

A further effect due to the absence of viscosity is the well-known formation of superfluid films with siphon effects [42–49].

For the properties of ^{3}He in the millikelvin range, see ref. [36,50–68].

References

[1] P. Debye: Ann. Phys. **81**, 1154 (1927)
[2] W.F. Giauque: J. Am. Chem. Soc. **49**, 1864 (1927)
[3] N. Kurti et al.: Nature **178**, 540 (1956)
[4] H. London: in Proc. Int. Conf. on Low Temp. Physics, p. 157 (1957)
[5] D.J. Cousins et al.: J. Low Temp. Phys. **114**, 547 (1999)
[6] A. Benoit, M. Caussignac, S. Pujol: Physica B **197**, 48 (1994)
[7] S. Triqueneaux et al.: Cryogenics **46**, 228 (2006)
[8] P.M. Berglund et al.: J. Low Temp. Phys. **6**, 357 (1972)
[9] G.R. Pickett: Physica B **280**, 467 (2000)
[10] J.T. Tuoriniemi, T.A. Knuuttila: Physica B **280**, 474 (2000)
[11] M. Naum, T.M. Eiles, J.M. Martinis: Appl. Phys. Lett. **65**, 3123 (1994)
[12] R.G. Scurlock ed.: *History and Origins of Cryogenics*, Clarendon Press, Oxford (1992)
[13] I.A. Tanaeva, A.T.A.M. de Waele: Cryogenics **45**, 578 (2005)
[14] P.V.E. Mc Clintock et al.: *Matter at Low Temperature*, Blackie, London (1984)
[15] B.A. Hand ed.: *Cryogenic Engineering*, Academic Press, London (1986)
[16] R.C. Reidel, J.M. Pruasnitz, T.K. Sherwood: *The Properties of Gases and Liquids*, McGraw Hill, New York (1977)
[17] J. White: *The Properties of Liquid and Solid Helium*, Clarendon, Oxford (1967)
[18] K. Mitizuki, T. Nagamiya: J. Phys. Soc. Jpn. **11**, 93 (1956)
[19] A.J. Berlinsky, N.W. Hardy: Phys. Rev. B **8**, 5013 (1973)
[20] P. Pedroni et al.: Solid State Commun. **14**, 279 (1974)
[21] W.R. Wampler, T. Schober, B. Lengeler: Phylos. Mag. **34**, 129 (1976)
[22] M. Schwark et al.: J. Low Temp. Phys. **53**, 685 (1983)
[23] M. Kolac, B.S. Neganov, S. Sahling: J. Low Temp. Phys **59**, 547 (1985)
[24] F. Pobell: *Matter and Methods at Low Temperatures*, Springer, Berlin (1991)
[25] N.W. Ashcroft, N.D. Mermin: *Solid State Physics*, Saunders College Publishing, Philadelphia (1976)
[26] D.S. Graywall: Phys. Rev. B **18**, 2127 (1978)
[27] D.S. Graywall: Phys. Rev. B **21**, 1329 (1979)
[28] D.S. Graywall: Phys. Rev. B **27**, 2747 (1983)
[29] D.S. Graywall: Phys. Rev. B **33**, 7520 (1986)
[30] S.W. Van Sciver: *Helium Cryogenics*, Plenum Press, New York (1986)
[31] X.Y. Li, Y.H. Huang, G.B. Chen, V.D. Arp: in *Proceedings of the Twentieth International Cryogenic Engineering Conference (ICEC20)*, Beijing 2004, Elsevier Science (2006)
[32] W.E. Keller: *Helium-Three and Helium-Four*, Plenum Press, New York (1969)

[33] M.J. Buckingham: in *Near Zero: New Frontiers of Physics*, ed. by J.D. Fairbank, B.S. Deaver Jr., C.W.F. Everitt and P.F. Michelson., p. 152, Published by W.H. Freeman and Co., New York (1988)
[34] R.A. Webb et al.: Phys. Rev. Lett. **30**, 210 (1973)
[35] J.C. Wheatley: Phys. Rev. Lett. **30**, 829 (1973)
[36] J.C. Wheatley: Rev. Mod. Phys. **47**, 415 (1975)
[37] W.P. Halperin et al.: Phys. Rev. B **13**, 2124 (1976)
[38] T.A. Alvesalo et al.: Phys Rev. Lett. **44**, 1076 (1980)
[39] T.A. Alvesalo et al.: Phys. Rev. Lett. **43**, 1509 (1979)
[40] D.S. Greywall: Phys. Rev. B **23**, 2152 (1981)
[41] D.M. Lee, H.A. Fairbank: Phys. Rev. **116**, 1359 (1959)
[42] J. Wilks: *The properties of liquid and solid helium*, Clarendon, Oxford (1967)
[43] R.J. Donnelly: *Experimental superfluiduty*, Chicago University Press, Chicago (1967)
[44] T. Tsuneto: *The Structure and Properties of Matter*, Springer, Berlin (1982)
[45] J. Wilks, D.S. Betts: *An Introduction to Liquid Helium*, Clarendon, Oxford (1987)
[46] D.R. Tilley, J. Tiley: *Superfluidity and Superconductivity*, Hilger, Bristol (1990)
[47] I.M. Khalatnikov: *An Introduction to the Theory of Superfluidity*, Benjamin, New York (1965)
[48] K.H. Bennemann, J.B. Ketterson eds.: *The Physics of Liquid and Solid Helium*, vol. 2, Wiley, New York (Part I, 1976, Part II, 1978)
[49] D.F. Brewer: in *The Physics of Liquid and Solid Helium*, ed. by K.H. Benneman and J.B. Keterson, part 2, p. 573, Wiley, New York (1978)
[50] D.D. Osheroff, R.C. Richardson, D.M. Lee: Phys. Rev. Lett. **28**, 885 (1972)
[51] D.D. Osheroff et al.: Phys. Rev. Lett. **29**, 920 (1972)
[52] A.J. Leggett: Rev. Mod. Phys. **47**, 331 (1975)
[53] J.C. Wheatley: Physica **69**, 218 (1973)
[54] J.C. Wheatley: in *Progress in Low temperature Physics*, ed. by D.F. Brewer, vol. 7, p. 1, North Holland, Amsterdam (1978)
[55] W.F. Brinkman, M.C. Cross: in *Progress in Low Temperature Physics*, ed. by D.F. Brewer, vol. 7, p. 105, North Holland, Amsterdam (1978)
[56] P. Woelfle: in *Progress in Low Temperature Physics*, ed. by D.F. Brewer, vol. 7, p. 19, North Holland, Amsterdam (1978)
[57] P.W. Anderson, W.F. Brikman: in *The Physics of Liquid and Solid Helium*, vol. 2, p. 177, Wiley, New York (1978)
[58] D.M. Lee, R.C. Richardson: in *The Physics of Liquid and Solid Helium*, ed. by K.H. Bennemann and J.B. Ketterson, vol. 2, p. 287, Wiley, New York (1978)
[59] D. Vollhard, P. Woelfle: *The Superfluid Phases of Helium-3*, Taylor & Francis, London (1990)
[60] W.P. Halperin, L.P. Pitaevskii eds: *Helium Three*, North Holland, Amsterdam (1990)
[61] L.D. Landau: Sov. Phys. JEPT **3**, 920 (1957); ibid. **5**, 101 (1957); ibid. **8**, 70 (1959)
[62] A.A. Abrikosov, I.M. Khalatnikov: Rep. Prog. Phys. **22**, 329 (1959)
[63] J.C. Wheatley: *Progress in Low Temperature Physics*, ed. by Gorter, vol. 6, p. 77, North Holland, Amsterdam (1970)
[64] G. Baym, C. Pethick: in *The Physics of Liquid and Solid Helium*, ed. by K.H. Benneman and J.B. Keterson, part 2, p. 1, Wiley, New York (1978)
[65] D.S. Graywall: Phys. Rev. B **29**, 4933 (1984)
[66] O.V. Lounasmaa: *Experimental Principles and Methods Below 1 K*, Academic Press, London (1974)
[67] R. Koenig, F. Pobell: Phys. Rev. Lett. **71**, 2761 (1993)
[68] R. Koenig, F. Pobell: J. Low Temp. Phys. **97**, 287 (1994)

3 Properties of Solids at Low Temperature

Contents

3.1 Introduction

The knowledge of the physical properties of materials is important in the design and construction of cryogenic equipments. Specific heat, thermal expansion, thermal conductivity, magnetic properties and mechanical moduli are relevant in the building of apparatus and in carrying out low-temperature experiments. On the other hand, the measurement of the low-temperature properties of a material and in particular their temperature dependence supplies precious information about the type and structure of the investigated sample. When the temperature is reduced, the properties of materials become in most cases easier to describe (solid ^{3}He is one of the few exceptions). In particular, the various constituents of a material can be sometimes studied independently: the nuclear spin system, for example, can be considered independent of the electron and phonon systems. Also the contributions of electrons and phonons to the specific heat of a metal can be simply added. This is not always true: the interaction of conduction electrons and phonons account for the temperature dependence of the electrical resistivity of a metal. In the following paragraphs, we shall recall the so-called 'thermal properties' of solid materials, i.e. specific heat, thermal expansion and thermal conductivity. Also the magnetic properties of solids will be discussed (Section 3.6) for their contribution to the specific heat. Dielectric properties will be considered in Section 9.8 in connection with dielectric constant thermometry.

3.2 Specific heat

The specific heat is the amount of heat required to change one mole of a substance by one degree in temperature. Therefore, unlike the extensive variable heat capacity, which depends on the quantity of material, specific heat is an intensive variable and has units of energy per number of moles (n) per degree.

The heat capacity of a substance can differ, depending on which are the variables held constant, with the quantity being held constant usually being denoted with a subscript. For example, the specific heat at constant pressure is commonly denoted c_P, while the specific heat at constant volume is commonly denoted c_V:

$$c_V = \frac{1}{n}\left(\frac{\mathrm{d}Q}{\mathrm{d}T}\right)_V; \quad c_P = \frac{1}{n}\left(\frac{\mathrm{d}Q}{\mathrm{d}T}\right)_P \tag{3.1}$$

To evaluate the specific heat of a material, the various excitations that take place in the material are to be considered. This is the reason why the specific heat gives plenty of information about the material.

The experimental data usually give the specific heat at constant pressure c_P. Theories usually refer to the specific heat at constant volume c_V. The specific heat c_P is greater than c_V by a factor $(1+\beta\ \gamma_G T)$, where β is the volumetric coefficient of thermal expansion and γ_G is the so-called Grüneisen parameter:

$$\gamma_G = \beta\frac{V}{\chi_T c_V} = \beta\frac{V}{\chi_S c_P} \tag{3.2}$$

where V is the molar volume and χ_T and χ_S are the isothermal and adiabatic compressibility coefficients, respectively. For most solids below room temperature, c_P exceeds c_V by less than 2% (about 1% at $T = \theta_D/2$, with θ_D the Debye temperature of the material).

Hence, we shall discuss how c_V depends on temperature and on materials, since theoretical models deal with c_V (see ref. [1–9]).

3.3 Lattice specific heat

At high temperatures (T > 100 K) c_V is due to the phonon contribution c_{ph} which approaches the classical Dulong and Petit value of $3\boldsymbol{n}R = 24.94\boldsymbol{n}$ [J/mol K], where $\boldsymbol{n}$ is the number of atoms in the molecule.

The Debye's elastic continuum model for solids [1–9] gives for c_{ph}:

$$c_{ph}(T) = 9\mathbf{n}N_A k_B\left(\frac{T}{\theta_D}\right)^3 \cdot \int_0^\infty x^4 e^x \cdot (e^x - 1)^{-2}\,\mathrm{d}x \tag{3.3}$$

The integral is tabulated in ref. [3] and yields:

$$c_{ph}(T) = \frac{12}{5}\pi^4 \mathbf{n}N_A k_B\left(\frac{T}{\theta_D}\right)^3 = 1944 \cdot \mathbf{n}\left(\frac{T}{\theta_D}\right)^3 \quad [\mathrm{J/mol\,K}] \tag{3.4}$$

for temperatures $T < \theta_D/10$.

From room temperature down to about 50 K, eq. (3.3) gives a quite good value of $c_{ph}(T)$ for most solids. Deviations from eq. (3.4) for $T < \theta_D/10$ depend on the material as shown in Fig. 3.1.

The cubic dependence of c_{ph} on temperature accounts for the small specific heat at low temperature of crystalline insulators. Specific heat of rare gas crystals is shown in Fig. 3.2.

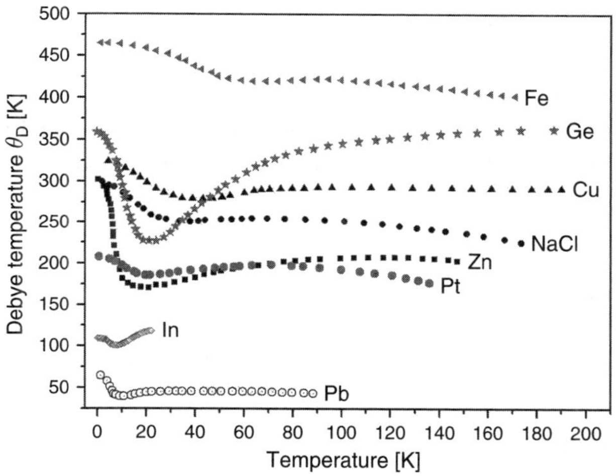

Fig. 3.1. Debye temperature versus temperature for some materials. The values of Debye temperatures are obtained by heat capacity measurements [10,11].

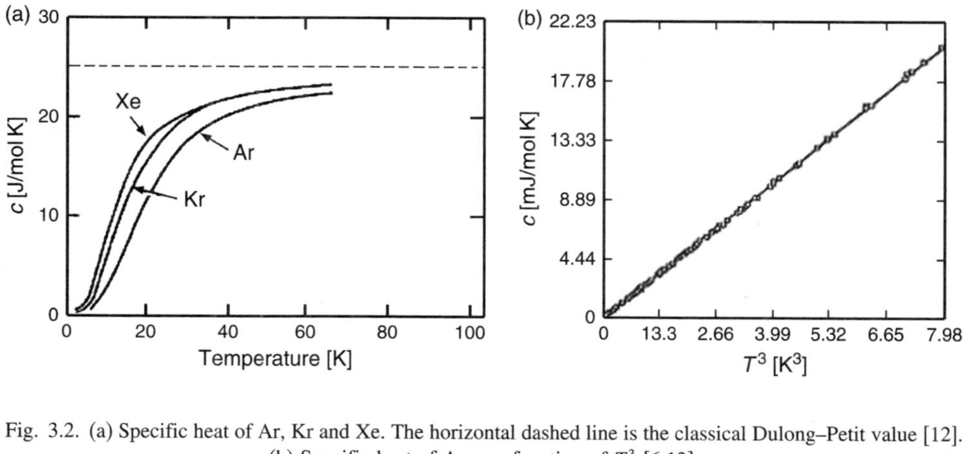

Fig. 3.2. (a) Specific heat of Ar, Kr and Xe. The horizontal dashed line is the classical Dulong–Petit value [12]. (b) Specific heat of Ar as a function of T^3 [6,13].

We wish to note that θ_D can also be calculated from the measured ultrasonic velocities, and data should be equal to those obtained from specific heat measurements. The Debye's temperature evaluated from data of ultrasonic velocity is (see e.g. [14,15]):

$$\theta_D = \frac{h}{k_B}\left(\frac{3N_A\rho}{4\pi A_m}\right)^{\frac{1}{3}} \cdot v_m \tag{3.5}$$

where h is the Planck's constant, N_A is Avogadro's number, A_m is the effective molar mass, ρ is the density and v_m is the mean integrated velocity given by:

$$v_m = \left(\frac{3}{1/v_L^3 + 2/v_T^3}\right)^{\frac{1}{3}} \tag{3.6}$$

where v_L and v_T are the velocities of the longitudinal and shear waves respectively.

The Debye's temperature is high for lattice made of strongly bound light atoms, as in diamond ($\theta_D = 2000\,K$) and low for lattice made of weakly bound heavy atoms as in lead ($\theta_D = 95\,K$).

3.4 Electronic specific heat

Free electrons contribute to specific heat with a term which, at least at low temperatures, is:

$$c_e(T) = \gamma \cdot T \tag{3.7}$$

The γ values (Sommerfeld constants) for some metals are reported in Table 3.1.

Thus, the specific heat of a metal is:

$$c = \gamma \cdot T + \beta \cdot T^3 \tag{3.8}$$

Table 3.1
Sommerfeld constant γ of some metals

Isotope compound	γ[mJ/mol K^2]
^{27}Al	1.35
^{63,65}Cu	0.691
^{93}Nb	7.79
^{107,109}Ag	0.640
^{113,115}In	1.69
^{117,119}Sn	1.78
^{195}Pt	6.49
^{197}Au	0.689
^{203,205}Tl	1.47
Au*In*$_2$	3.15
*Pr*Ni$_5$	40

For the two compounds AuIn$_2$ and PrNi$_5$, the data are for the element indicated in italics [16].

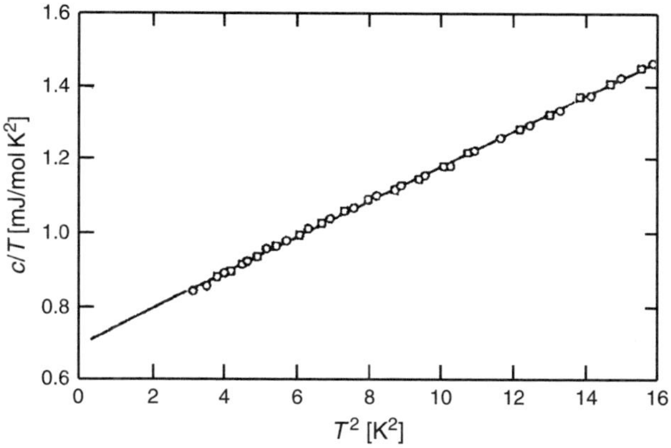

Fig. 3.3. Specific heat c of copper divided by the temperature T as a function of T^2 [8].

In practice, the linear term in eq. (3.8) becomes important only for $T \leq 10\,\mathrm{K}$.

Formula (3.8) is in good agreement with experimental results as shown in Fig. 3.3 for copper. The experimental check of the validity of eq. (3.8) is done by plotting T^2 versus c/T as in Fig. 3.3.

3.5 Electronic specific heat in superconducting materials

Materials that exhibit the phenomenon of superconductivity enter into a new state below a critical temperature T_C (see Table 8.11).

In this 'superconducting state', the electrical resistivity of the material becomes zero, and its thermal properties change. T_C is a function of the material as well as of its purity and of the applied magnetic field.

In particular, if the magnetic field is strong enough, the material does not enter the superconducting state. The latter property is shown in Fig. 3.4(a) where the specific heat of Al was measured [17] in no-field and in a moderate field.

We notice from Fig. 3.4(b) that the lattice specific heat c_{ph} is not modified by the superconducting transition: $c_{ph} = \beta T^3$ with the same β of the normal state.

We can conclude that the new behaviour of the superconducting material is due to a new state for the electrons: in fact, at the critical temperature, there is a jump of the electronic specific heat. In no external magnetic field, it is a second-order transition, which does not involve latent heat.

The jump in c_e is due to the fact that the superconducting metal has a new degree of freedom, i.e. the possibility of entering the superconducting state. For 'simple' superconductors, such as Al and Sn, the Bardeen–Cooper–Schrieffer (BCS) theory [18–22] gives:

$$\Delta c_e = 1.43 \cdot \gamma T_C \tag{3.9}$$

where γT_C is the electronic specific heat at T_C in the normal state.

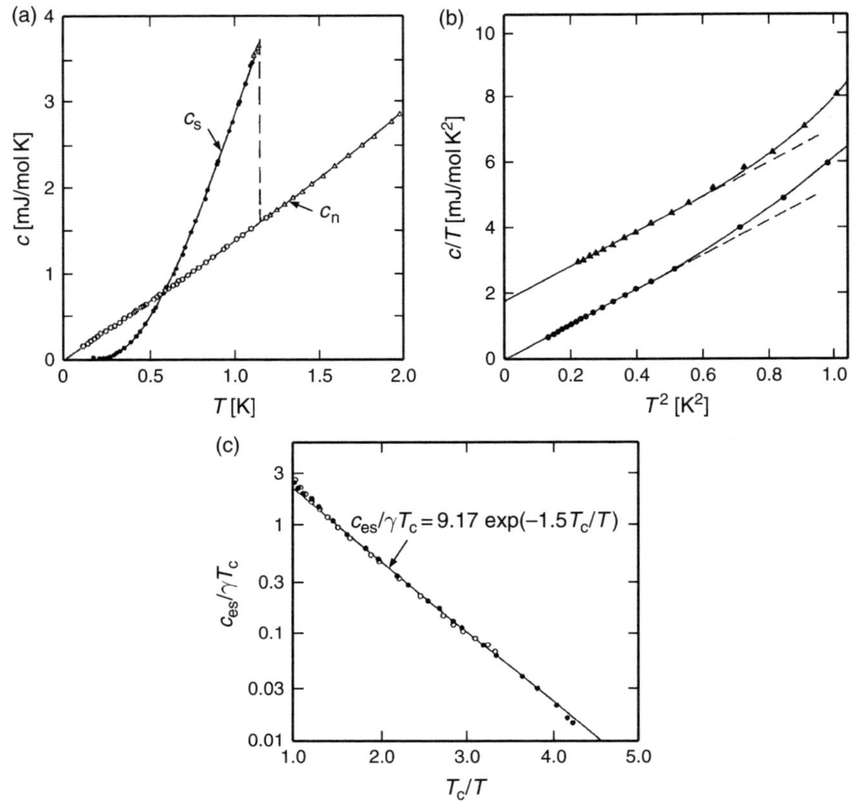

Fig. 3.4. (a) Specific heat of Al in the superconducting state (c_s) and in the normal state (c_n) [17]. (b) Specific heat c of Hg in the normal (▲) and in the superconducting (•) state. The straight lines correspond to eq. (3.11) with $\theta_D = 72\,\text{K}$ and $\gamma = 1.82\,\text{mJ}/(\text{mol K}^2)$ [24]. (c) Electronic specific heat c_{es} of superconducting V (•) and Sn (∘). The full line represents formula (3.10) [25].

Below T_C, the electronic specific heat c_{es} decreases with the temperature as (see Fig. 3.4 (c)):

$$c_{es} \propto \exp\left(-b\frac{T_C}{T}\right) \tag{3.10}$$

Thermodynamic arguments [23] indicate that the transition from the normal to the superconducting state at zero field does not involve a latent heat and therefore must be a higher-order transition. Experimental evidence indicates that it is second-order transition.

The observed specific heat of single-crystal vanadium of Fig. 3.4(c), reported in Fig. 3.5 at $H = 0$, shows the form of the theoretically predicted curve. When the transition takes

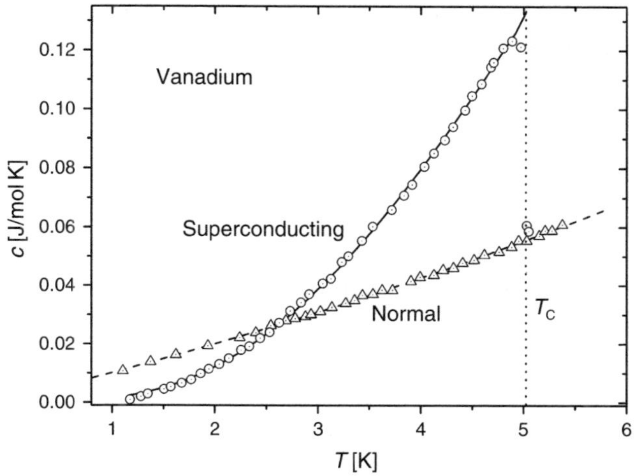

Fig. 3.5. Specific heat of vanadium. Data from [26]. The solid line represents data taken at $H=0$ and illustrates the second-order transition. The dashed curve for normal vanadium was obtained by applying a magnetic field such that $H > H_c$.

place in the presence of a magnetic field, a latent heat is observed and the transition is of the first order. This latent heat is given by:

$$L = T\,(S_n - S_s) = -\left(\frac{VH_cT}{4\pi}\right)\frac{dH_c}{dT} \tag{3.11}$$

which gives:

$$(S_n - S_s) = -\left(\frac{VH_c}{4\pi}\right)\frac{dH_c}{dT} \tag{3.12}$$

where H_c is the temperature-dependent critical field, V is the volume and S_n and S_s are the entropies in the normal and superconducting states.

Since:

$$c = T\frac{dS}{dT} \tag{3.13}$$

$$(c_s - c_n) = \left(\frac{VT}{4\pi}\right)\frac{d}{dT}\left(H_c\frac{dH_c}{dT}\right) = \left(\frac{VH_cT}{4\pi}\right)\frac{d^2H_c}{dT^2} + \left(\frac{VT}{4\pi}\right)\left(\frac{dH_c}{dT}\right)^2 \tag{3.14}$$

At $T = T_C$ and $H_c = 0$:

$$(c_s - c_n) = \left(\frac{VT_C}{4\pi}\right)\left(\frac{dH_c}{dT}\right)^2 \tag{3.15}$$

At very low temperatures, the second derivative term in eq. (3.14) dominates $(d^2H_c/dT^2 < 0)$ and $c_s - c_n < 0$. The difference $(c_s - c_n)$ at T_C has been carefully determined and is 9.75, 10.6 and 41.5 mJ/mol K for indium, tin and tantalum respectively.

Corresponding values computed from eq. (3.15) using values of $\mathrm{d}H/\mathrm{d}T$ from ref. [27] are 9.52, 10.56 and 41.6 mJ/mol K. This change in specific heat is significant, for example 44% in Sn.

3.6 Magnetic specific heat

When a material capable of magnetization is considered, an additional term appears in the thermodynamic expression for the change in total energy of the system. The change is given by:

$$\mathrm{d}Q = T\mathrm{d}S = \mathrm{d}E + p\mathrm{d}V + M\mathrm{d}H \tag{3.16}$$

where the magnetization $M = \chi H$ is the vector sum of the magnetic dipole moments, H is the magnetic field strength and χ is the magnetic susceptibility.

Fundamental definitions for the two primary magnetic heat capacities may be derived [3] and are:

$$\begin{aligned} C_H &= \left(\frac{\mathrm{d}Q}{\mathrm{d}T}\right)_\mathrm{H} = T\left(\frac{\mathrm{d}S}{\mathrm{d}T}\right)_\mathrm{H} \\ C_M &= \left(\frac{\mathrm{d}Q}{\mathrm{d}T}\right)_\mathrm{M} = T\left(\frac{\mathrm{d}S}{\mathrm{d}T}\right)_\mathrm{M} \end{aligned} \tag{3.17}$$

The heat capacity at constant magnetization is independent of an external magnetic field and, as was the case for C_V, appears in most theoretical developments. Heat capacity at constant field strength, C_H, is the quantity generally determined experimentally (corresponds to C_p in the nonmagnetic case). The difference between the heat capacity contribution at low temperatures from ferromagnetic and ferrimagnetic alignments, C_M, and C_H is given by [3]:

$$C_H - C_M = T\left(\frac{\partial M}{\partial T}\right)_H^2 \left(\frac{\partial M}{\partial H}\right)_T^{-1} \tag{3.18}$$

There are several types of magnetic behaviour that affect the specific heat of a material: paramagnetism, ferromagnetism, antiferromagnetism and ferrimagnetism. Diamagnetism, being independent of temperature, causes no specific heat contribution and is not considered.

A material is paramagnetic when a magnetic field tends to align the electronic magnetic moments parallel to the field. The degree of alignment is proportional to the field strength, and the alignment is destroyed by thermal agitation when the external field is removed.

In centimetre-gram-second units (CGS), the susceptibility $\chi = M/H$ is positive for paramagnetic materials. The magnetic moments of ferromagnetic materials spontaneously and cooperatively assume a regular order when the temperature is low enough. The magnetization of these materials is finite even for $H = 0$; this is possible because there is a singularity in the susceptibility at the Curie temperature, T_c [6]. Above T_c, the spontaneous parallel alignment vanishes, leaving the disordered paramagnetic state.

In the presence of a magnetic field, the magnetization of ferromagnetic materials increases rapidly with increasing H and saturates at a very high level compared with materials in the paramagnetic state. A lower energy state exits for some materials when the magnetic moments or spins are aligned antiparallel instead of parallel. The spontaneous antiparallel alignment occurs through cooperative interaction for temperatures below the Néel temperature. Materials that undergo antiparallel alignment of magnetic moments are called antiferromagnetic if the all moments are of the same magnitude and ferrimagnetic if they are of different magnitudes. Macroscopic behaviour shows no net magnetization for antiferromagnetic materials, whereas ferrimagnetic materials behave in a manner similar to ferromagnets. The regularity of the spin alignments mentioned above is destroyed by thermal agitation. It can be demonstrated [28,29] that for ferromagnetic and antiferrimagnetic materials:

$$C_M \propto T^{3/2} \tag{3.19}$$

and for antiferromagnetic alignments:

$$C_M \propto T^3 \tag{3.20}$$

The general behaviour of the specific heat near a magnetic transition is illustrated in Fig. 3.6 for $MnCO_3$. This carbonate undergoes a paramagnetic to antiferromagnetic transition at 24.9 K as the temperature is lowered.

The difference between the $MnCO_3$ and $CaCO_3$ data is because $MnCO_3$ undergoes a transition and $CaCO_3$ does not.

These phenomena are theoretically described by several models. Theories differ both for the type of interaction and for the 'geometry' and dimensionality of the magnetic

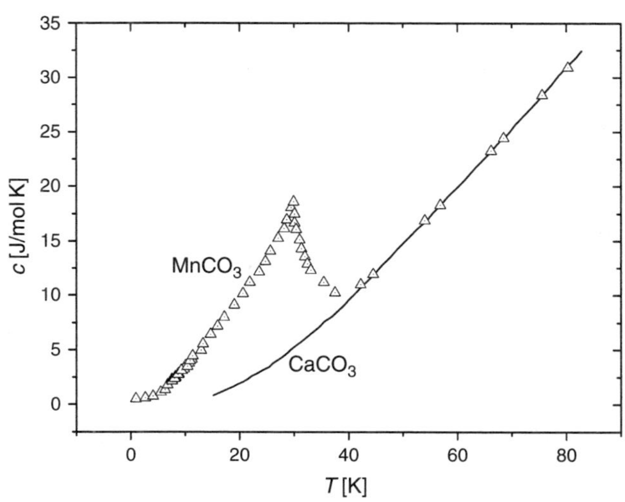

Fig. 3.6. Specific heats of $MnCO_3$ and $CaCO_3$ illustrate the general effect of a magnetic transition. Data from [30]. The Néel temperature for $MnCO_3$ is 29.4 K [30].

lattice (one, two or three dimensions (see ref. [29]). Changes in the dimensionality of the 'magnetic lattice' have a dramatic effect upon the thermodynamic properties.

The magnetic contributions to specific heat discussed above are the result of cooperative interactions and are initiated by lambda-type transitions. One more magnetic effect occurs, which is due to non-cooperative changes in order. This phenomenon is known as the Schottky effect and is observed in paramagnetic salts and in some ferromagnetic metals. It is a result of electronic energy degeneracies in the paramagnetic materials and nuclear energy degeneracies in the ferromagnetic materials.

The observed Schottky-specific heat is due to changes in internal energy that occur when nearly adjacent energy levels are occupied. The degenerate energy levels may be caused by external or internal magnetic fields.

If atoms having magnetic moments are in a magnetic field, the effect of the crystal field is to produce a set of orbital levels for the single magnetic ion. An atom with a magnetic moment M has $(2M+1)$ possibilities of orientation of the magnetic moments with respect to the magnetic field. At a given temperature, only the ground state and the excited states lower than $k_B T$ will be occupied. For the magnetic properties, only the levels with energies not much larger than $k_B T_c$ must be considered. This gives an additional contribution to the specific heat. In the simplest case of spin $I = 1/2$, there are two possible spin orientations with equal degeneracy and energetic splitting ΔE. The contribution to the specific heat (Schottky anomaly) is given by [1,7,31]

$$c_m = k_B N_0 \left(\frac{\Delta E}{k_B T}\right)^2 \frac{\exp\left(\frac{\Delta E}{k_B T}\right)}{\left(1+\exp\left(\frac{\Delta E}{k_B T}\right)\right)^2} \qquad (3.21)$$

and shown in Fig. 3.7.

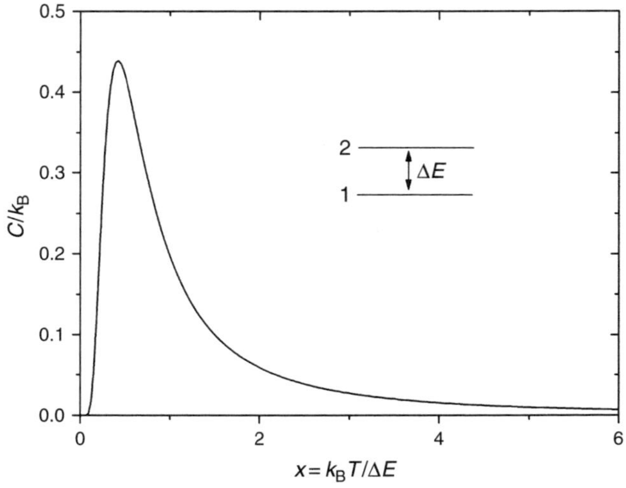

Fig. 3.7. Specific heat c of a two-level system with energy separation ΔE.

From formula (3.21), we see that the temperature at which the maximum of the magnetic contribution to the specific heat occurs is determined by the energy splitting ΔE of the levels:

$$\left(\frac{\Delta E}{k_B T}+2\right)=\left(\frac{\Delta E}{k_B T}-2\right)\exp\left(\frac{\Delta E}{k_B T}\right) \tag{3.22}$$

Very often ΔE is small in comparison with the thermal energy $k_B T$ and from formula (3.21) we obtain:

$$c_m = k_B N_0 \left(\frac{\Delta E}{2k_B T}\right)^2 \quad \text{for } T \gg \frac{\Delta E}{k_B} \tag{3.23}$$

corresponding to the right part of Fig. 3.7 well beyond the maximum.

In the case of a metal, for $\Delta E/k_B = T = \theta_D$:

$$c_m = \gamma T + \delta T^{-2} \tag{3.24}$$

For example, the specific heat of stainless steel 304, between 70 and 700 mK, is [32] (see Fig. 3.8):

$$c = \left(465 \cdot T + 0.56 \cdot T^{-2}\right) \quad [\mu\text{J/g K}] \tag{3.25}$$

The magnetic specific heats of some alloys containing paramagnetic atoms together with copper for comparison are shown in Fig. 3.8. Note that below 0.1 K, magnetic materials as manganin have a specific heat 10^3 higher than copper.

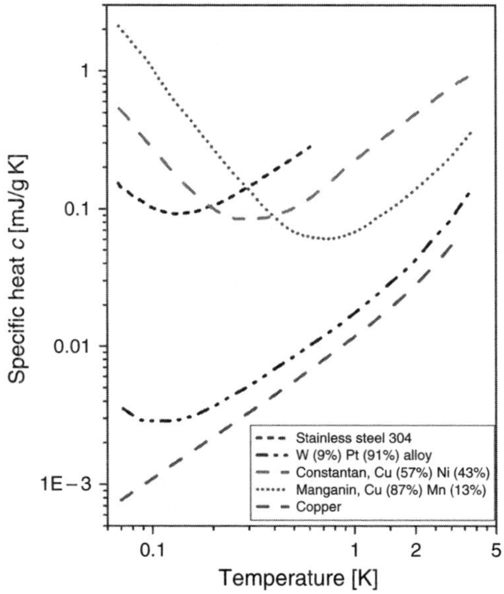

Fig. 3.8. Specific heat of: stainless steel 304 [32], costantan (57% Cu, 43% Ni), manganin (87% Cu, 13% Mn), 9% W, 91% Pt alloy [33] and copper [8].

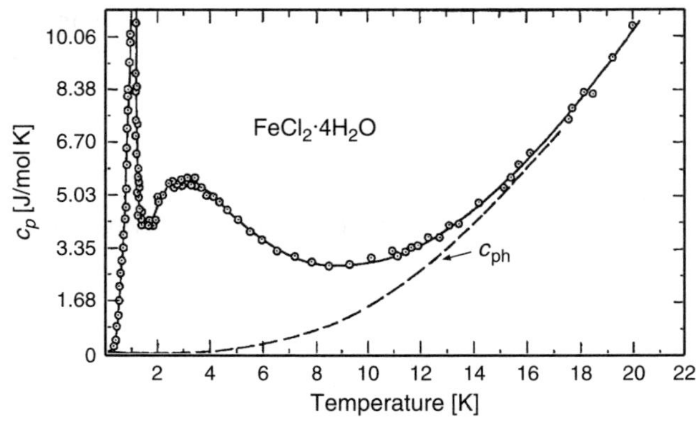

Fig. 3.9. Specific heat of $FeCl_2 \cdot 4H_2O$ drawn from data obtained by Friedberg et al. [35] and Raquet and Friedberg [36]. The peak near 1 K is only partly shown, the highest value of the specific heat measured being above 6 cal/mol K [29].

Data of Fig. 3.8 show that the use of wires of manganin or constantan is to be avoided in some applications as wiring of detectors (see Chapter 15).

In general, the spin may be larger than $^1/_2$, and there are more than two levels, but results are qualitatively similar [34]. An useful discussion about the nuclear magnetic moment contribution can be found in ref. [34].

An example of magnetic contributions to the specific heat is reported in Fig. 3.9 that shows the specific heat of $FeCl_2\,4H_2O$, drawn from data of ref. [35,36]. Here the Schottky anomaly, having its maximum at 3K, could be clearly resolved from the lattice specific heat as well as from the sharp peak at ~1K, which is due to a transition to antiferromagnetic order (lambda peak).

As we can see from Fig. 3.9, the magnitude of the Schottky effect may be several orders greater than that due to lattice and electronic contributions even at liquid helium temperatures.

Contribution of various origins is described in ref. [29].

3.7 Specific heat due to the amorphous state

In the non-crystalline or amorphous solids (glasses and polymers) an arrangement of atoms in a lattice with a periodic order does not exist.

Figure 3.10 shows the typical dependence on temperature of the specific heat of an amorphous and a crystalline polymer. For both materials, the specific heat has a steep dependence on temperature, but the behaviour is more complex in the case of the amorphous material.

In general, for temperatures below about 80 K, the shape of the $c(T)$ curves for the various amorphous polymers do not differ very much and are at a first approximation independent of the chemical composition. Until the beginning of the 1970s, an explanation

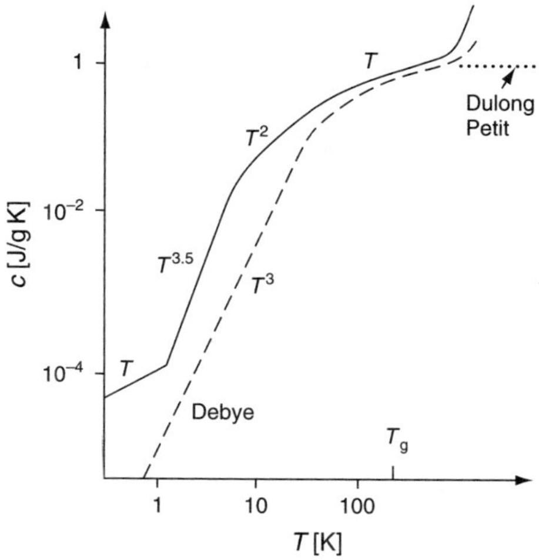

Fig. 3.10. Typical temperature dependence of the specific heat for an amorphous (——), and a crystalline polymer (- - - -).

of how the difference in the microscopic structure could produce such a difference in the material properties did not exist. It was believed that the Debye theory should explain also the properties of the amorphous solids since at low temperature, the wavelength and mean free path of phonons increase; as a consequence, defects and irregularity in the structure of the solid should not play an important role. The first systematic work of measure and collection of data both of specific heat and of thermal conductivity of non-crystalline solids was due (1971) to Zeller and Pohl [37]. Their revolutionary hypothesis was that the excess in specific heat observed below 1 K in amorphous polymers and glasses was to be attributed to low-energy localized excitations (e.g. vibrations of atoms or of groups of atoms) capable of producing scattering centres for phonons [37]. Only the measurements of thermal conductivity carried out by Zaitlin and Anderson in 1975 [38] demonstrated that below 1K, the acoustic phonons are the main responsible for the heat transfer. Instead, the excitations that produce the excess (the linear contribution) of specific heat cannot carry thermal energy because they are to be considered as localized excitations.

These excitations may be represented by two-level systems (TLS) or more generally by strongly anharmonic oscillators. The linear contribution to the specific heat is to be attributed to the thermal excitation of the TLS. A physical explanation is given in the frame of the tunnelling model proposed independently by Anderson et al. [39] and Phillips [40] in 1972, with the aim of explaining the measured thermal and acoustic properties of amorphous materials. According to this theory, because of the structural disorder, groups of atoms have more than one possible position corresponding to a small energy difference. The typical excitation energy of TLS is of the order of $E \leq 10^{-4}$ eV, and the quantum tunnelling transition between the two levels can take place only with absorption or emission of phonons in order to conserve the energy.

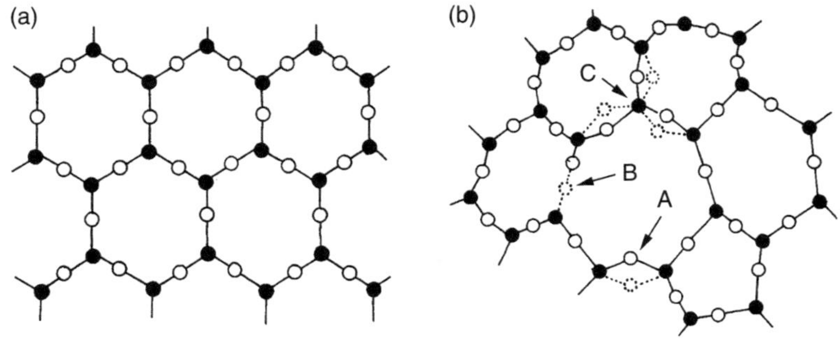

Fig. 3.11. Schematic two-dimension representation of the structure of cristobalite (a crystalline form of SiO_2) and of vitreous SiO_2. Si atoms are represented by full circles, oxygen by open circles. A, B and C represent three cases of double possible equilibrium positions for the atoms of the material in the amorphous state: A, transversal displacement; B, longitudinal displacement; C, small-angle rotation of the SiO_4 tetrahedron.

As an example, in Fig. 3.11, a schematic two-dimension representation of the structure of cristobalite (a crystalline form of SiO_2) and of vitreous SiO_2 is shown. A, B and C represent three cases of double possible equilibrium positions for the atoms of the material in the amorphous state [41]. Atoms can tunnel from one position to another. The thermal excitation of TLS is responsible for the linear contribution to the specific heat of amorphous solids.

The model proposed by Anderson and Phillips gives a phenomenological explanation of the properties of the amorphous materials without supplying a detailed microscopic description [42]. Low-temperature measurements of the specific heat of amorphous solids have however shown that instead of a linear contribution as expected from the TLS theory, the best representation of data is obtained with an overlinear term of the type [43,44]:

$$c = a \cdot T^{1+\delta} \quad \text{with } 0.1 < \delta < 0.5 \tag{3.26}$$

This does not mean that the TLS theory is wrong, but only some approximations are to be revised. In fact, the TLS theory does not take into account the processes of absorption and emission of a phonon by a TLS which lead to relaxation phenomena in the tunnelling levels.

The speed and time by which the perturbed TLS systems relax to the equilibrium thermal populations depend on the TLS characteristics and in particular on the coupling energy of tunnelling states.

Even Anderson et al. [39] pointed out that an important consequence of the tunnelling model was the (logarithmic) dependence of the measured specific heat on the time needed for the measurement of c. The latter phenomenon was due to the large energy spread and relaxation time of TLS. In 1978, Black [45], by a critic revision of the tunnelling theory, has been able to explain the time dependence of the low-temperature specific heat.

We wish also to mention the discovery of an unusual magnetic field dependence of specific heat in some multicomponent glasses in the 0.3–4 K range [46]. A theoretical explanation of the phenomenon can be found in ref. [[47] pp. 17–46].

In the case of polymers, the complex chain structure allows for a variety of vibrational motions which give a contribution to the specific heat and characterize other properties of the material. Vibrations taking place along a chain or among the chains are called stretching or longitudinal vibrations, whereas the bending or transversal vibrations, characteristic of the specific polymer, depend on the intrinsic bending stiffness of the polymeric chains [48]. The latter ones are responsible for peculiar temperature dependences of the specific heat and of the thermal expansion of some polymeric materials (e.g. they may produce negative expansion coefficients). Even these types of vibrations can be described as phonons whose density distribution depends on the vibration mode. For each vibration mode, a state density function can be obtained together with the respective contribution to the specific heat. Each mode has its own Debye temperature θ_{Di}. Above θ_{Di}, the specific heat contribution tends to the Dulong–Petit limit. In fact, for each mode, there are two limiting cases [49–52]:

$$\begin{cases} T \gg \theta_{Di} \rightarrow c_i = \text{constant} \\ T \ll \theta_{Di} \rightarrow c_i \propto T^{n+1} \end{cases} \tag{3.27}$$

with $0 < n < 2$; n depends on the vibrational modes that contribute to the specific heat.

The overall specific heat of a polymer is given by a combination of the various contributions to the specific heat of longitudinal and transversal phonons. At temperatures below 1 K, the linear contribution due to the TLS must be added.

All amorphous materials, in summary, show a specific heat with a cubic and an overlinear contribution:

$$c = \beta \cdot T^3 + a \cdot T^{1+\delta} \tag{3.28}$$

The almost linear contribution and the heat release must be carefully considered when an amorphous material is used at very low temperatures.

3.8 Data of specific heat

Data of specific heat can be found in ref. [53–58]. Some of these data are shown in Fig. 3.12.

From Fig. 3.12, we see that at fixed (low) temperature, the specific heat of various solids range over many orders of magnitude. Low specific heat materials (high Debye temperature) are for example very important in the realization of detectors (see Chapter 15). High specific heat materials are essential as regenerators in cryocoolers (see Chapter 5 and Fig. 3.13).

Useful data of specific heat of technical solids above 25 K are reported in Table 3.2.

A special attention is to be devoted to copper, which is very often used in a cryogenic apparatus. The low-temperature specific heat of copper is usually considered as given by: $c = 10^{-5}\,T$ [J/g K]. However, an excess of specific heat has been measured, as reported in the literature [59–69]. For $0.03\,\text{K} < T < 2\,\text{K}$, this increase is due to hydrogen or oxygen impurities, magnetic impurities (usually Fe and Mn) and lattice defects [59–66]. The increase of copper specific heat observed in the millikelvin temperature range is usually attributed to a Schottky contribution due to the nuclear quadrupole moment of copper [67,68].

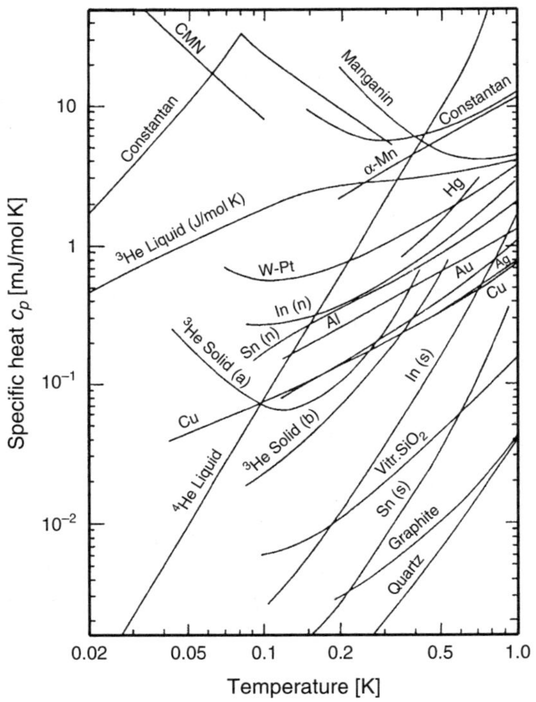

Fig. 3.12. Specific heat of some materials below 1 K [55].

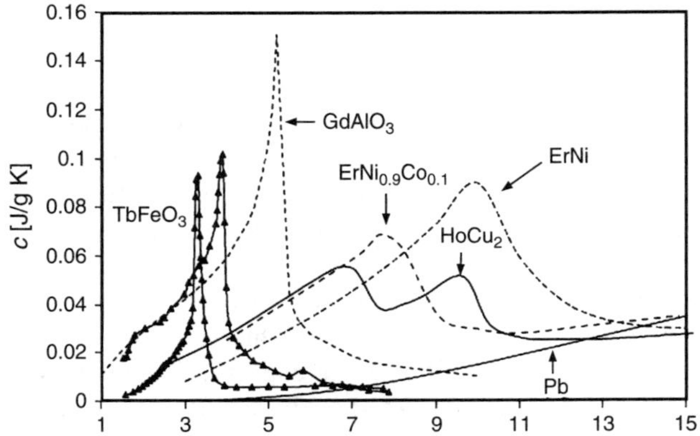

Fig. 3.13. Specific heat of some materials used in regenerators for cryocoolers [70].

Table 3.2
Specific heat c_P [J/g/K] of technical solids: various solids [56]; metals, alloy and non-metals [58]; metals [71]; polymers [47]

Material	25 K	50 K	75 K	100 K	150 K	200 K	293 K
Metals							
Al	0.0175	0.142	0.322	0.481	0.683	0.797	0.897
Cu	0.015	0.097	0.187	0.252	0.322	0.355	0.383
Fe	0.0080	0.051	0.136	0.216	0.324	0.384	0.444
Nb	0.020	0.085	0.147	0.188	0.230	0.248	0.262
Ni	0.0098	0.069	0.156	0.232	0.329	0.383	0.435
Si	0.0085	0.078	0.170	0.260	0.425	0.557	0.694
Ti	0.0137	0.098	0.210	0.300	0.408	0.466	0.518
W	0.0041	0.032	0.064	0.087	0.112	0.123	0.133
Alloys							
Al2024	—	—	—	0.46	0.65	0.73	0.84
CuZn (65/35)	0.022	0.118	0.21	0.27	0.33	0.36	0.377
Constantan	0.013	0.08	0.17	0.24	0.32	0.36	0.41
Inconel 718	—	0.07	0.16	0.27	0.36	0.40	0.43
Nb-38Ti	0.03	0.11	0.24	—	—	—	—
SnPb (50/50)	0.062	0.116	0.140	0.152	0.163	0.170	0.178
S.S. 304/316	0.019	0.092	0.19	0.28	0.35	0.42	0.47
Ti-6Al-4V	—	—	0.21	—	0.40	0.49	0.55
Non-metals							
Sapphire	0.0014	0.0148	0.0558	0.126	0.313	0.501	0.763
MgO	0.0019	0.0207	0.085	0.195	0.0449	0.661	0.916
Pyrex	0.043	—	—	0.28	0.406	0.533	0.72
Silica	0.038	0.111	0.188	0.268	0.420	0.546	0.728
ZrO_2	0.009	0.041	0.095	0.15	0.26	0.35	0.45
Polymers							
Epoxy	0.13	0.27	0.39	0.48	—	1.0	1.3
G10 (GFRP)	—	0.3	0.4	0.5	—	1.0	1.5
Nylon 6	—	—	0.47	—	0.81	1.01	1.5
Stycast	0.032	0.088	0.15	0.22	—	—	—
Teflon	0.10	0.21	0.29	0.39	0.56	0.72	1.0

3.9 Thermal expansion

For an isotropic material, the linear coefficient of thermal expansion α at constant pressure is:

$$\alpha_p = \frac{1}{L}\left(\frac{\mathrm{d}L}{\mathrm{d}T}\right)_p \tag{3.29}$$

As in the case of c_P and c_V, α_p does not differ significantly from α_V at low temperatures.

For anisotropic solids, there are two or three (depending on the symmetry of the crystal) principal linear coefficients. For isotropic solids, the volumetric thermal expansion is $\beta = 3\alpha$.

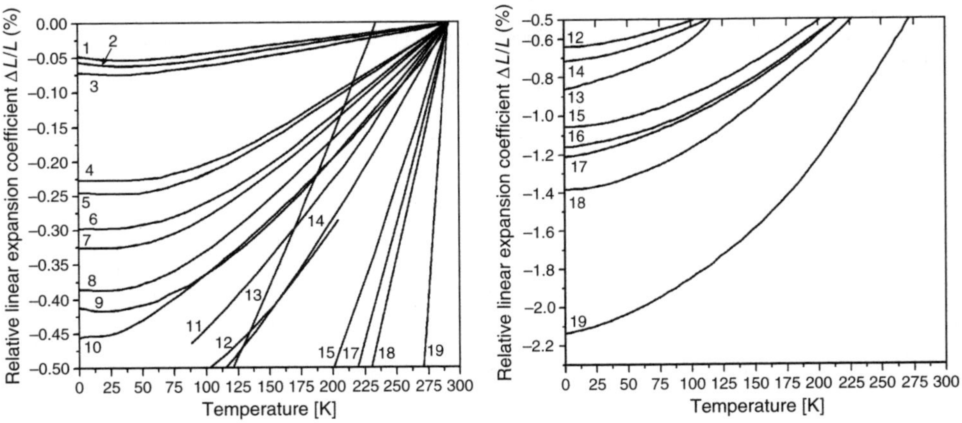

Fig. 3.14. Relative linear thermal expansion coefficient of: **(1)** Invar, **(2)** Pyrex, **(3)** W, **(4)** Ni, **(5)** $Cu_{0.7}Ni_{0.3}$, **(6)** stainless steel, **(7)** Cu, **(8)** brass, **(9)** Al, **(10)** Torlon, **(11)** soft solder, **(12)** Vespel SP-22, **(13)** Hg, **(14)** In, **(15)** Araldite, **(16)** Stycast 1266, **(17)** PMMA, **(18)** Nylon, **(19)** Teflon [60]. Some additional data are: Ag between **(8)** and **(9)**; Stycast 2850 GT slightly larger than **(9)**. The integral contraction between 300 and 4 K is $10^3 \Delta L/L = 11.5$, 4.2, 6.3 and 5.7 for Stycast 1266, Stycast 2850 GT, Vespel SP-22 and solders respectively [69,72,73].

Thermal contraction is due, as we know, to the anharmonicity of the potential an atom experiences in a crystal and vanishes for $T \to 0$ (e.g. $\alpha_p = 3 \times 10^{-10}$/K for the Cu at 1 K [74]). For the microscopic theory of thermal contraction, see e.g. ref. [1]. The thermal contraction below room temperature of several materials is shown in Fig. 3.14.

We can approximately divide the materials into three groups: (1) commercial alloys and glasses with an extremely small expansion coefficient; (2) metals that exhibit an integrated contraction of 0.2–0.4% between 300 and 4 K; (3) organic materials with integrated contraction of 1–2%. A fourth group not shown in Fig. 3.14 includes more recently developed composite materials (e.g. carbon fibres in epoxy) which may show extremely small, and even negative [75,76], expansion coefficients as Kevlar fibres. These recent composites, together with other organic materials, are of particular interest for cryogenics because of their very low thermal conductivity. It is also worth noting that some polymeric materials as Torlon show thermal contraction coefficients as low as metals [77].

Joining materials with different thermal expansion in a cryogenic structure which must stand repeated thermal cycles needs a careful attention to avoid stresses and breaking of the joint. In particular, the vacuum tightness at low temperature may be jeopardized.

The measurement of the linear expansion coefficient can be carried out by several methods (see Chapter 13), for example, by means of an interferometric dilatometer [77], a capacitance dilatometer [78] or a SQUID dilatometer [74]. The latter can achieve resolutions as small as 2×10^{-14} m.

Data of linear thermal contraction coefficients are reported in ref. [34,79,80] and in Table 3.3.

Table 3.3
Linear thermal contraction $\Delta L/L$ (%) from 293 K, and a at 293 K (10^{-6} K^{-1}) [79,81–85]

	$\Delta L/L$ (%)						$a(10^{-6}\,K^{-1})$
Material	4 K	40 K	80 K	100 K	150 K	200 K	293 K
Metals							
Al	0.414	0.412	0.390	0.369	0.294	0.201	22.9
Cu	0.326	0.323	0.302	0.283	0.221	0.149	16.65
Fe	0.204	0.202	0.195	0.185	0.149	0.102	11.8
Ni	0.229	0.227	0.217	0.206	0.165	0.112	12.8
Si	0.022	0.022	0.023	0.024	0.024	0.019	2.56
Ti (pc)	0.151	0.150	0.142	0.134	0.107	0.073	8.6
W	0.088	0.087	0.081	0.076	0.059	0.040	4.42
Alloys							
Al2024	0.396	0.394	0.372	0.351	0.278	0.190	21.2
Al5083	0.415	0.413	0.390	0.368	0.294	0.201	22.8
Berylco 25	0.316	0.315	0.296	0.277	0.218	0.151	17.9
Brass (65/35)	0.384	0.380	0.350	0.326	0.253	0.169	19.6
$Fe_{64}Ni_{36}$	0.045	0.048	0.048	0.045	0.030	0.020	~1
Hastelloy C	0.218	0.216	0.204	0.193	0.154	0.105	12.8
Inconel X	0.229	0.228	0.217	0.205	0.164	0.112	13.0
Nb-45Ti	0.188	0.184	0.167	0.156	0.117	0.078	≈10
Cu/NbTi	0.265	0.262	0.245	0.231	0.178	0.117	≈12
S.S. 304/316	0.296	0.296	0.278	0.260	0.203	0.138	15.8
S.S. 310	0.273	0.270	0.252	0.237	0.187	0.127	14.5
Ti-6Al-4V	0.173	0.171	0.162	0.154	0.118	0.078	8
Non-metals							
Sapphire (‖)	0.0715	0.0715	0.0705	0.069	0.061	0.045	5.80
Sapphire (⊥)	0.0605	0.0605	0.0595	0.0585	0.052	0.039	5.06
MgO	0.139	0.0139	0.137	0.133	0.114	0.083	10.3
Pyrex	0.056	0.057	0.054	0.050	0.0395	0.027	3.0
Silica	−0.0005	−0.0003	0.0001	0.0013	0.0030	0.0031	0.45
ZrO_2 (stab)	0.131	0.130	0.124	0.118	0.098	0.068	8.0
YBCO-123(ab)	0.15	0.15	—	0.14	0.115	0.08	10
YBCO-123(c)	0.36	0.35	0.33	0.31	0.25	0.17	17
Polymers							
Torlon	0.448	0.434	0.387	0.358	0.279	0.191	~24
Araldite	1.06	1.02	0.935	0.88	0.71	0.505	~60
Nylon	1.39	1.35	1.25	1.17	0.95	0.67	~80
Stycast	0.44	0.43	0.40	0.38	0.32	0.225	30
Teflon	2.14	2.06	1.93	1.85	1.60	1.25	~200
G10 (‖)	0.705	0.69	0.64	0.60	0.49	0.35	~40
G10 (⊥)	0.24	0.235	0.21	0.20	0.155	0.11	~12

3.10 Thermal conductivity

The heat transfer in a solid is due both to lattice vibrations (phonons) and to conduction electrons. Experiments show that in reasonably pure metals, nearly all the heat is carried by the electrons. In impure metals, alloys and semiconductors, however, an appreciable proportion of the thermal conductivity is due to lattice conduction. In dielectric crystals and amorphous insulating solids, of course, all the heat is transported by the phonons.

These carriers of heat do not move 'balistically' from the hotter part of the material to the colder one. They are scattered by other electrons, phonons, defects of the lattice and impurities. The result is a diffusive process which, in the simplest form, can be described as a gas diffusing through the material. Hence, the thermal conductivity k can be written as:

$$k = \frac{1}{3}\frac{c}{V_{\mathrm{m}}} \cdot v\lambda \tag{3.30}$$

where λ is the mean free path, v is the velocity of the carriers, c is the specific heat and V_{m} is the molar volume. Typical values of the phonon velocity is the sound speed $3 \div 5 \times 10^5$ cm/s; for the electrons, the velocity is $10^7 \div 10^8$ cm/s. Both velocities are independent of temperature at low temperatures. Thus, for the evaluation of k in (3.30), it is necessary to determine λ.

The main scattering processes limiting the thermal conductivity are phonon–phonon (which is absent in the harmonic approximation), phonon defect, electron–phonon, electron impurity or point defects and more rare electron–electron. For both heat carriers, the thermal resistivity contributions due to the various scattering processes are additive. For the phonons:

$$\frac{1}{k_{\mathrm{ph}}} = R_{\mathrm{pp}} + R_{\mathrm{pe}} + R_{\mathrm{pi}} + R_{\mathrm{pb}} + R_{\mathrm{pd}} \tag{3.31}$$

where R_{pp} is caused by phonon–phonon scattering (umklapp process), R_{pe} by scattering phonon–electron (present only in metals and electrical conductors), R_{pi} by scattering from impurities or point defects, R_{pb} by scattering from boundaries and R_{pd} by scattering from dislocations.

In the case of metals and electrical conductors, for the electrons, the thermal resistivity is:

$$\frac{1}{k_{\mathrm{e}}} = R_{\mathrm{ei}} + R_{\mathrm{ep}} \tag{3.32}$$

where R_{ei} is caused by impurity scattering and R_{ep} by scattering from phonons.

Since the number of phonons increases with temperature, the electron–phonon and phonon–phonon scattering are temperature dependent. The number of defects is temperature independent and correspondingly, the mean free path for phonon defect and electron defect scattering does not depend on temperature.

Each of the two terms in formula (3.32) depends on temperature, and its contribution has a different importance depending on the temperature range of interest.

In the following, we shall describe separately the temperature dependence of the contributions to the thermal conductivity for the two 'heat carriers'. In the case of phonons, the Debye temperature θ_{D} will be taken as a reference in analysing the temperature dependence of the thermal conductivity.

3.10.1 Phonons

At temperatures well below the Debye's temperature, the lattice conductivity is given by:

$$k_{\text{ph}} = \frac{1}{3}\frac{c_{\text{ph}}}{V_{\text{m}}} v_{\text{s}} \lambda_{\text{ph}} \propto T^3 \lambda_{\text{ph}}(T) \quad \text{for } T \leq \frac{\theta_{\text{D}}}{10} \tag{3.33}$$

(a) Intermediate temperatures, approximately: $\theta_{\text{D}}/30 \leq T \leq \theta_{\text{D}}/10$

In this temperature range, the phonon–phonon scattering dominates and λ_{ph} decreases with increasing temperature because the number of phonon increases [1,6–9,31,86] (see Figs 3.15 and 3.16). Hence, in this temperature range, the thermal conductivity decreases with increasing temperature. The thermal conductivity of several materials at $T > 2\,\text{K}$ is shown in Fig. 3.16.

(b) Low temperatures: $T < \theta_{\text{D}}/30$

In this temperature range, the number of phonons is small, and their scattering is due to lattice defects or to crystal boundaries. Of the two processes of scattering, the latter is of more importance since, at low temperatures, the dominant phonon wavelength is larger than the size of the lattice imperfections. As a consequence λ_{ph} is usually temperature independent. Hence, the temperature dependence of the thermal conductivity is that of the specific heat:

$$k_{\text{ph}} \propto c_{\text{ph}} \propto T^3 \tag{3.34}$$

The result of the considerations made in (a) and (b) is that the phonon thermal conductivity goes through a maximum as illustrated in Figs 3.15 and 3.16. It is to be noted that,

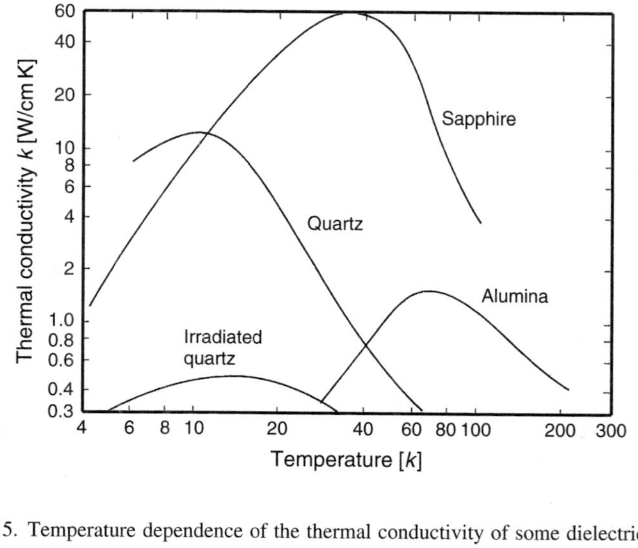

Fig. 3.15. Temperature dependence of the thermal conductivity of some dielectric solids.

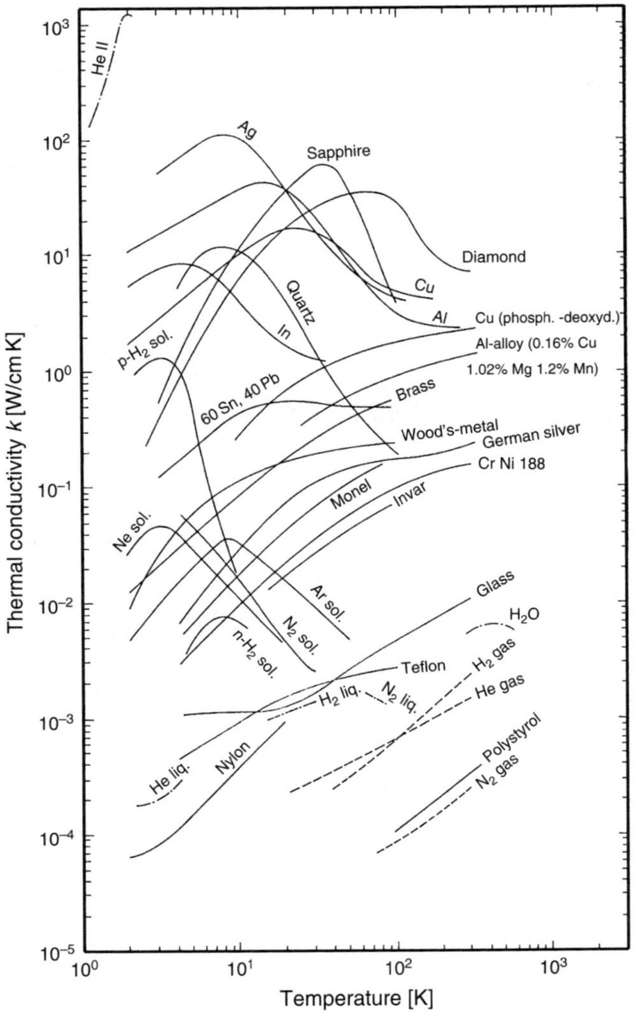

Fig. 3.16. Thermal conductivity of various materials at $T > 2\,K$ [87–91].

owing to the difference in the number of defects, the thermal conductivity of nominally identical samples can vary considerably. Note also that the thermal conductivity of a crystal with very few defects and impurities can be rather high (100 W/cm K), as shown in Figs 3.15 and 3.16, comparable with that of metals as Cu or Al.

On the other hand, for an amorphous insulator, λ_{ph} is determined by the scattering of phonons with defects and can be very small, even approaching atomic distances.

In Fig. 3.17, we see the thermal conductivity of $(BaF_2)_{1-x}(LaF_3)_x$ [92]. The pure BaF_2 ($x = 50$) has a thermal conductivity typical of perfect crystals. The addition of LaF_3 leads to a thermal conductivity approaching that of an amorphous solid (a-SiO_2).

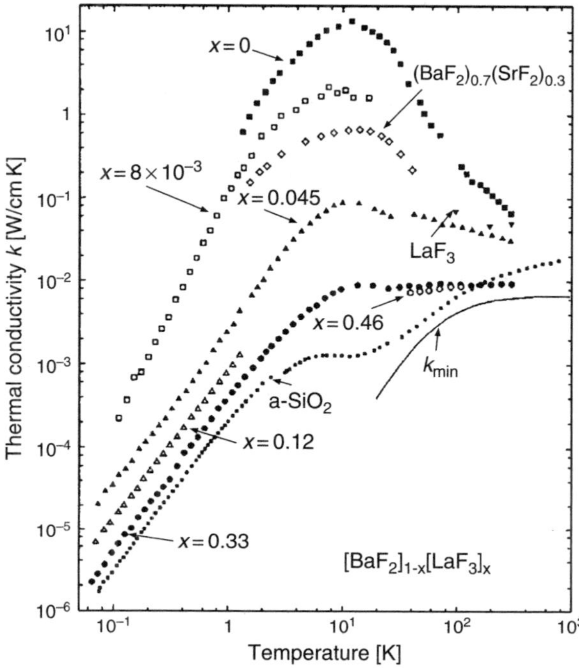

Fig. 3.17. Thermal conductivity of $(BaF_2)_{12x}(LaF_3)_x$ [92,93].

Note that the conductivities for $x = 50.33$ and 0.46 are very close, approaching the minimum thermal conductivity k_{min} at high temperatures [92]. Undoped LaF_3 again has a large thermal conductivity. The addition of a large concentration of (divalent) SrF_2 to BaF_2 has a relatively small effect on the thermal conductivity. A glass-like thermal conductivity is not reached.

In some cases (amorphous materials as polymers or glasses), the scattering of phonons with tunnelling states is also relevant. This process leads to an approximate [40–42,94,95] dependence as T^2 of k (see Table 3.4) and to a plateau in the $2\,K < T < 20\,K$ range (see Fig. 3.18).

3.10.2 Electron thermal conductivity

Usually in a metal, this contribution is much larger than the lattice contribution due to the strong difference in the velocity of heat carriers. Again we wish to point out that there are two phenomena which produce the scattering of the electrons as heat carriers:

1. The lattice may be distorted because of several reasons as vacancies, interstitials, dislocations and impurities. These lattice defects cause the so-called impurity scattering which produces the term R_{ei}. At low temperatures, R_{ei} is the constant electronic thermal resistance typical of metals.
2. The other contribution to R_{ep} is due to phonons.

Table 3.4
Thermal conductivity $k = a \cdot T^n$ [W/cm K]

Material	Temperature range	a	n
Nylon [96]	0.2–0.8 K	2.6×10^{-5}	1.75
Nylon [90]	2–20 K	$2\ 10^{-5}$	1.31
Polyetheretherketone (PEEK) [97]	0.1–1 K	1.87×10^{-5}	1.47
Polymethylmethacrylate (PMMA) [98]	0.3–0.7 K	2.8×10^{-4}	1.84
Polymethylmethacrylate (PMMA) [99]	0.1–2 K	1.9×10^{-4}	1.65
Polymethylmethacrylate (PMMA) [100]	0.4–1 K	3.3×10^{-4}	1.81
Polymethylmethacrylate (PMMA) [101]	0.05–0.5 K	2.9×10^{-4}	1.77
Polypropylene (PP) [102]	0.1–4 K	2.74×10^{-5}	1.28
Polystyrene (PS) [98]	0.3–0.7 K	1.9×10^{-4}	1.93
Polystyrene (PS) [100]	0.4–1 K	2×10^{-4}	1.87
Polystyrene (PS) [101]	0.05–0.5 K	1.56×10^{-4}	1.78
Teflon [103]	0.17–4 K	3×10^{-5}	2.0
Teflon [96]	0.3–0.7 K	3.8×10^{-5}	2.4
Torlon 4203 [104]	0.1–1 K	7.72×10^{-5}	2.23
Torlon 4203 [78]	0.1–5 K	6.13×10^{-5}	2.18
Vespel SP1 [94]	0.1–1 K	1.8×10^{-5}	1.2
Vespel SP22 [105]	0.07–2 K	1.7×10^{-5}	1.85
Vespel SP22 [94]	0.1–2 K	1.7×10^{-5}	2.0
Vespel SP5 [91]	0.1–1 K	3×10^{-5}	1.73

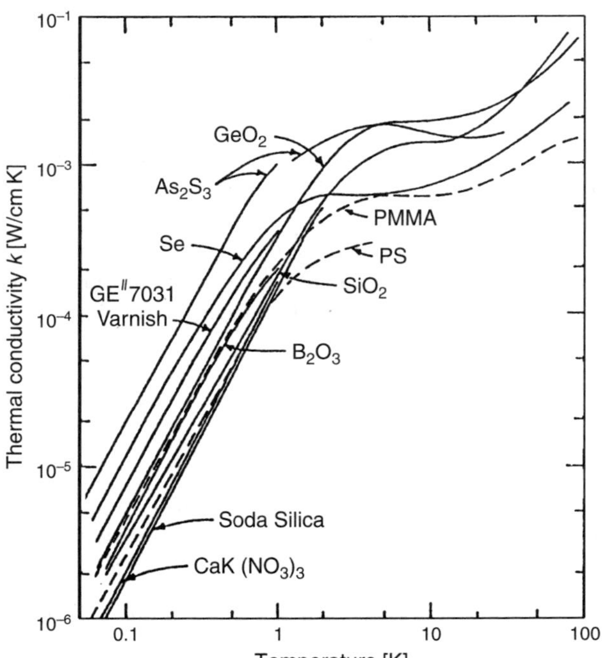

Fig. 3.18. Thermal conductivity of some amorphous solids [100].

The two contributions give rise to the electron thermal conductivity k_e (eq. (3.32)).

In the scattering of electrons by impurities, the mean free path of electrons does not change with temperature ($\lambda =$ constant in eq. (3.30)) and $R_{ei} \propto 1/c$ and $R_{ei} = a/T$.

About R_{ep}, it decreases as temperature decreases, due to the fact that the number of phonons decreases. A full treatment of the problem, however, can only be obtained by solving the Boltzmann transport equation, which has only been solved for the case of quasi-free electrons. Further information and approximate solutions can be found in ref. [7,106,107]. The general result of these calculations shows that at low temperature ($T < \theta_D/10$), the thermal resistance R_{ep} is of the form $b \cdot T^2$.

Thus, the total electronic thermal resistance is of the form:

$$\frac{1}{k_e} = R_{ei} + R_{ep} = \frac{a}{T} + b \cdot T^2 \tag{3.35}$$

At low temperature ($T \leq 20\,\text{K}$):

$$k_e \propto T \tag{3.36}$$

At high temperatures ($T \geq 20\text{K}$), the electron–phonon scattering is dominant and k decreases with T. Hence, we find a maximum of thermal conductivity (see Figs 3.16, 3.19 and 3.20) which is around 10 K for pure metals and 40 K for alloys. For example, in the case of Al1050 Al alloy, where the thermal conduction is mainly due to electrons:

$$k(T) = \frac{T}{(a + b \cdot T^3)} \tag{3.37}$$

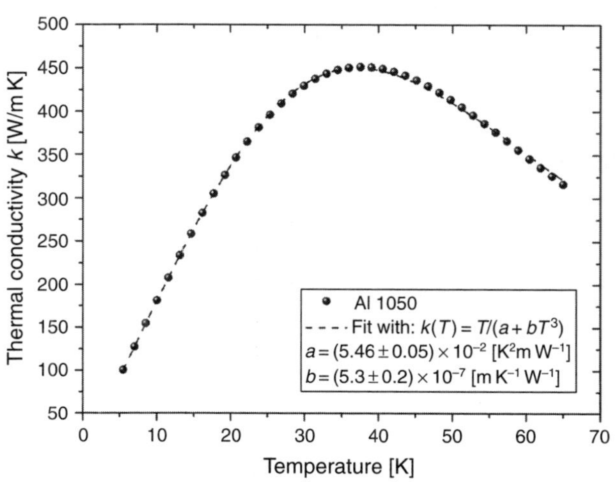

Fig. 3.19. Thermal conductivity of Al 1050 as a function of T [108].

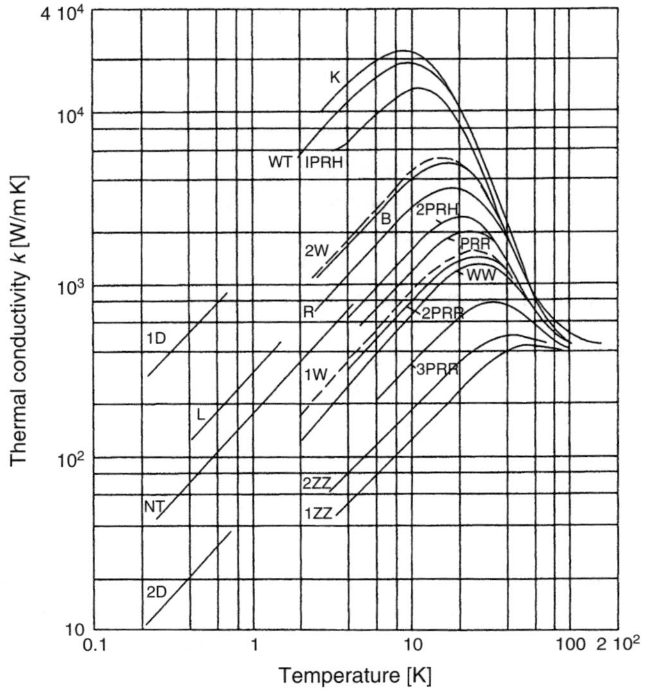

Fig. 3.20. Thermal conductivity of copper samples with residual resistivity ratio (RRR) ranging from 3000 down to about 10 [109].

(see Fig. 3.19). Figure 3.20 shows instead the thermal conductivity of copper of varying purity, expressed by residual resistivity ratio (RRR) (see Section 3.13) ranging from 3000 (high purity) down to about 10 (low purity).

3.11 Superconducting metals

In a superconducting metal, some of the electrons are paired into the so-called 'cooper pairs' which are all in the same zero entropy state and do not carry heat. Heat is carried only by unpaired electrons which are in energy states separated from cooper pairs by an energy gap $\Delta E(T)$. The number of unpaired electrons varies as $\exp(-\Delta E/k_B T)$. Hence:

$$k_{e,s} \propto T \exp\left(-\frac{\Delta E}{k_B T}\right) \tag{3.38}$$

with $\Delta E = 1.76\, k_B T_C$ for most of the superconducting elements [20–24,105,110]. The result is that at low temperature (below $T_C/10$), the total thermal conductivity of a

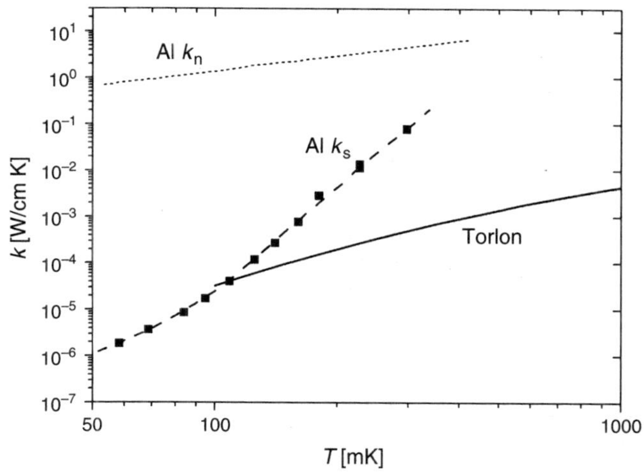

Fig. 3.21. Thermal conductivity k of Al in the normal conducting state and in the superconducting state compared to k of the dielectric Torlon (see Section 11.5.2) at $T > 50$ mK.

superconducting metal approaches the thermal conductivity of an insulator ($k \propto T^3$). As an example, in Fig. 3.21, the thermal conductivity of Al in the normal and in the superconducting state is shown.

Since a suitable magnetic field prevents superconductivity, it is possible to build 'thermal switches', as we shall see in Section 4.3.2.

3.12 Data of low-temperature thermal conductivity

Major compilations of data on thermal conductivity are:

- In general [89].
- Metals and alloys [109]; metals, alloys and non-metals [90]; metals and alloys [111] (short review). For Al alloys, see ref. [112].
- Non-crystalline solids extending down to 0.05K [37] and see also [100].
- Polymers [47].
- More data on polymers and other fibre-reinforced polymers can be found in the proceedings of various ICMC (International Cryogenic Materials Conference) meetings devoted to non-metallic materials and composites at low temperatures [113–117] and some special issues of the journal *Cryogenics* [118–120].

In Table 3.5, the low-temperature thermal conductivity k of some materials of cryogenic interest is reported together with an analytic fit.

Table 3.5
Low-temperature thermal conductivity k in W/cm K of some materials of cryogenic interest

Material	Temperature range	a	n
Glass fibre + nylon [121]	0.2–5K	2.42×10^{-5}	1.77
Kapton [122]	5–20 K	1.451×10^{-5}	1.177
Kapton [123]	0.02–4 K	1.273×10^{-5}	0.9817
Kapton H [124]	10–300 K	2.5×10^{-5}	1
Kapton HN [125]	0.2–5 K	0.65×10^{-4}	1.00
Kapton HN [126]	0.5–5 K	4.638×10^{-5}	0.5678
Kapton HN [127]	4–300 K	2.996×10^{-5}	0.9794
Kapton MT [128]	5–300 K	5.24×10^{-5}	1.02
Kel-F [96]	0.2–0.8 K	0.66×10^{-4}	2.0
Kevlar [129]	0.1–2.5 K	3.9×10^{-5}	1.71
Polyethylene (PET SH0)[a] [130]	0.09–0.4 K	0.732×10^{-5}	1.116
Polyethylene (PET SH0) [130]	0.4–0.65 K	1.054×10^{-5}	1.481
Polyethylene (PET SH0) [130]	0.65–2 K	1.242×10^{-5}	1.885
Polyethylene (PET SH2)[b] [130]	0.08–0.2 K	5.357×10^{-5}	1.634
Polyethylene (PET SH2) [130]	0.2–0.9 K	2.222×10^{-5}	1.07
Polyethylene (PET SH2) [130]	0.9–2 K	2.21×10^{-5}	1.187
Polyethylene (PET SH3)[c] [130]	0.08–0.2 K	0.6097×10^{-4}	1.805
Polyethylene (PET SH3) [130]	0.2–0.9 K	2.194×10^{-5}	1.142
Polyethylene (PET SH3) [130]	0.9–2 K	2.167×10^{-5}	1.238
Upilex R [130]	0.1–9 K	1.8×10^{-5}	1.15
Upilex R [131]	0.1–1 K	1.50×10^{-5}	1.00
Crystal Si_3 N_4 [132]	~0.1–0.3 K	1.62×10^{-4}	2.5
Manganin [133]	1–4 K	0.94×10^{-3}	1.2
Al/Si 1% alloy (n) [134]	0.05–1.2 K	$(38.03\ T + 0.68\ T^2) \times 10^{-3}$	
Al/Si 1% alloy (s) [134]	0.07–0.2 K	2.0×10^{-3}	1.92
Stycast 1266 plain [105]	0.05–1 K	0.39×10^{-3}	1.9
Stycast 1266 plain [135]	0.05–0.5 K	0.49×10^{-3}	1.98
Stycast (44%weight)+$CaCO_3$ grain Ø1-5 μm [105]	0.05–1K	0.33×10^{-3}	2.65
Stycast 6551 MM [105]	0.05–1K	1.7×10^{-4}	2.65
Stycast 2850 FT [105]	0.05–1K	0.92×10^{-4}	2.65
Stycast 2850 GT [133]	1–4K	0.78×10^{-4}	1.8
Stycast 2850 FT [136]	2–10K	0.53×10^{-4}	1.8
CuNi (70% Cu by weight) [105]	0.1–2K	0.65×10^{-3}	1.1
CuNi (55% Cu by weight) [105]	0.09–0.9K	0.39×10^{-3}	1.2
CuNi (55% Cu by weight) [137]	1.3–100K	$(4.2\ T - 0.02\ T^2) \times 10^{-3}$	
CuNi 70/30 by weight [138]	0.3–4K	0.93×10^{-3}	1.23
CuNi 70/30 by weight [139]	0.05–0.2K	0.64×10^{-3}	1
Nb-Ti [91]	4–9K	0.75×10^{-4}	1.85
NbTi rod Ø2.7 mm 45% Nb by weight [105]	0.1–7K	0.27×10^{-3}	2
NbTi wire Ø0.125 mm 45% Nb by weight [105]	0.1–2K	0.15×10^{-3}	2
SPAM [105]	0.06–2K	0.81×10^{-4}	2.25
Oak wood [140]	0.02–1K	0.93×10^{-5}	2.7
AGOT graphite [141]	0.3–3K	0.49×10^{-5}	1.86

Table 3.5
Continued

Material	Temperature range	a	n
AGOT graphite [94]	0.1–2 K	0.51×10^{-5}	1.76
Nuclear graphite [94]	0.1–2 K	1.5×10^{-5}	1.13
Macor [142]	0.4–1.1 K	0.58×10^{-4}	2.24
Al_2O_3 [143]	2–8 K	2.7×10^{-3}	2.5
Al_2O_3 [94]	0.1–2 K	2.9×10^{-4}	2.7
Pyrex [96]	0.18–0.8 K	1.5×10^{-4}	1.75
Pyrolytic Boron Nitride [144]	0.1–1 K	See reference	
Varnish GE7031 [100]	0.05–2 K	0.54×10^{-3}	1.87
Cu-Be C17510 [145]	0.1–1 K	2.39×10^{-2}	0.99
SiO_2 [100]	0.4–1 K	2.4×10^{-4}	1.87
3 $SiO_2 \times Na_2O$ [100]	0.4–1 K	1.7×10^{-4}	1.92
Corning Code No. 7740 [100]	0.4–1 K	2.1×10^{-4}	1.92
GeO_2 [100]	0.4–1 K	3.7×10^{-4}	1.91
As_2S_3 [100]	0.4–1 K	1.7×10^{-5}	1.92
B_2O_3 [100]	0.4–1 K	3.5×10^{-4}	1.96
Se [100]	0.4–1 K	0.79×10^{-3}	1.81
PMMA [100]	0.4–1 K	3.3×10^{-4}	1.81
PS [100]	0.4–1 K	2.0×10^{-4}	1.87
$CaK(NO_3)_3$ [100]	0.4–1 K	1.6×10^{-4}	1.90
Glycerol [146]	1.5–100 K	See reference	
Torlon [85,147]	0.1–300 K	See reference	

[a] Unextruded sample;
[b] Sample with extrusion ratio of 3.85;
[c] Sample with extrusion ration of 4.4.

3.13 The Wiedemann–Franz law

As we shall see in Section 11.4, the measurement of the low-temperature conductivity k of a metal can be very difficult. On the contrary, the measurement of the electrical conductivity σ is quite easy.

Since both conductivities mainly depend on electrons and are limited by the same scattering processes, it is possible to find a relation between k and σ.

This relation is the Wiedemann–Franz (W–F) law:

$$\frac{k}{\sigma} = L_0 T \tag{3.39}$$

where $L_0 = 2.443 \times 10^{-8}\ \mathrm{V^2/K^2}$ is the Lorentz constant.

The W–F law is fairly well verified at $T << \theta_D$. At intermediate temperatures, the measured thermal conductivity is smaller than that calculated from the electrical conductivity by means of the W–F law.

Lorentz ratios for metallic elements are reported in ref. [109] and for technically relevant alloys in [148].

In the literature, we often find the RRR as a measure of the purity of a metal (see Fig. 3.20). This is the ratio of the electrical conductivity at low temperature (usually 4.2 K) to the room temperature electrical conductivity:

$$\mathrm{RRR} = \frac{\sigma_{4.2\mathrm{K}}}{\sigma_{300\mathrm{K}}} \tag{3.40}$$

Since at low temperatures, the electrical conductivity is determined by the scattering on defects, whereas at room temperature it depends on the scattering on phonons, the RRR is a measure of the limiting defect scattering. It indicates how pure a material is. The RRR of particularly pure materials can be as high as 10^4.

For Cu, an useful relation between the RRR and the thermal conductivity is:

$$k \approx \left(\frac{\mathrm{RRR}}{76}\right) \cdot T \ [\mathrm{W/cm\,K}] \tag{3.41}$$

For Ag:

$$k \approx \left(\frac{\mathrm{RRR}}{55}\right) \cdot T \ [\mathrm{W/cm\,K}] \tag{3.42}$$

References

[1] N.W. Ashcroft, N.D. Mermin: *Solid State Physics*, Holt Rinehart and Winston, New York (1976)
[2] H.M. Rosenberg: *The Solid State*, Clarendon Press, Oxford (1984)
[3] E.S.R. Gopal: *Specific Heats at Low Temperatures*, Plenum Press, New York (1966)
[4] A. Cezairliyan: *Specific Heat of Solids*, ed. by C.Y.O. Ho and A. Cezairliyan, Hemisphere Publishing Corp., New York (1988)
[5] T.H.K. Barron, G.K. White: *Heat Capacity and Thermal Expansion at Low Temperatures*, Plenum Press, New York (1999)
[6] C. Kittel: *Introduction to Solid State Physics*, 6th ed., Wiley, New York (1986)
[7] J.M. Ziman: *Electrons and Phonons*, Clarendon Press, Oxford (1972)
[8] H. Ibach, H. Lüth: *Solid-State Physics, an Introduction to Theory and Experiment*, Springer-Verlag, Berlin, Heidelberg (1991)
[9] P.V.E. McClintock, D.J. Meredith, J.K. Wigmore: *Matter at Low Temperatures*, Blackie, London (1984)
[10] J.R. Clement, E.H. Quinnell: Phys. Rev. **92**, 258 (1953)
[11] G.K. White, P.J. Meeson: *Experimental Techniques in Low-Temperature Physics*, 4th ed., Clarendon Press, Oxford (2002)
[12] M.L. Klein, G.K. Horton, J.L. Feldman: Phys. Rev. **184**, 968 (1969)
[13] L. Finegold, N.E. Philips: Phys. Rev. **177**, 1383 (1969)
[14] N.E. Phillips: Crit. Rev. Solid State Sci. **2**, 467 (1971)
[15] O.L. Anderson: J. Phys. Chem. Solids **24**, 909 (1963)
[16] D.L. Martin: Phys. Rev. **38**, 5357 (1973); ibid. B **17**, 1670 (1978)
[17] N.E. Philips: Phys. Rev. **114**, 676 (1959)
[18] D.R. Tilley, J. Tilley: *Superfluidity and Superconductivity*, 3rd ed., Hilger, Bristol (1990)
[19] M. Tinkham: *Introduction to Superconductivity*, McGraw-Hill, New York (1975)
[20] A.C. Rose-Innes, E.H. Rhoderick: *Introduction to Superconductivity*, Pergamon Press, London (1977)
[21] W. Buckel: *Supraleitung*, 4 Aufl., VCH, Weinheim (1990)
[22] J. Bardeen, L.N. Cooper, J.R. Schrieffer: Phys. Rev. **108**, 1175 (1957)
[23] M.W. Zemansky: *Heat and Thermodynamics*, McGraw Hill, New York (1968)
[24] N.E. Philips, M.H. Lambert, W.R. Gardner: Rev. Mod. Phys. **36**, 131 (1964)
[25] M.A. Biondi et al.: Rev. Mod. Phys. **30**, 1109 (1958)

[26] W.S. Corak et al.: Phys. Rev. **102**, 656 (1956)
[27] D. Schoenberg: *Superconductivity*, Cambridge University Press, Cambridge (1952)
[28] J. van Kranendonk, J.H. van Vleck: Rev. Mod. Phys. **30**, 1 (1958)
[29] L.J. de Jongh, A.R. Miedema: Adv. Phys.**50**, 947 (2001)
[30] I.N. Kalinkina: Sov. Phys. JEPT **16**, 1434 (1963)
[31] R. Berman: *Thermal Conduction in Solids*, Clarendon, Oxford (1976)
[32] C. Hagmann, P.L. Richards: Cryogenics **35**, 345 (1995)
[33] J. H. Ho, N. E. Philips: Rev. Sci. Instrum. **35**, 1382 (1985)
[34] F. Pobell: *Matter and Methods at Low Temperature*, p. 174, 2nd ed., Springer-Verlag, Berlin (1991)
[35] S.A. Friedberg, A.F. Cohen, J.H. Schelleng: *Proc. Int. Conf., Magn. Crystallogr.*, Kyoto, J. Phys. Soc. Japan **17**, suppl. B-1, p. 515 (1961)
[36] S.A. Friedberg, C.A. Raquet: J. Appl. Phys. **39**, 1132 (1968)
[37] R.C. Zeller, R.O. Pohl: Phys. Rev. B **4**, 2029 (1971)
[38] M.P. Zaitlin, A.C. Anderson: Phys. Rev. B **12**, 4475 (1975)
[39] P.W. Anderson, B.I. Halperin, C.M. Varma: The Philosophical Magazine **25**, 1 (1792)
[40] W.A. Phillips: J. Low Temp. Phys. **7**, 351 (1972)
[41] S. Hunklinger: in *Ultrasonic Symposium Proceedings*, p. 493, IEEE, New York (1974)
[42] S. Hunklinger: Cryogenics **28**, 224 (1988)
[43] J.C. Lasjaunias, A. Ravex, M. Vandorpe: Solid State Communications **17**, 1045 (1975)
[44] M. Meissner, K. Spitzmann: Phys. Rev. Lett. **46**, 265 (1981)
[45] J.L. Black: Phys. Rev. B **17**(6), 2740 (1978)
[46] M. Meissner, S. Abens, P. Strelow: *Hahn-Meitner Institute Report*, Berlin (2000) www.hmi.de/bensh/sample-env/home.html
[47] G. Jug: Philos. Mag. **84**, 3599 (2004)
[48] G. Hartwig: *Polymer Properties at Room and Cryogenic Temperatures*, Plenum Press, New York (1994)
[49] H. Baur: Kolloid Z. u. Z. Polymere **241**, 1057 (1970)
[50] H. Baur: Kolloid Z. u. Z. Polymere **244**, 293 (1971)
[51] H. Baur: Kolloid Z. u. Z. Polymere **250**, 1015 (1972)
[52] M. Meissner:*Nonmetallic Materials and Composites at Low Temperatures*, vol. 2, Plenum Press, New York and London (1982)
[53] M. Barucci et al.: J. Low Temp. Phys. **123**, 303 (2001)
[54] G.K. White: *Experimental Techniques in Low Temperature Physics*, 3rd ed., Clarendon, Oxford (1979)
[55] O.V. Lounasmaa: *Experimental Principles and Methods Below 1 K*, Academic, London (1974)
[56] R.J. Corruccini, J.J. Gniewek: *Specific Heat an Henthalpies of Technical Solids at Low Temperatures*, NBS Monograph **21**, US Govt. Print. Office, Washington DC (1960)
[57] K. Schäfer, E. Lax: *Kalorische Zustandsgrößen*, Landolt-Börnstein, 2. Band, 4.Teil, 6. Aufl., Spinger Berlin, Heidelberg (1961)
[58] Y.S. Touloukian, E.H. Buyco: *Thermophysical Properties of Matter (Specific Heat)*; vol. 4, *Metallic Elements and Alloy*, vol. 5, *Nonmetallic Solids*, Plenum Press, New York (1970)
[59] D.L. Martin: Rev. Sci. Instrum. **38**, 1738 (1967)
[60] G. Ahlers: Rev. Sci. Instrum. **37**, 477 (1966)
[61] N. Waterhouse: Can. J. Phys. **47**, 1485 (1969)
[62] Y. Hiki, T. Maruyama, Y. Kogure: J. Phys. Soc. Jpn. **34**, 723 (1973)
[63] J. Bevk: Philos. Mag. **28**, 1379 (1973)
[64] G.J. Sellers, A.C. Anderson: Rev. Sci. Instrum. **45**, 1256 (1974)
[65] D.S. Greywall: Phys. Rev. B **18**, 2127 (1978)
[66] E.J. Cotts, A.C. Anderson: J. Low Temp. Phys. **43**, 437 (1980)
[67] D.J. Bradley et al.: J. Low Temp. Phys. **57**, 359 (1984)
[68] K. Gloos et al.: J. Low Temp. Phys. **73**, 101 (1988)
[69] R.F. Seligmann, R.E. Sarwinski: Cryogenics **12**, 239 (1972)
[70] I.A. Tanaeva et al.: Cryogenics **43**, 441 (2003)
[71] G.T. Furukawa, T.B. Douglas, N. Pearlman: *The American Institute of Physics Handbook*, 3rd ed., Chapter 4, McGraw-Hill, New York (1972)
[72] Leybold-Heraeus: *Kryotechnisches Arbeitsblatt* Nr. 5, Leybold AG, D-5000 Köln 1, FR Germany

[73] G.W. Swift, R.E. Packard: Cryogenics **19**, 362 (1979)
[74] D.A. Ackerman, A.C. Anderson: Rev. Sci. Instrum. **53**, 1657 (1982)
[75] G. Hartwig: Cryogenics **35**, 717 (1995)
[76] G. Hartwig, K. Endres, O. Haider: Adv. Cryog. Eng. **40B**, 1107 (1994)
[77] G. Ventura et al.: Cryogenics **39**, 481 (1999)
[78] P. Roth, E. Gmelin: Rev. Sci. Instrum. **63**, 2051 (1992)
[79] R.J. Corruccini, J.J. Gniewek: *Thermal Expansion of Technical Solids at Low Temperatures*, NBS Monograph 29, US Govt. Print. Office, Washington, DC (1961)
[80] Y.S. Touloukian, P.K. Kirby, R.E. Taylor, P.D. Desai, T.Y.R. Lee: *The Thermophysical Properties of Matter (Thermal Expansion)*, vols. 12 (1975), 13 (1977), Plenum Press, New York
[81] R.K. Kirby, T.A. Hahn, R.D. Rothrock: *The American Institute of Physics Handbook*, 3rd ed., Chapter 4, McGraw-Hill, New York (1972)
[82] A.F. Clark: Cryogenics **8**, 282 (1968)
[83] A.F. Clark: *Materials at Low Temperatures*, R.P. Reed and A.F. Clark eds., Chapter 3, *American Society of Metals*, Metals Park, Ohio (1983)
[84] A.F. Clark, G. Fujii, M.A. Ranney: *IEEE Trans. Magn.* MAG-**17**, 2316 (1981)
[85] G. Ventura et al.: Cryogenics **39**, 481 (1999)
[86] P.G. Klemens: *Solid State Physics* 7, p. 1, Academic Press, New York (1958)
[87] Leybold-Heraeus: *Kryotechnisches Arbeitsblatt* Nr. 1, Leybold AG, D-5000 Köln 1, FR Germany
[88] R.L. Powell, W.A. Blaupied: *Thermal Conductivity of Metal and Alloys at Low Temperatures*, Nat. Bureau of Standards Circular 556, US Govt. Print. Office, Washington, DC (1954)
[89] G.E. Childs, R.L. Ericks, R.L. Powell: *Thermal Conductivity of Solids*, NBS Monograph 131, US Govt. Print. Office, Washington, DC (1973)
[90] Y.S. Touloukian, R.L. Powell, C.Y. Ho, P.G. Klemens: *The Thermophysical Properties of Matter (Thermal Conductivity)*, vol. 1 (1970), vol. 2 (1971), Plenum Press, New York
[91] C.Y. Ho, R.L. Powell, P.E. Liley: J. Phys. Chem. Ref. Data **1**, 279 (1972)
[92] D.G. Cahill, S.K. Watson, R.O. Pohl: Phys. Rev. B **46**, 6131 (1992)
[93] R.O. Pohl, X. Liu, E.J. Thompson: Rev. of Mod. Phys. **74**, 991 (2002)
[94] M. Locatelli, D. Arnand, M. Routin: Cryogenics **16**, 374 (1976)
[95] J.J. Freeman, A.C. Anderson: Phys. Rev. B **34**, 5684 (1986)
[96] A.C. Anderson, W. Reese, J.C. Wheatley: Rev. Sci. Instr. **34**, 1386 (1963)
[97] G. Ventura et al.: *Proc. of the 8th International Symposium on Temperature and Thermal Measurements in Industry and Science* (TEMPMEKO '01), ed. by B. Fellmuth, J. Seidel and G. Scholz, VDE Verlag, Berlin (2001)
[98] A. Nittke et al.: J. Low Temp. Phys. **98**, 517 (1995)
[99] C.L. Choy: *Polymer* **18**, 984 (1977)
[100] R.B. Stephens: Phys. Rev. B **8**, 2896 (1973)
[101] R.B. Stephens: Phys. Lett. A **38**, 215 (1972)
[102] M. Barucci et al.: Cryogenics **42**, 551 (2002)
[103] T. Scott, M. Giles: Phys. Rev. Lett. **29**, 642 (1972)
[104] G. Ventura et al.: Nucl. Phys. B (Proc. Suppl.) **78**, 571 (1999)
[105] J.R. Olson: Cryogenics **33**,729 (1993)
[106] P.G. Klemens: *Handbuch der Physik*, vol. 14, p. 198, Springer, Berlin (1956)
[107] A.H. Wilson: *Theory of Metals*, 2nd ed., Cambridge University Press., London and New York (1953)
[108] M. Barucci et al.: *Proceedings of 8th International Conference Advanced Technology and Particle Physics*, p. 541, ed. by M. Barone et al., Como 2003, World Scientific (2003)
[109] O. Madelung, G.K. White (eds): *Landolt-Börnstein Vol. III/15c*, Springer, Berlin (1991)
[110] W. Odoni, P. Fuchs, H.R. Ott: Phys. Rev. B **28**, 1314 (1983)
[111] P.G. Klemens, R.K. Williams: Internat. Metals Rev. **31**, 197 (1986)
[112] A.L. Woodcraft: Cryogenics **45**,626 (2005)
[113] *Advances in Cryogenic Engineering Transactions of the International Cryogenic Materials Conference* – ICMC, vol. **51–52**, Keystone, Colorado, 29 August–2 September 2005, ed. by Balachandran, U. (Balu), Springer, New York (2006)

[114] *Advances in Cryogenic Engineering Transactions of the International Cryogenic Materials Conference* – ICMC, vol. **49–50**, Anchorage, Alaska, 22–26 September 2003, ed. by Balachandran, U. (Balu), Springer, New York (2004)
[115] *Advances in Cryogenic Engineering Proceedings of the International Cryogenic Materials Conference* – ICMC, vol. **47–48**, Madison, Wisconsin, USA 16–20 July 2001, ed. by Balachandran, Balu, Gubser, Donald, Hartwig, K. Ted, Springer, New York (2002)
[116] *Advances in Cryogenic Engineering Materials*, vol. **45–46** *Proceedings of the Cryogenic Engineering Conference & International Cryogenic Materials Conference CEC/ICMC '99*, 12–16 July 1999, Montreal, Quebec, Canada, Kluwer Academic/Plenum Publishers (2000)
[117] *Advances in Cryogenic Engineering Materials*, vol. **43–44**, *Proceedings of the Cryogenic Engineering Conference And International Cryogenic Materials Conference (CEC/ICMC 97)* 27 July–1 August 1997, Portland, Orgeon, Plenum Press (1998)
[118] *Cryogenics*, vol. **35**, Elsevier (1995)
[119] *Cryogenics*, vol. **31**, Elsevier (1991)
[120] *Cryogenics*, vol.**28**, Elsevier (1988)
[121] M. Barucci et al.: Cryogenics **40**, 465 (2000)
[122] S.L. Wipf: *Proc. of the Ninth Intern. Conf. on Magnet Technology*, Villiigen. Swiss Inst. Nucl. Res., p. 692 (1985)
[123] R. Radebaugh, N.. Frederick, J.D. Siegwarth: Cryogenics **13**, 41 (1973)
[124] H. Yokoyama: Cryogenics **35**, 799 (1995)
[125] M. Barucci et al.: Cryogenics **40**, 147 (2000)
[126] J. Lawrence, A.B. Patel, J.G. Brisson: Cryogenics **40**, 203 (2000)
[127] D.L. Rule, D.R. Smith, L.L. Sparks: *NISTIR 1990*, p. 3948 (1990)
[128] D.J. Benford, T.J. Powers, S.H. Moseley: Cryogenics **39**, 93 (1999)
[129] G. Ventura et al.: Cryogenics **40**, 489 (2000)
[130] D.M. Finlayson, P. Mason: J. Phys. C **18**, 1777 (1985)
[131] E. Gottardi et al.: Nucl. Phys. B **78**, 581 (1999)
[132] M. Nahum, T.M. Eiles, J.M. Martinis: Appl. Phys. Lett. **65**, 3123 (1994)
[133] D.T. Corzett, A.M. Keller, P. Seligmann: Cryogenics **16**, 505 (1976)
[134] G. Ventura et al.: Nucl. Phys. B **61**, 576 (1998)
[135] G. Armstrong, A.S. Greenberg, J.R. Sites: Rev. Sci. Instrum. **49**, 345 (1978)
[136] C.L. Tsai, H. Weinstock, W.C. Overton Jr: Cryogenics **18**, 562 (1978)
[137] D.G. Cahill et al.: Phys. Rev. B **40**, 3254 (1989)
[138] H.A. Fairbank, D.M. Lee: Rev. Sci. Instrum. **31**, 660 (1960)
[139] D.S. Greywall: Phys. Rev. B **29**, 4933 (1984)
[140] Th. Wagner, S. Götz, G. Eska: Cryogenics **34**, 655 (1994)
[141] D.O. Edwards et al.: Cryogenics **8**, 392 (1968)
[142] D.A. Zych: Cryogenics **29**, 758 (1989)
[143] K. Gloos et al.: Cryogenics **30**, 14 (1990)
[144] T.C. Hsieh, C. Anderson: Rev. Sci. Instrum. **52**, 1919 (1981)
[145] A. Woodcraft, R.V. Sudiwala, R.S. Bhatia: Cryogenics **41**, 603 (2001)
[146] C. Talón et al.: Phys Rev. B **65**, 012203 (2001)
[147] M. Barucci et al.: Cryogenics **45**, 295 (2005)
[148] J.G. Hust, L.L. Sparks: *Lorenz Ratios of Technically Important Metals and Alloys*, NBS Tech. Note 634, US Dept. of Commerce, Washington, DC (1973)

4 Heat Transfer and Thermal Isolation

Contents

4.1 Introduction

The transfer of heat and the thermal isolation are extremely important in low-temperature apparatuses and experiments. These problems become more serious as the temperature decreases.

The thermal conductivity of materials has been examined in Chapter 2 and Chapter 3. As we shall see in this chapter, in many cases, at very low temperatures, the heat conduction is not limited by the bulk thermal resistivity of the material but by the contact thermal resistance appearing at the interface of two materials. This is a particularly severe problem, below 1K, in the case of the heat transfer between liquid He and a solid (see Section 4.3). Heat transfer by radiation will be considered in Section 5.3.2.2.

4.2 Selection of materials of appropriate thermal conductivity

The low-temperature thermal conductivity of different materials may differ by many orders of magnitude (see Fig. 3.16). Moreover, the thermal conductivity of a single material, as we have seen, may heavily change because of impurities or defects (see Section 11.4). In cryogenic applications, the choice of a material obviously depends not only on its thermal conductivity but also on other characteristics of the material, such as the specific heat, the thermal contraction and the electrical and mechanical properties [1]. For a good thermal conductivity, Cu, Ag and Al (above 1K) are the best metals. Anyway, they all are quite soft especially if annealed. In case of high-purity aluminium [2] and copper (see Section 11.4.3), the thermal conductivities are $k \sim 10\,T$ [W/cm K] and $k \sim T$ [W/cm K], respectively.

For a good thermal insulation, the following are often used: plastic materials or graphite (pay attention to the various types), Al_2O_3, stainless steel, or $Cu_{0.70}Ni_{0.30}$ easier to solder. The last two materials can be slightly magnetic at low temperatures.

For the electric wires that carry the signals from low to room temperature, low thermal conductivity electric conductors of suitable dimensions (typical diameters from 25 up to 200 μm) are used. Because of the low thermal conductivity and its weak dependence on temperature, the most commonly used materials are: constantan ($\rho_{300K} = 52.5\,\mu\Omega\,cm$, $\rho_{4K} = 44\,\mu\Omega\,cm$), manganin ($\rho_{300K} = 48\,\mu\Omega\,cm$, $\rho_{4K} = 43\,\mu\Omega\,cm$) and less frequently phosphor bronze. Pay attention to the magnetic contribution to the specific heat (see Section 3.6). The coating of these wires with a thin layer of superconducting solder alloy ($T_C \approx 7\,K$) produces a non-insulated conductor of zero electrical resistance. Below 1K, it is better to use NbTi wires with or without Cu or CuNi matrix.

The matrix can be taken away by HNO_3 except at the ends of the wire, to allow easy soldering. Below 1K, the electrical contact can be done also by squeezing the wires to be connected inside a thin Al tube (see, for example, Section 12.4). For low-level signals, shielded twisted pairs or cryogenic coaxial cables are used. When thousands of wires are necessary, as in large experiments, Kapton strips with deposited conductors [3] or woven wires [4] are used.

Table 4.1
Electrical resistivity ρ of some elements and alloys in units of $10^{-8}\,\Omega\,m$ [5–11].

T [K]	20	50	75	100	150	200	250	295
Metals								
Ag	0.0038	0.11	0.26	0.42	0.73	1.04	1.34	1.61
Al	0.00065	0.048	0.20	0.47	1.06	1.60	2.20	2.74
Au	0.0125	0.20	0.43	0.63	1.04	1.44	1.83	2.21
Cu	0.0008	0.050	0.182	0.345	0.70	1.044	1.385	1.68
Fe	0.0061	0.125	0.540	1.26	3.14	5.17	7.42	9.70
Nb	0.08	1.0	2.3	3.95	6.9	9.8	12.3	14.4
Pb	0.56	2.75	4.6	6.5	10.2	13.9	17.6	21.0
Pt	0.0367	0.734	1.74	2.80	4.89	6.92	8.91	10.6
Ta	0.046	1.08	2.27	3.54	6.09	8.55	11.0	13.1
Ti(pc)	0.02	1.4	4.1	7.9	16.7	25.7	34.8	42.7
W	0.0019	0.141	0.51	1.02	2.10	3.18	4.30	5.28
T [K]	4	20	75	100	200	295		
Alloys								
Al2024	1.6	1.6	1.9	—	3.3	4.5		
Al5083	3.0	3.0	3.3	—	4.8	5.9		
Ti6Al4V	—	147	150	—	162	169		
$Cu_{70}Zn_{30}$	4.3	—	4.7	5.1	—	7.2		
$Cu_{70}Zn_{30}$	36.4	36.5	36.7	—	38.0	38.5		
Hastelloy C	123	123	124	—	126	128		
Inconel X	121	120	121	—	123	125		
Invar	50	50.5	54	—	69	81		
Berylco 25	8.2	8.2	8.5	—	9.7	10.5		
Stainless steel 304	49.5	49	51	—	63	71		
Stainless steel 310	68.5	69	72	—	82	88		
Manganin	43	—	45	—	—	48		
Evanohm	133	—	133	—	—	134		
Pt10Rh	9.8	—	11.5	—	—	18.2		

Table 4.2
Thermal conductivity in W/m/K for some technical materials [9–13]

T [K]	0.4	1	4	10	40	80	150	300
Metals								
Al5083	—	0.7	3	8	34	56	80	120
Berylco	—	—	1.9	5	21.5	37	—	—
CuZn (70/30)	0.2	0.7	3	10	37	65	85	120
CuNi (60/40)	0.03	0.1	0.8	3	14	20	~25	30
Inconel (ann)	—	0.05	0.45	1.7	8	12	13	15
Inconel (drawn)	—	—	0.27	0.9	5	9	—	—
Manganin	0.02	0.06	0.5	2	7	13	16	22
Stainless steel 304/316	0.03	0.08	0.3	0.7	5	8	11	15
Solder (60Sn40Pb)	—	—	16	42	52	52	—	~50
Non-metals								
Alumina (sintered) (5-30 μm)	—	—	0.5	4.5	80	150	70	—
Macor	0.0003	0.003	0.06	0.3	—	~2	—	—
Pyrex	0.003	0.015	0.09	0.13	—	0.45	0.8	1.1
Silica	—	—	0.1	0.1	0.3	0.5	0.8	1.4
Zirconia	—	—	—	—	—	—	0.9	1.5
Polymers								
Nylon	0.0006	0.003	0.01	0.04	—	—	—	—
PMMA	0.005	0.02	0.05	0.07	—	0.15	0.18	0.2
PVC	—	—	0.03	0.04	—	0.12	—	0.14
HDPE	—	0.003	0.03	0.09	—	0.4	—	0.4
Teflon	0.0004	0.004	0.05	0.1	0.2	0.2	—	—
Epoxy	0.0007	0.007	0.06	0.06	—	—	—	—
Torlon	0.001	0.004	0.013	0.015	0.07	0.1	0.15	0.26

If large electrical currents, as in the case of superconducting magnets, must be carried to low temperature, Cu wires are in most cases used to carry currents from room temperature to 4 K. In Tables 4.1 and 4.2, data of electrical and thermal conductivity of some commonly used materials are reported.

The mechanical and electrical contact between two metals may be made by soldering them together. Unfortunately most soft solder become superconducting at low temperature (see Table 4.3).

4.3 Heat switches

4.3.1 Gas heat switches

The simplest and most commonly used method of thermally connecting the various parts of a cryogenic apparatus is the use of a gas (usually He) at a pressure low enough to avoid condensation at the working temperature. The switch is opened by pumping the

Table 4.3
Melting temperatures T_m and superconducting transition temperatures T_C of some solders [14–21]

Solder	T_m [°C]	T_C [K]
12÷14% Sn, 25÷27% Pb, 50% Bi, 10÷13% Cd (Wood's metal)	70	8÷9
50 ÷ 52% In, 50 ÷ 48% Sn	120	7.1÷7.5
30 ÷ 60% Sn, 70 ÷ 40% Pb	257÷183	7.1÷7.8
97% Sn, 3% Ag	240	3.7
95.5% Sn, 3.5% Ag, 1% Cd	220	3.05
26% Sn, 54% Bi, 20% Cd	103	3.7
43% Sn, 57% Bi	140	2.25
82.5% Cd, 17.5% Zn	265	1÷1.6
70% Au, 30% Sn (eutectic)	280	1.17
Ga	30	1.1
60% Bi, 40% Cd	—	<0.8
40% Ag, 19% Cu, 20% Cd, 21% Zn	610	<0.064
56% Ag, 22% Cu, 17% Zn, 5% Sn	650	<0.064
60% Ag, 30% Cu, 10% Sn	(700)	<0.057
50% Ag, 15.5% Cu, 16.5% Sn, 18% Cd	630	<0.057

gas. This method is used in pre-cooling an experimental apparatus under vacuum to 77 K or to 4.2 K. A gas pressure of 0.1 torr is enough for a satisfactory heat transfer.

A special care is to be devoted to the control that all the parts of the apparatus have reached the desired temperature: when parts remain at higher temperature, due to the high value of the specific heat, the cooling only by radiative exchange is usually impossible. To open a gas heat switch, several hours of pumping are usually necessary to reduce the pressure to a value suitable for the thermal isolation. An insufficient pumping leads to a time-dependent heat leak due to desorption and condensation of the residual gas at the coldest surfaces.

For $T < 2.2\,K$, ^4He can also form a superfluid film which contributes to the heat transfer. H_2 can be used as exchange gas: the advantage is that it can be condensed when ^4He is transferred into the cryostat and does not need to be pumped. However, the ortho-para conversion produces heating (see Section 2.2). ^3He, with a high vapour pressure, no exothermic reactions and no superfluidity in the kelvin temperature range is the best solution except when its residual radioactivity cannot be tolerated (see Section 16.5). Examples of gas switches are reported in ref. [22–27].

4.3.2 *Superconducting heat switches*

In Section 3.10, we have seen that the thermal conductivity of a superconducting metal can become very small at temperatures well below the transition temperature T_C (about 1/10 of T_C); it can be orders of magnitude smaller than the thermal conductivity of the same material in the normal state. Some metals can be easily switched from the superconducting to the normal state by applying a moderate magnetic field (10^{-2}–10^{-1} T).

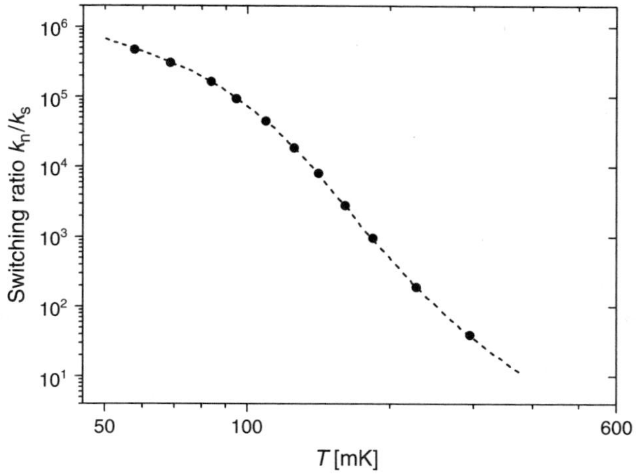

Fig. 4.1. Switching ratio k_n/k_s of the thermal conductivity of Al.

This switch, of course, can be used only when a magnetic field does not interfere with the running experiment. In small refrigerators, the available space may not be enough for the switch. Hence, it is possible to build a superconductor heat switch; the ratio k_n/k_s (switching ratio) can be very high (see Figs 3.21 and 4.1). The heat produced by the eddy currents during the switching is little if the switch is well designed.

From eqs (3.36) and (3.38):

$$k_n/k_s = a \cdot T^2 \quad \text{for } T < 0.1 \cdot T_C \tag{4.1}$$

with $a = 10^2$–10^3 for a good switch.

There are several tricks described in the literature [28–39], which must be kept in mind in projecting a good superconducting thermal switch. High-purity metals are used in the form of thin foils or wires in order to increase k_n and to reduce eddy current heating. An useful discussion of the constructive details of thermal switches is given in ref. [40].

4.3.3 Other heat switches

Other types of thermal switches have been built, e.g. mechanical switches which can be turned on and off easily, but which produce heat in the switching and are therefore used at $T > 1$K. Thermal contact is made by pressing together metallic surfaces. Conductance of a few mW/K can be obtained in the 'on' state [14,30,41–45]. Another disadvantage of these switches are the large forces (typically 10 kg) needed to produce a good contact. Nowadays, the use of these switches is not common.

Liquid He II in an isolating tube has also been used as a switch [46]. Thermal switches can also be produced using materials, like beryllium, graphite or stainless steel, whose thermal conductivity varies rapidly with temperature [47–49].

4.4 Contact thermal resistance

One of the main problems encountered in low-temperature experiments is the realization of good thermal contacts between the parts of a cryogenic system.

Around 1936, the existence of a 'very considerable' thermal resistance at the interface between liquid helium and solids was discovered (Kapitza resistance R_k) [50]. A similar effect (contact resistance R_C) is present in any contact between two materials. In the presence of a heat flux across the boundary, this thermal resistance causes a temperature discontinuity (see Fig. 4.2).

If $\dot{Q}$ is the power flowing through a surface A of two materials in thermal contact (see Fig. 4.2), a temperature drop ΔT is observed at the discontinuity [51]:

$$\Delta T = \frac{R_C}{A} \cdot \dot{Q} \tag{4.2}$$

where R_C is the thermal contact resistance. We use the word resistance, instead of resistivity, for historical reasons, because resistivity usually describes a bulk property. However, the term resistivity would be more appropriate because the thermal boundary resistance is a geometry-independent property.

Experimental data show that the thermal boundary resistance between solids is poorly reproducible [52–53]. The experiments in fact demonstrated that the physical and chemical condition of the interfaces is a critical factor determining the thermal boundary resistance. For this reason, the study of the contact resistance has been carried out on evaporated surfaces in order to reduce the irregularities and make R_C more reproducible.

In 1941, Kapitza [54] reported his measurements of the temperature drop at the boundary between helium and a solid (bronze) when heat flows across the boundary. More than ten years later, Khalatnikov (1952) presented a model, an approximation to what is now known as the 'acoustic mismatch model', to explain that a thermal resistance R_k (thermal boundary resistance) occurs at boundaries with helium [55].

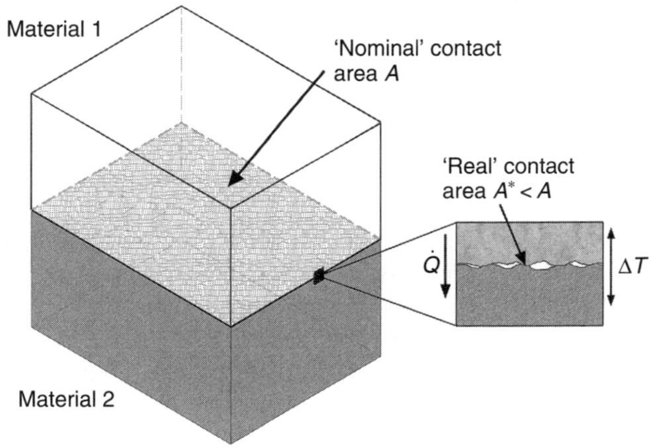

Fig. 4.2. Schematics of the contact between two solids.

The acoustic mismatch model, essentially in its modern form, was first written down by Mazo and Onsager (1955) [56], independently of Khalatnikov's earlier version. The model is currently known as Mazo and Onsager model. The citation "Mazo (1995)" is a post-doc thesis and L. Onsager is the supervisor.

In the acoustic mismatch model, the only essential simplifying assumption is that the phonons are governed by continuum acoustics and the interface is treated as a plane. Phonons are treated as plane waves, and the materials in which the phonons propagate are treated as continua (no lattice). For phonons with wavelength much greater than typical inter-atomic spacing, this continuum approximation might be expected to be accurate. Under this strong assumption, there are only a few results possible when a phonon is incident on the interface: the phonon can specularly reflect, reflect and mode convert, refract or refract and mode convert. The transmission probability is the total fraction of the energy transmitted across the interface.

The thermal boundary resistance to helium predicted by this theory is of the order of 10^3 [K cm^2/W] at 1 K and varies as T^{-3}. This magnitude would be disastrous because it would severely limit attainable cooling powers and increase time constants for reaching thermal equilibrium, thus complicating the quest for experimentation at ever-lower temperatures. Fortunately, the experimentally observed Kapitza resistance was often about two orders of magnitude smaller, near 1K, than that predicted using the acoustic mismatch model. It was originally thought that the measured Kapitza resistance was lower than that expected from acoustic mismatch theory only at interfaces to liquid ^{4}He. Then, it was found that the same enhanced transport existed at interfaces between copper and liquid or solid ^{3}He and solid ^{4}He. Even in this case, the acoustic mismatch model overestimates the contact resistance, but it is able to predict the dependence of resistance on the liquid helium pressure. Such dependence can be experimentally measured only at temperatures below 0.1K [57]. For $T < 1$K, a good thermal contact with liquid helium is obtained increasing the contact surface, using heat exchanger made of sintered metallic powders (see Fig. 4.3 and Section 6.3).

At milli-kelvin temperatures, the problem of contact resistance between helium and solids becomes more complex. Thermal transfer phenomena take place involving spins and thermal resistance of sintered materials. The understanding of the thermal transport at very low temperature is of the utmost importance, also from a technical point of view, since helium is the working substance in dilution refrigerators (see Chapter 6).

A review of the measurements below 100 mK has been done by Harrison in 1979 [58], whereas a study of the heat transfer between helium and sintered metals is due to Rutherford et al. in 1984 [59]. A measurement of Kapitza resistance between Mylar and helium around 2 K is reported in ref. [60].

As can be noted in Fig. 4.4, even if for $0.01\,\mathrm{K} \leq T \leq 0.2\,\mathrm{K}$, experimental data confirm the theory of acoustic mismatch with $A \cdot R_k \cdot T^3 \approx 10^{-2}\,[\mathrm{m^2\,K^4/W}]$; the good thermal contact at higher temperatures is not yet explained, in spite of the studies of the problem [57,61–68]. Referring to Fig. 4.4, data of R_C between solids are only indicative of the temperature dependence, since the pressure on the contact is not known.

In 1959, Little [69] extended the acoustic mismatch model to interfaces between solids. The experiments have revealed that in this case, the thermal contact resistance between solids is higher than that evaluated from the model and that data are less reproducible.

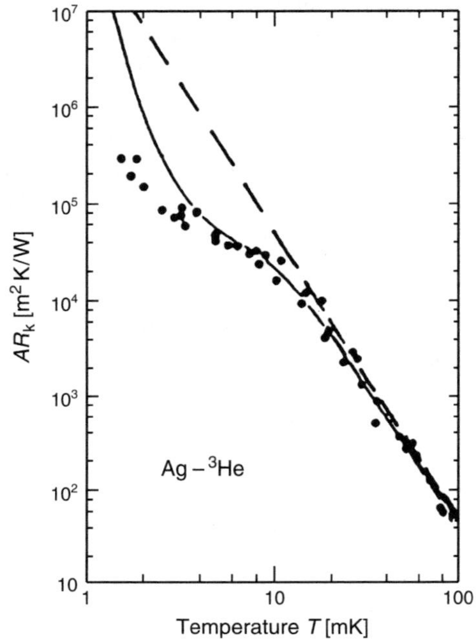

Fig. 4.3. Thermal resistance R_k multiplied by surface area A between ^{3}He and Ag sintered powder. Experimental data from [70]. The dashed line represents the prediction of the acoustic mismatch theory for bulk Ag. The full line is the prediction for the coupling of (zero) sound modes of liquid ^{3}He with modes with a characteristic energy of 15 mK k_B of the Ag sinter [60].

Only at very low temperatures ($T < 100$ mK) [71], the measurements of contact resistance are in good agreement with the mismatch theory.

The study of the contact resistance R_C between solids can be carried out over a temperature range (from milli-kelvin up to hundreds kelvin) much larger than in the case of the Kapitza resistance R_k (below 4 K). It is hence possible to evidence effects, such as dispersion and inelastic phonon scattering, which are absent at low temperatures.

The different behaviour of contact resistance in the two cases can be examined through the two models: the just described acoustic mismatch model and the diffuse mismatch model which suppose that all the phonons are scattered at the interface. Hence these two models define two limits in the behaviour of phonons at a discontinuity.

In 1987, Swartz [73] measured the thermal boundary resistance between metal films and the dielectric substrates onto which the films were deposited, in the range 0.6–200 K. A typical example is the measurement of the thermal contact resistance between indium and sapphire [72]. To minimize the dependence on surface irregularities, indium was vacuum deposited onto the sapphire rods; the two surfaces were then pressed together and annealed. Analogous measurements have been carried out also with lead and aluminium. In all these cases, it has been clear that the contact resistance was strongly dependent on the sample preparation. In particular, obtained data suggest that the contact between the two materials was not complete.

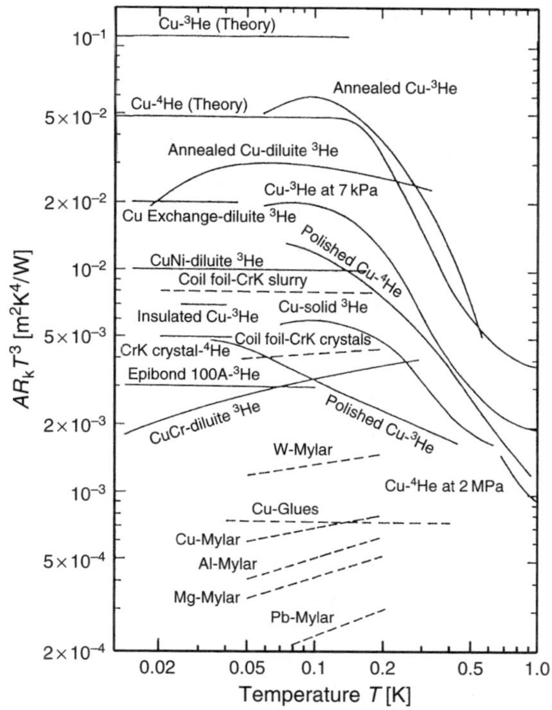

Fig. 4.4. Thermal contact resistance R_k multiplied by $A \cdot T^3$ between liquid helium and various solids and also R_C between metals and dielectrics [30].

In the case of contact between metals and epoxy resins, a sandwich of the epoxy between two flat and well-cleaned metal (copper or gold) surfaces was realized. A steady-state technique was used in most cases.

Some experiments outlined the frequency dependence of phonon scattering on surfaces [74]. Thus, Swartz made the hypothesis that a similar phenomenon could take place at the interface between solids and proposed the 'diffuse mismatch model' [72]. The latter model represents the theoretic limit in which all phonons are heavily scattered at the interface, whereas the basic assumption in the acoustic mismatch model is that no scattering phenomenon takes place at the interface of the two materials. In the reality, phonons may be scattered at the interface with a clear reduction of the contact resistance value as calculated by the acoustic model.

In the diffuse mismatch model, the scattering destroys the correlation between the wave vector of the impinging phonon and that of the diffused one. In other words, the scattering probability is the same independent of which of the two materials the phonon comes from. This probability is proportional to the phonon state density in the material (Fermi golden rule).

Contact resistance values calculated by the diffuse scattering model are higher than those obtained experimentally in the case that the two materials are similar (in density and

Debye phonon velocity) and lower in the case of very dissimilar materials. For example, the estimated Kapitza resistance is smaller by about an order of magnitude due to the great difference in the characteristics of helium and any solid. On the other hand, for a solid–solid interface, the estimated resistance is quite close (30%) to the value given by the mismatch model. The agreement with experimental data is not the best in many cases. This is probably due to many phenomena such as surface irregularities, presence of oxides and bulk disorder close to the surfaces. Since the physical condition of a contact is hardly reproducible, measurements give, in the best case, the temperature dependence of R_C.

At higher temperatures, scattering processes other than diffuse elastic scattering are effective which account for the differences with experimental data for the contact of two solids [75–76].

Instead, when two solids are pressed together, due to surface irregularity, the real contact area A^* (see Fig. 4.2) can be much smaller (for metal a factor $\sim 10^{-6}$) than the nominal contact area A. An applied force results in a strong pressure on the contact area A, and a plastic or permanent deformation occurs. The deformation changes the area A: hence the thermal conductance of the contact is proportional to the force. A drawback of the deformation of the lattice is the reduction of the bulk conductivity of the material.

If the two solids are of the same (or similar) materials and the depth of surface impurities (e.g. oxides) is thin in comparison with the heat carrier wavelength, the expected contact thermal resistance is $R_C \propto T^{-1}$ (see eq. 3.36) for metals, and $R_C \propto T^{-3}$ (see eq. 3.33) for dielectric material. For a dirty contact between metals (heat conduction by phonons only) $R_C \propto T^{-2}$ (see eq. 3.35). These dependences of R_C have been observed experimentally.

The way to make good a thermal contact has been described in the literature [14,30,77–86] for metals. In this case, a good thermal and electrical contact can be realized by soldering the two parts together. Unluckily, most solder alloys (especially those with a low melting temperature) become superconducting at low temperature (see Table 4.3). For further information about soldering of cryogenic components see for example ref. [87].

As we said before, the contact resistance between normal metals should theoretically vary as T^{-1}. Experimental results show dependence as T^{-1} [88], but also with larger exponents. For example, in the contact between Au/Au and Cu/Cu the contact resistance varies as $T^{-1.3}$ [89]. It is to be remarked that the Au/Au contact shows a thermal resistance 20 times smaller than that of the Cu/Cu contact (probably because of the presence of copper oxides).

For a review of the problems in thermal contacts between 4 and 300 K see ref. [90]. For a survey on the theories and measurements of contact resistance at temperatures above 0.1 K see ref. [72].

As we can see in Fig. 4.4, in many cases $R_k \propto T^{-3}$, while $R_C \propto T^{-n}$ with $n < 3$.

The use of data of contact resistance between solid materials of Fig. 4.4 must be very cautious: in fact the contact thermal resistance vary with the applied force. In Fig. 4.5, the change of R_C of a $Cu/Sn/TeO_2$ contact is shown for an approximate doubling of the force.

The contact thermal resistance has sometimes advantageous applications in the realization of very effective thermal insulations [91–93].

A system reaches the thermal equilibrium in a time that depends on its heat capacity, on the thermal conductance of the various parts and on the thermal contact resistance. The latter contribution becomes more important as the temperature decreases. The ideal

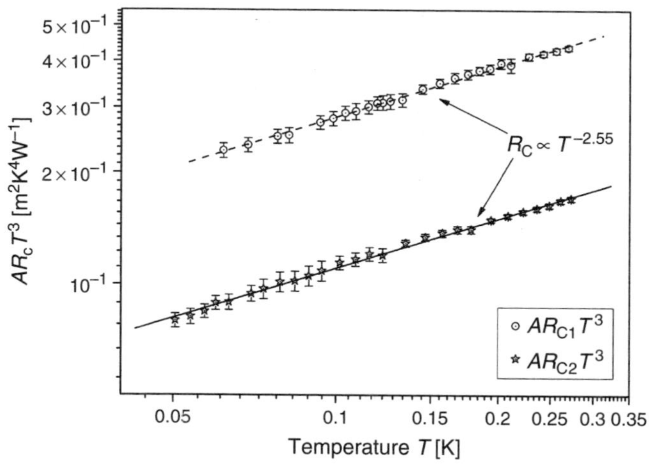

Fig. 4.5. Change of the thermal resistance R_C of a contact Cu/Sn/TeO_2 for an approximate doubling of the applied force.

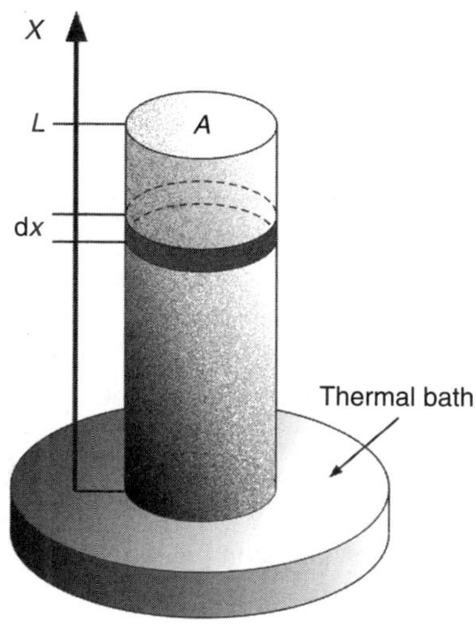

Fig. 4.6. Scheme for the calculation of the thermal time constant of a sample connected to a thermal bath by a thermal resistance of negligible value (see text).

case is that of a sample of heat capacity C and infinite thermal conductivity which is connected to the thermal bath through a pure thermal conductance G (of negligible heat capacity): this is analogous to a lumped constant electrical circuit (see e.g. [94]) and the thermalization time constant is simply $\tau = C/G$.

In reality we cannot usually neglect the internal thermal time constant τ_i of the sample and the contact resistance R_C. Let us consider the example of Fig. 4.6 that shows a cylindrical sample connected to a thermal bath by a thermal resistance of negligible value.

If c is the specific heat of the material [J/K cm^3], k the thermal conductivity, A the cross-sectional area of the sample and L the length of the sample, a simplified calculation gives:

$$\tau_i = \int_0^L \frac{c}{k} \cdot x \cdot dx = \frac{c \cdot L^2}{2k} \tag{4.3}$$

(a more exact calculation would give a constant $4/\pi^2$ instead of 1/2).

If we evaluate, e.g., formula (4.3) for a sample of Teflon with $L = 10$ cm at 80 K, from Tables 3.2 and 4.2 we get: $\tau_i \sim 1.3 \times 10^4$ s; at 200 mK, from Table 3.4 and [95] we get: $\tau_i \sim 45$ s.

Example: We want to cool a cylinder ($A = 1\,\text{cm}^2$, $L = 0.5$ cm) of stainless steel 304 from 80 to 40 mK.

First case: Good thermal contact with the thermal bath.

The specific heat of stainless steel is given by [96]:

$$c_A = \left(465 \cdot T + 0.56 \cdot T^{-2}\right) \times 10^{-6} \text{ [J/g K]} \tag{4.4}$$

With $\bar{k}$ (at 60 mK) $= 6 \times 10^{-3}$ [W/m K], eq. (4.3) gives $\tau_i \sim 2.5$ s.

If the contact resistance is $R_C = 5 \times 10^3\, T^{-1}$ [K/W], the time constant $\tau = c_A \times R_C$, due to the contact at the mean temperature of 60 mK, is ~120 s. Hence the sample temperature lags the bath temperature with a delay of about 120 s.

Second case: Same sample of first case, but with a thin isolating layer between the stainless steel cylinder and thermal bath (contact resistance $R_C = 2 \times 10^4\, T^{-3}$ [K/W]).

The time constant $\tau = c_A \times R_C$ at 60 mK is now about 1.3×10^5 s, i.e., the sample temperature lags the bath temperature with a delay of ~36 h.

Recent developments of materials and devices with structures in nanometer length scales have created new opportunities and challenges in the science of thermal transport. Interfaces play a particularly important role in the properties of nanoscale structures and nanostructured materials [97–98]. This is why a renewed interest for contact resistance arose in recent years with studies of nanocomposite, semicrystalline and polycrystalline materials where contact resistances has a controlling role to determine the bulk thermal conductivity of the material [99–100].

References

[1] P. Richard, Reed and Alan F.Clark (eds): *Materials at low temperature*, American Society for Metals, Metals Park, PH (1983)

[2] A.L. Woodcraft: Cryogenics **45**, 626 (2005)

[3] Minco Corporate Headquarters: Minco, 7300 Commerce Lane, Minneapolis, MN 55432, U.S.A., E-mail: Info@minco.com

[4] Tekdata's Cryoconnect division, Tekdata Limited, Westport House, Federation Road, Burslem, Stoke-on-Trent, Staffordshire, ST6 4HY, E-mail: cables@tekdata-interconnect.co.uk
[5] J. Bass: *Landolt-Börnstein volume III/15a*, p. 70, Springer Verlag, Berlin (1982)
[6] J. Babiskin, J.R. Anderson: In *The American Institute of Physics handbook*, 3rd edn, ch. 9, Mc Graw Hill, New York (1972)
[7] G.T. Meaden: *Electrical resistance of metals*, Heywood, London (1966)
[8] A.F. Clark, G.E. Childs, G.H. Wallace: Cryogenics **10**, 295 (1970)
[9] G.E. Childs, L.J. Ericks, R.L. Powell: *Thermal conductivity of solid at room temperature and below*, NBS Monograph 131 U. S. Govt. Printing Office, Washington, DC (1973)
[10] O. Madelung, G.K. White: *Landolt-Börnstein volume III/15a*, p. 358, Springer, Berlin (1991)
[11] Y.S. Touloukian, R.W. Powell, C.Y. Ho, P.G. Klemens: *Thermophysical properties of matter*, vols. **1–2**, *Thermal conductivity*, Plenum Press, New York (1970)
[12] G. Hartwig: *Polymer properties at room and cryogenic temperatures*, Plenum Press, New York (1994)
[13] M. Barucci et al.: Cryogenics **45**, 295 (2005)
[14] G.K. White: *Experimental techniques in low temperature physics*, 3rd ed., Clarendon, Oxford (1979)
[15] J. Landau, R. Rosenbaum: Rev. Sci. Instrum. **43**, 1540 (1972)
[16] H.C. Meijer, G.J.C. Bots, G.M. Coops: *Proc. 6th Int. Cryogenic Engineering Conf.*, Grenoble (1976)
[17] W.H. Warren Jr., W.G. Bader: Rev. Sci. Instrum. **40**, 180 (1969)
[18] R.G. Gylling: Acta Polytechn. Scand. Physics **81**, 1 (1971)
[19] H.C. Meijer, C. Beduz, T. Mathu: J. Phys. E: Sci. Instrum. **7**, 424 (1974)
[20] D. Dummer, P. Anderson, W. Weyhmann: Cryogenics **31**, 388 (1991)
[21] C.L. Reynolds, A.C. Anderson: Cryogenics **16**, 687 (1976)
[22] E.N. Smith, J.M. Parpia, J.R. Beamish: Physica B **284**, 2026, Part 2 (2000)
[23] J.P. Torre, G. Chanin: Rev. Sci. Instrum. **55**, 213 (1984)
[24] H.M. Chang, H.J. Kim: Cryogenics **40**, 769 (2000)
[25] S.A.J. Wiegers, P.E. Wolf, L. Puech: Physica B **165–166** , 139 (1990)
[26] G.F.M. Wiegerinck et al.: Cryogenics **46**, 9 (2006)
[27] S. Jeong et al.: Cryogenics **46**, 705 (2006)
[28] R.M. Mueller et al.: Rev. Sci. Instum. **49**, 515 (1978)
[29] R.M. Mueller et al.: Cryogenics**20**, 395 (1980)
[30] O.V. Lounasmaa: *Experimental principles and methods below 1 K*, Academic, London (1974)
[31] V.P. Peshkov, A.Yu. Parshin: Sov. Phys.-JETP **21**, 258 (1965)
[32] R.H. March, O.G. Symko: *Proc. Grenoble Conf. Int. Inst. Refrigeration*, Annexe 2, p. 57, Paris (1965)
[33] J.M. Cotignola et al.: Rev. Sci. Instrum. **38**, 87 (1967)
[34] Yu.M. Bunkov: Cryogenics **29**, 938 (1989)
[35] N.S. Lawson: Cryogenics **22**, 667 (1982)
[36] K.W. Wittekers et al.: Cryogenics **29**, 904 (1989)
[37] E. Schuberth: Rev. Sci. Instrum. **55**, 1486 (1984)
[38] M. Krusius, D.N. Paulson, J.C. Wheatley: Rev. Sci. Instrum. **49**, 396 (1978)
[39] K. Gloos et al.: Cryogenics **30**, 14 (1990)
[40] F. Pobell: *Matter and methods at low temperature*, pp. 75–79, 2nd ed., Springer-Verlag, Berlin (1991)
[41] N.E. Philips: *Heat capacity of aluminum between 0.1°K and 4.0°K*, Phys. Rev. **114**, 676 (1959)
[42] R.W. Hill, G.R. Pickett: *Annales Academiae Scientiarum Fennicae*, Series A6: Physica **210**, 40 (1966)
[43] J.H. Colwell: Rev. Sci. Instrum.**40**, 1182 (1969)
[44] J.A. Birch: J. Phys. C: Solid State Phys. **8**, 2043 (1975)
[45] P.R. Roach et al.: Rev. Sci. Instrum. **46**, 207 (1975)
[46] F.J. Shore: Rev. Sci. Instrum. **31**, 966 (1960)
[47] B. Marland, D. Bugby, C. Stouffer: Cryogenics **44**, 413 (2004)
[48] M. DiPirro et al.: Cryogenics **44**, 559 (2004)
[49] B. Marland, D. Bugby: Cryocoolers **11** , 729 (2001)
[50] W.H. Keesom, A.P. Keesom: Physica **3**, 359 (1936)
[51] A. Iwamoto, R. Maekawa, T. Mito: Cryogenics **41**, 367 (2001)
[52] O. Weis: Z. Angew. Phys **26**, 325 (1969)
[53] T. Klitsner, R.O. Pohl: Phys. Rev. B **36**, 6551 (1987)

[54] P.L. Kapitza: Zh. Eksp. Teor. Fiz **11**, 1 (1941) [J. Phys. (USSR) 4, 181 (1941)], also in *Collected Papers of P.L. Kapitza*, vol. **2**, ed. by D. ter Haar Pergamon Oxford, p. 581 (1965)
[55] I.M. Khalatnikov: Zh. Eksp. Teor. Fiz.**22**, 687 (1952)
[56] R.M. Mazo: *Theoretical studies on low temperature phenomena*, Ph.D. thesis, Yale University (1955)
[57] A.C. Anderson, J.J. Connolly, J.C. Wheatley: Phys. Rev. **135**, A910 (1964)
[58] J.P. Harrison: J. Low Temp. Phys. **37**, 467 (1979)
[59] A.R. Rutherford, J.P. Harrison, M.J. Stott: J. Low Temp. Phys. **55**, 157 (1984)
[60] G. Hattenberger et al.: Cryogenics **45**, 404 (2005)
[61] T. Nakayama: In *Progress in low temperature physics*, vol. **7**, ed. by D.F. Brewer North-Holland, Amsterdam (1989)
[62] A.C. Anderson, W.L. Johnson: J. Low Temp. Phys. **7**, 1 (1972)
[63] J.T. Folinsbee, A.C. Anderson: J. Low Temp. Phys. **17**, 409 (1974)
[64] J.D. Siegwarth, R. Radebaugh: In *Proc. 13th Int. Conf. Low Temp. Phys.*, p. 398 and 401, ed. by K.D. Timmerhaus, W.J. O'Sullivan, E.F. Hammel, Plenum Press, New York (1973)
[65] W.E. Braun ed.: *Non-equilibrium phonon dynamics*, Plenum Press, New York (1985)
[66] Y. Sun, P. Wolfle, S.K. Yip: Phys Rev Lett. **63**, 1613 (1989)
[67] M. Grabinski, M. Liu: Phys. Rev. B Condens Matter. **40**, 8720 (1989)
[68] J. Amrit, M.X. François: J. Low Temp. Phys. **128**, 113 (2002)
[69] W.A. Little: Can. J. Phys. **37**, 334 (1959)
[70] K. Andres, W. Sprenger: In *Proc. 14th Int. Conf. Low Temp. Phys.*, vol. **1**, p. 123, ed. by M. Krusius, M. Vuorio, North-Holland, Amsterdam (1975)
[71] D.S. Matsumoto et al.: Phys. Rev. B**16**, 3303 (1977)
[72] E.T. Swartz, R.O. Pohl: Rev. Mod. Phys. **61**, 605 (1989)
[73] E.T. Swartz: *Solid–solid thermal boundary resistance*, Ph.D. Thesis, Cornell University, Ithaca, New York (1987)
[74] W. Eisenmenger: in *Phonon scattering in condensed matter V*, p. 204, ed. by A.C. Anderson and J.P. Wolfe, Springer, Berlin (1986)
[75] R.J. Stoner, H.J. Maris: Phys. Rev. B **48**, 16373 (1993)
[76] R.J. Stevens, A.N. Smith, P.M. Norris: J. Heat Transfer **127**(3), 315 (2005)
[77] C. Schmidt: Phys. Rev. B **15**, 4187 (1977)
[78] T. Okamoto et al.: Rev. Sci. Instrum. **61**, 1332 (1990)
[79] T. Mamiya et al.: Rev. Sci. Instrum. **59**, 1428 (1988)
[80] K.A. Corbett, A. Sawada, E.N. Smith: Physica B **194–196**, 1211 (1994)
[81] K.M. Lau, W. Jr Zimmerman: Rev. Sci. Instrum. **50** , 254 (1979)
[82] L.J. Salerno, P. Kittel, A.L. Spivak: Cryogenics **34**, 649 (1994)
[83] M. Deutsch: Cryogenics **19**, 273 (1979)
[84] K. Muething, G.G. Ihas, J. Landau: Rev. Sci. Instrum. **48**, 906 (1977)
[85] R.I. Boughton, N.R. Brubaker, R.J. Sarwinski: Rev. Sci. Instrum. **38**, 1177 (1967)
[86] M. Suomi, A.C. Anderson, B. Holmström: Physica **38**, 67 (1968)
[87] W. Hand, J. Tallis: *Cryogenic Engineering ed by Hand*, pp. 293–302, Academic Press, London (1986)
[88] I. Didschuns et al.: Cryogenics **44**, 293 (2004)
[89] R. Berman, C.F. Mate: Nature **182,** 1661 (1958)
[90] E. Gmelin et al.: J. Phys. D-Appl. Phys. **32** (6), R19 (1999)
[91] G. Ventura et al.: Cryogenics **37**, 877 (1997)
[92] M. Barucci et al.: J. Low Temp. Phys. **123**, 303 (2001)
[93] Ch. Buzzi et al.: in *Proc. 7th Int. Workshop on Low Temp. Detectors LTD-7*, p. 126, 27 July–2 August 1997, Munich, Germany, pub. by MPI Physik,
[94] B.J. Ley, S.G. Lutz, C.F. Rehberg: *Linear circuit analysis*, McGraw-Hill, New York (1959)
[95] A. Nittke et al.: Eur. Phys. J.B - Condens. Matter **8**, 19 (1998)
[96] C. Hagmann, P.L. Richards: Cryogenics **35**, 345 (1995)
[97] H.K. Lyeo, D.G. Cahill: Phys. Rev. B**73** (14), Art. No. 144301 (2006)
[98] D.G. Cahill et al.: J. Appl. Phys. **93**, 793 (2003)
[99] J. Amrit: J. Phys. D **39**, 4472 (2006)
[100] M.A. Angadi et al.: J. Appl. Phys. **99**(11), Art. No. 114301 (2006)

PART III

5 Cooling Down to 0.3K

Contents

5.1 Introduction

Here and in the two following chapters, we will describe the equipments needed to obtain low temperatures.

In this chapter, some low-temperature refrigeration techniques (down to 0.3K) using ^{4}He or ^{3}He will be discussed.

Until few years ago, the production of temperatures below 10K without using liquid ^{4}He was a real challenge [1]. Nowadays commercial mechanical refrigerators, even of modest dimensions, allow for temperatures below 4K without using cryoliquids (see Section 5.9).

We will first describe 'classic' cryogenic techniques in which either liquid ^{4}He or ^{3}He is evaporated (see e.g. ref [2]). In this case, an important practical aspect is the limitation in the consumption of cryogens, in particular of ^{4}He which is expensive and sometimes not recovered.

In a dewar, the thermal isolation of cold parts is mainly realized by means of vacuum. Hence a good knowledge of vacuum techniques is needed and also skillness in the localization and elimination of leaks. A cryogenist should know how to solder, weld and glue.

These technologies are very important since low-temperature structures experience large mechanical stresses due to temperature gradients and different thermal expansion coefficients of various materials.

A typical and stressing problem of cryogenics is that defects, leak or cracks develop when the system is at low temperature and in vacuum, but the repair is only possible at room temperature.

Also disappointing is the fact that in some cases defects appearing at low temperatures disappear when the system is warmed up. Hence their localization is very difficult.

Last, we must remember that cold surfaces adsorb gases. Should a small leak to the atmosphere be present, air will condense at helium-cooled surfaces. If air contains a certain amount of He, which does not condense at the walls, the pressure in the vacuum space raises and the thermal isolation is lost.

All dewars of cryogenic liquids must be equipped with safety valves, so that pressure cannot raise to dangerous values in case of an accident.

5.2 Transport and storage vessels

Some confusion exists in the terminology regarding liquid helium (LHe) containers. In the following, we shall call 'dewars' the storage and the transport containers, and cryostats the cryogen containers used in experiments. Because of their large capacity (typically

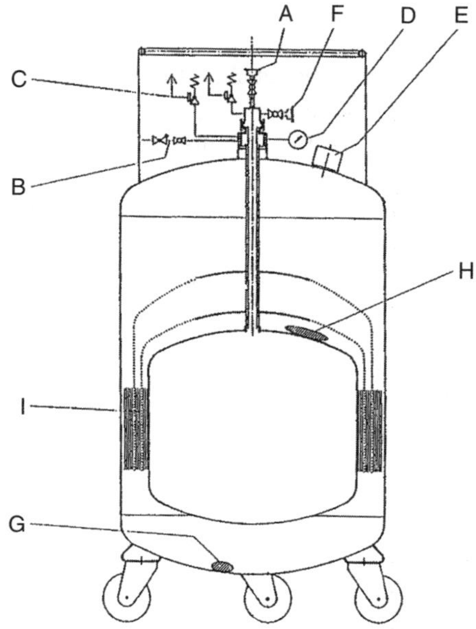

Fig. 5.1. Commercial storage vessel for liquid ^{4}He (A: connection for transfer tube, B: overflow valve, C: safety valve, D: manometer, E: vacuum and safety valves, F: gas valve, G: getter material, H: adsorbent material, I: superinsulation).

100–500 l), dewars are not pumped, and the liquid is kept at a pressure slightly higher than the atmospheric pressure, in order to minimize the vaporization rate. Note that the temperature of the liquid ^{4}He can be well lower than 4.2K, depending on the altitude (observatories).

The typical volume of a dewar is 100 l. Figure 5.1 shows the design of a commercial vacuum-superinsulated dewar for liquid ^{4}He. Such dewars are made from aluminium or stainless steel. The evaporation rate of a good dewar is about 1% per day.

Commercial dewar are equipped with several safety features. Note that there are usually two overpressure safety valves, adjusted for about 1.5 atm and 6 atm respectively. The former safety valve can be usually excluded by a small lever valve which allows liquid transfer operations. If it is forgot closed, the pressure reaches 6 atm and, due to the higher conduction through the gas, the liquid consumption can become as high as 5% per day or more.

5.3 Liquid ^{4}He in the cryostats

There are two phases in the use of a dewar or a cryostat: the cool-down period and the constant temperature period.

5.3.1 Cool-down period

The heat of evaporation of ^{4}He is about 2.6 kJ/l at its normal boiling temperature of 4.2 K. The enthalpy of helium gas between 4.2 and 77 K (300K) is 64 kJ/l (200 kJ/l). Hence, it is important to make use of the enthalpy, letting the gas leave the cryostat to a temperature close to room temperature.

A precooling of dewar with liquid nitrogen (LN_2) from 300 to 77 K saves a large amount of LHe: liquid N_2 has a vaporization heat 60 times that of LHe and a cost 20 times less. Moreover, the enthalpy of materials (es. Al) between 77 and 300 K (162 J/g) is much higher than that between 4 and 77 K (9 J/g).

5.3.2 Constant temperature period

When the cold system has reached a stationary state, the heat transferred from external sources or produced by the experiment must be compensated by the cooling power provided by the cryoliquid.

The power load is usually due to four sources:

1. Conduction by the walls of the cryostat and by electrical wires,
2. Radiation,
3. Convection of the residual gas,
4. Thermoacoustic oscillations.

5.3.2.1 Heat conduction

Heat conduction takes mainly place along solid materials as electric wires or along the walls of the cryostat, by radiative exchanges, and through gases.

As regards the heat conduction through the solid parts of a cryostat, in the choice of the structural materials a compromise is sought for a low thermal conductivity and suitable mechanical properties. When possible, disordered materials are used; in the case of metals, low-conductivity alloys are used as Cu-Ni or stainless steel, in the form of thin-walled tubes. In the evaluation of the heat conduction, the most useful data are the thermal conductivity integrals shown in Fig. 5.2 for some structural materials. The thermal conductivity integral between two temperatures T_L and T_H is defined as:

$$\int_{T_L}^{T_H} k \cdot dT \tag{5.1}$$

5.3.2.2 Heat radiation

For a body of surface A, at a temperature T, the total power emitted is:

$$P = \varepsilon \cdot \sigma \cdot A \cdot T^4 \tag{5.2}$$

with $\sigma = 5.67 \times 10^{-8}$ [$W/m^2 K^4$] is the Stefan–Boltzmann constant.

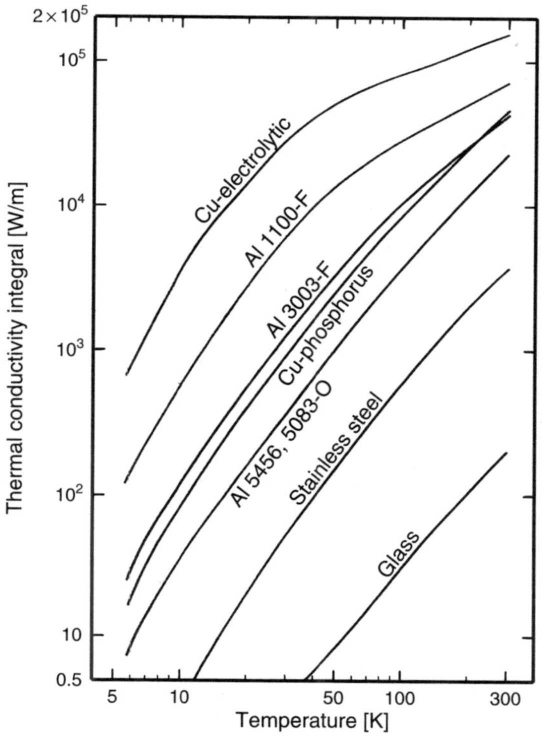

Fig. 5.2. Thermal conductivity integrals, between 4.2 and 300 K, for some structural materials [3].

The radiated power is distributed over a range of wavelengths λ with a spectral distribution given by:

$$P(\lambda) = \frac{2\pi hc^2}{\lambda^5 \left(\exp\left(\frac{hc}{\lambda K_B T} \right) - 1 \right)} \tag{5.3}$$

The function $P(\lambda)$ has a maximum given by the Wien's law:

$$\lambda_{max} \cdot T = \text{constant} = 2.9\,\text{mm K} \tag{5.4}$$

Hence, in the case of a cryogenic apparatus, $\varepsilon(\lambda)$ ($\varepsilon = 1$ for a black body) must be known in the far infrared and is different from the value in the visible region. However, well-polished metals are usually good reflectors even in the far infrared.

To get a high reflectivity, metal surfaces should be gold plated to avoid oxidation.

From eq. (5.2) we see that the total power emitted by $1\,\text{cm}^2$ with $\varepsilon = 1$ at 300 K is 45 mW corresponding to an evaporation of $70\,\text{cm}^3/\text{h}$ of ^{4}He. At 77 K, a surface of $1\,\text{cm}^2$ emits about 0.2 mW, with a ^{4}He consumption of $0.3\,\text{cm}^3/\text{h}$. Hence the part of the dewar cooled at helium temperature is surrounded by radiation shields or baffles at intermediate temperatures. The latter are either gas cooled or thermally anchored to a LN_2 reservoir.

Also shields are inserted in the dewar neck to prevent liquid ^{4}He from 'seeing' the room temperature.

The emissivity of some solids at three temperatures are reported in [3]. Other data for total and spectral emissivity can be found in ref. [4,5].

For two plane parallel surfaces, both of area A and of emissivities ε_1, ε_2 at respective temperatures T_1 and T_2, the radiative power transfer is:

$$P = \sigma \left(T_1^4 - T_2^4\right) \frac{\varepsilon_1 \varepsilon_2}{\varepsilon_1 + \varepsilon_2 - \varepsilon_1 \varepsilon_2} \tag{5.5}$$

To reduce the radiation power input, several solutions are possible: inserting n thermally floating shields between the cold and hot surfaces, the transmitted power is reduced by a factor $(n+1)$. The practical realization is the so-called superinsulation used in the dewar of Fig. 5.1: a few layers of a thin metallized insulating foil about $4\,\mu$m thick is used. To prevent thermal contact between adjacent layers, the material is often corrugated or a thin layer of fibreglass cloth is inserted between layers.

In the 'powder insulation', the vacuum space of the dewar is filled with a finely grained ($\sim 10\,\mu$m) insulating powder. The powder insulation relies on the fact that the contact area between neighbouring granules is so small that the effective thermal paths are long and tortuous. The powders used have very poor thermal conductivities; however, they have surprisingly large emissivities. The successful attenuation of thermal radiation is achieved by the large number of scatterings created by the powder. The scheme is analogous to the attenuation of a light beam by a dense fog.

Both types of insulation act to suppress thermal radiation by the intermediate shield principle. The insulation also acts to reduce the effective cell size for any residual gas in the vacuum space, thereby suppressing the thermal conductivity of the gas. In a typical commercial superinsulated dewar, there are about 50 layers of superinsulation, corresponding to a thickness of about one inch. The first few layers are the most effective in the attenuation of thermal radiation; however the subsequent layers are important for the suppression of thermal conductivity of any residual gas. One can define an effective thermal conductivity for these insulations, which in the case of superinsulation is about 10^{-6} W/(cm K) between 300 and 4 K.

Although the use of most insulation can render the dewar almost serviceable even with 10^{-3} mtorr of residual pressure in the vacuum space, the jacket should be evacuated to at least 10^{-5} mtorr for practical use. To help matters along, it is standard practice to incorporate some sort of 'getter' (see Section 1.6.4) into the vacuum jacket.

5.3.2.3 Conduction by gas particles

Surfaces at 4 K act as cryopump for all gases except He. In this case, the power transferred by the gas is [3]:

$$P = 0.02 \cdot a \cdot A\ [\mathrm{cm}^2] \cdot p\ [\mathrm{mbar}] \cdot \Delta T\ [\mathrm{K}] \tag{5.6}$$

if the mean free path of the gas is larger than dimensions of the container, as usually happens.

In eq. (5.6) 'a' is the 'accommodation coefficient' of the gas on the walls. 'a' is a very uncertain parameter which in the case of He and very clean surface can be as low as 0.02.

It is worth noting that the p in formula (5.6) is the pressure inside the dewar (e.g. the vacuum jacket) which is different from the pressure measured by a gauge at room temperature connected to the vacuum space at low temperature. The power transferred by the gas in the sub-kelvin range is usually negligible.

To improve the cryopumping action, a getter pump can be attached to the cold wall of a cryostat.

5.3.2.4 Thermoacoustic oscillations

Depending on the geometry of the dewar neck and of the other gas-filled tubes connecting cold with warm parts, the excitement of thermoacoustic oscillation is possible above the helium bath.

These oscillations may cause an additional heat flux into the dewar and vibrations which disturb delicate experiments. Often the geometry is changed by the insertion of an additional volume or adding a damping valve.

5.4 ^{4}He cryostats

Because of their brittleness, glass dewars have nowadays only an historical interest [3]. We shall hereafter describe only metal dewars (or cryostat).

Some of them use two reservoirs (liquid N_2 and ^{4}He, see below), but most metallic dewars do no longer use LN_2 (which produces vibrations in boiling) to cool radiative shields. Instead superinsulation is used.

Figure 5.3 shows a metallic cryostat with LN_2 shield used for optical measurements in the far infrared (an optical filter in this case is inserted on the 77K to block the visible radiation).

The He bath can be pumped for cooling down to about 1.3K.

It is important to note that some types of dewar necks are made of plastic materials which are permeable to gases and in particular to He. The permeation phenomenon has a strong dependence on temperature and is negligible at 4K (see e.g. ref. [6]). If the dewar remains for a long time at room temperature in an atmosphere containing He gas, the vacuum space is slowly filled with He which must be pumped before the filling with cryogenic liquids.

5.4.1 Cryostats for $T > 4.2$ K

To obtain temperatures above 4.2K, one can use the main ^{4}He bath and regulate the experiment temperature by means of a thermal impedance and a heater. The method is expensive, in terms of helium consumption, especially for $T >> 4$K.

It is more convenient to use the enthalpy of the cold gas evaporating from the LHe in a continuous gas flow cryostat (Fig. 5.4). The flux and hence the temperature of the experiment can be regulated by a heater or a needle valve.

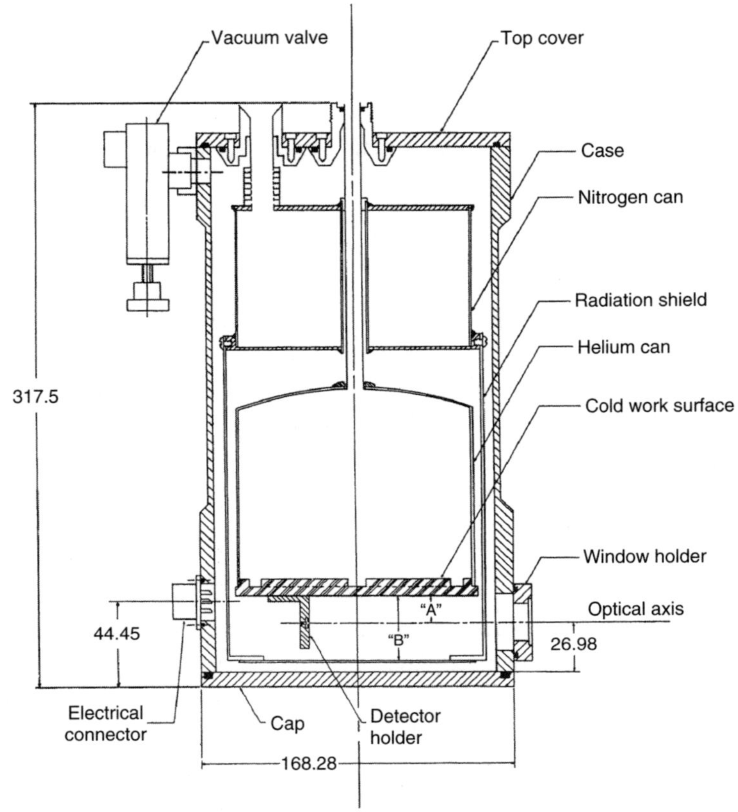

Fig. 5.3. Scheme of a metallic cryostat with LN_2 shield for optical measurements (dimensions are in mm).

Very practical for measurement between 4.2 K and room temperature are the so-called 'dipstick' cryostats, designed for insertion into storage dewars (typically 5 cm neck diameter). Construction details of an insertable cryostat are reported in ref. [7].

Sometimes, to get data from 4 K up to room temperature, the simplest and more economical way is to let the cryostat warm: if the thermal insulation is good and the vacuum chamber is kept under pumping, the warm-up time can be several days. If the experiment thermal time constant is much shorter, data at practically constant temperature can be recorded.

5.4.2 *Cryostats for* 1.3 K < T < 4.2 K

By lowering the pressure above the surface of the ^{4}He bath, temperatures below 4.2 K can be achieved. The practical limit is usually 1.3 K, with an evaporation of 40% of the liquid.

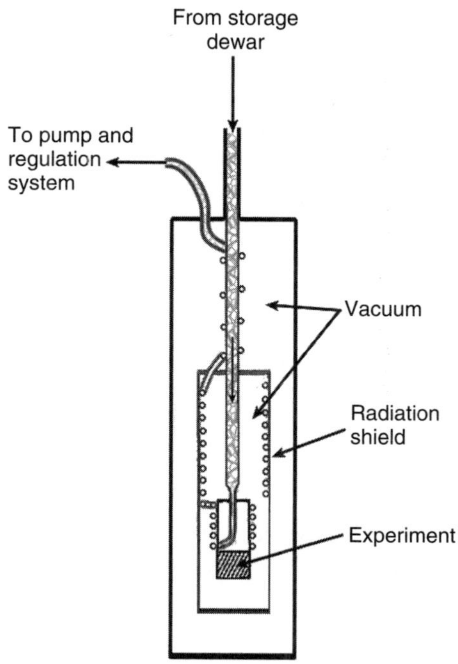

Fig. 5.4. Evaporation cryostat for temperatures $T > 4.2\,\mathrm{K}$. A mixture of gaseous and liquid ^{4}He is transferred from the storage dewar to the cryostat by means of a transfer tube. The mixture is pumped through a spiral tube which is first thermally anchored to the experiment chamber. The temperature of the experiment can be changed by a heater.

Lower temperatures (0.7 K) have been reached with a great consumption of LHe. However the specific heat of most solids (as Cu) is much lower than that of ^{4}He in this temperature range; hence it is convenient to keep most of the liquids at 4.2 K and pump on a small quantity to decrease the temperature only of the experiment [8].

The process of evaporation of a mass of liquid ^{4}He can be described by:

$$L(T) \cdot \mathrm{d}m = m \cdot c(T)\,\mathrm{d}T \tag{5.7}$$

where $L(T)$ is the latent heat of vaporization and $c(T)$ is the liquid specific heat.

The pumping process reduces the mass of liquid in the reservoir from m_i to m and the temperature from T_i to T:

$$\ln\left(m/m_i\right) = \int_{T_i}^{T} c(T)/L(T) \cdot \mathrm{d}T \tag{5.8}$$

The remaining mass of liquid ^{4}He is shown in Fig. 5.5.

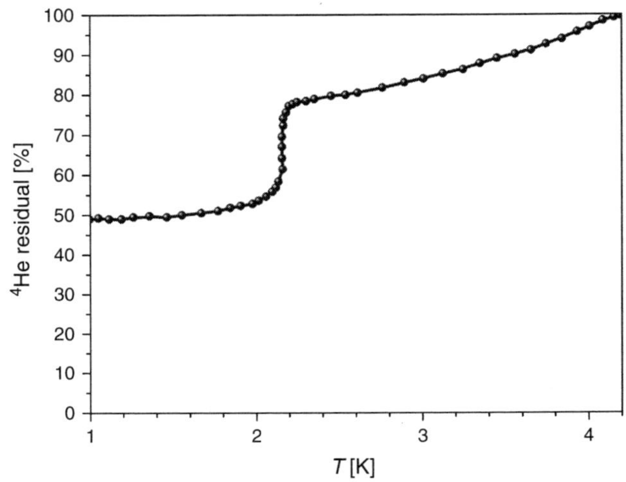

Fig. 5.5. Per cent of remaining liquid ^{4}He after pumping the bath down to the temperature T. Note the step around 2.2 K due to the transition to the superfluid phase with a peak in the specific heat (see Fig. 2.11).

5.5 ^{3}He cryostats

The range of temperature (4.2–1.3 K) accessible by pumping on a liquid ^{4}He bath can be extended to about 0.3 K if ^{3}He is used. The main reason is the difference of vapour pressure between the two isotopes at the same temperature. The ratio p_3/p_4 is 74 at 1 K and about 10^4 at 0.5 K. Moreover in cooling ^{3}He from 1.5 to 0.3 K by pumping on the liquid, only 20% is lost.

There are two disadvantages in the use of ^{3}He: its heat of vaporization is smaller than that of ^{4}He; its price is much higher (about 10^4 the price of ^{4}He). Due to its high price, ^{3}He is used only in a closed gas handling and cryogenic systems.

^{3}He is liquefied in a ^{4}He pumped cryostat at a temperature of about 1.3 K. The critical temperature is 3.324 K. At lower temperatures, the gas condenses on the walls if the pressure is sufficiently high. Since a ^{3}He refrigerator needs the support of a ^{4}He cryostat, it is always enclosed in a thermal shield at the temperature of the ^{4}He bath. Also all other thermal inputs are loaded onto the ^{4}He bath.

The description of some ^{3}He refrigerators is reported in ref. [9–22].

In most applications, ^{3}He refrigerators are now replaced by dilution refrigerators (see Chapter 6). They are still used in the range 0.3–1.3 K, when a very compact vibration-free system is required.

^{3}He refrigerators fall into two categories: (1) refrigerators with external adsorption pump and (2) refrigerators with internal adsorption pump. Hereafter, we will describe a refrigerator of the second type.

5.5.1 ^{3}He refrigerator with internal pump

This type of ^{3}He refrigerator was developed for astrophysical measurements in the far infrared on board of stratospheric balloons, rockets [23] and aeroplanes and also for ground-based telescopes [24].

Referring to Fig. 5.6, the ^{3}He refrigerator [25] contains a pump P and an evaporator E. They are connected by a stainless steel tube T internal to the copper support C. The latter is in good thermal contact with the working plane B of a pumped ^{4}He cryostat (for example that of Fig. 5.3) not shown in figure. The tube is connected to a charcoal cryopump P linked to the ^{4}He bath by a thermal connection L. A thermometer Th monitors the temperature of the pump. A thermal shield (not shown), at the temperature of the ^{4}He bath, surrounds the refrigerator.

The refrigerator (parts P, E and T) contains ^{3}He (see Table 5.1).

The cooling cycle starts when all parts of the refrigerator are at about 1.3K. At this temperature, the ^{3}He is completely adsorbed by the pump. The pump temperature is now raised to about 25 K by means of an heater. At 25 K, the ^{3}He is desorbed, and its pressure increases over the saturation pressure at 1.3 K. Consequently, ^{3}He condenses in the part of the tube T internal to the copper support C and drops down into the evaporator E. In this phase, the latent heat of condensation and the enthalpy variation are delivered to the ^{4}He bath. The cooling phase starts when all the ^{3}He is condensed in E and the power on the pump heater is switched off. The pump starts cooling towards the bath temperature, reducing the pressure on liquid ^{3}He in E. The adsorption heat of the ^{3}He vapour is delivered to the ^{4}He bath by L.

The refrigerator characteristics are summarized in Table 5.2.

In Fig. 5.7 the cooling power of the ^{3}He refrigerator as a function of temperature is shown for bath temperatures of 1.5 and 4.2 K. In fact, once ^{3}He is liquefied, but is possible

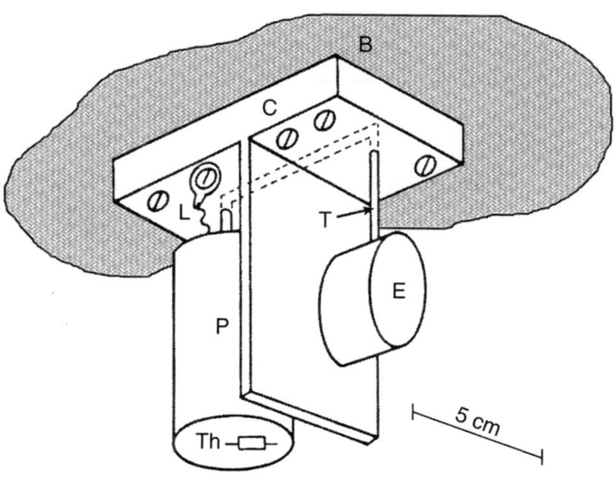

Fig. 5.6. Schematic drawing of a miniature ^{3}He refrigerator.

Table 5.1
Emissivity ε for some materials at three temperatures

Material	T [K]	ε_n
Stainless steel	300	0.20
	80	0.12
	4	0.10
Aluminium (unpolished)	300	0.25
	80	0.12
	4	0.07
Aluminium (polished mechanically)	300	0.20
	80	0.10
	4	0.06
Aluminium (polished electronically)	300	0.15
	80	0.08
	4	0.04
Chromium	300	0.08
Copper (polished mechanically)	300	0.03
	80	0.019
	4	0.015
Tin	300	0.05
	80	0.013
	4	0.012
Brass (polished)	300	0.03
	80	0.029
	4	0.018
Nickel	300	0.045
	80	0.022
Gold	300	0.20
	80	0.10
Silver	300	0.022
	80	0.008
	4	0.005

Table 5.2
Miniature ^{3}He refrigerator characteristics

^{3}He charge	~1.5 l STP
Pressure at room temperature	100 atm
Refrigeration capacity/cycle	1.8 J
Size	70 × 55 × 46 mm
Weight	290 g
Charcoal capacity	8 g
Operating temperature (bath 1.5K)	0.294 ± 0.002 K
Operating temperature (bath 4.2K)	0.333 ± 0.002 K
Stability of operating temperature	< ±0.001 K
Time to stabilize at operating temperature	30 min
Time to condense	10 min
Running time (bath at 4.2K)	24 h

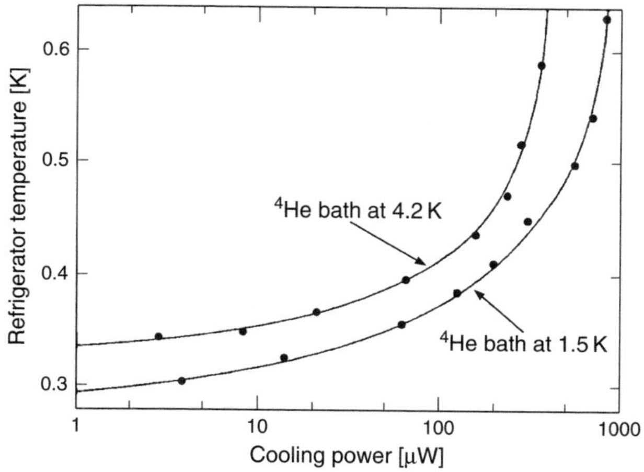

Fig. 5.7. Working temperature of miniature refrigerator versus cooling power for two temperatures of the ^{4}He bath.

to remove the pumping on the ^{4}He bath, the cooling power is slightly lower, but the system becomes movable and free of vibrations.

In ^{3}He refrigerators, it is also possible to avoid pumping on ^{4}He, taking advantage of an adiabatic expansion as proposed in ref. [26].

5.6 Accessories

5.6.1 N_2 *transfer tubes*

For the transfer of LN_2, a simple thick-walled rubber tube or a low-conductivity metallic tube which can be bent to the required shape may be used. In the latter case, a convenient covering, e.g. a flexible foamed neoprene, is needed. Less frequent is the use of double-walled metallic tubes. In this case, the vacuum enclosure must contain some charcoal active as a cryopump.

It is common to use compressed gas to force the liquid N_2 through the transfer tube. If transfers are frequent, pressurization and depressurization of the container leads to waste and contamination of the liquid. To avoid these drawbacks, one can leave the container sealed by a blowoff valve (release for an overpressure of about 0.2 atm). A valve in the transfer tube allows liquid N_2 to be drawn under this pressure when desired.

5.6.2 ^{4}*He transfer tubes*

^{4}He has a latent heat of vaporization much smaller than LN_2 and a higher cost. A double-walled transfer tube is always used.

Example 5.1. If the inner wall of a transfer tube is stainless steel of 4 mm outer diameter, 0.25 mm wall thickness and 1.5 m length, the room temperature enthalpy of the tube is about 2700 J. Should only the latent heat be used to precool the tube, about 1 l of liquid ^{4}He would be needed. However, the heat capacity of the evaporated gas contributes to the cooling. The room temperature enthalpy of the gas produced by 1 cc of liquid is about 190 J. If half of this enthalpy is used (slow initial transfer), then only about 30 cc of liquid is needed for precooling the transfer line.

To transfer the liquid, a small overpressure is created in the dewar from which the liquid is to be transferred. A special care must be taken when transferring LHe into either a warm dewar or a partially filled dewar to avoid waste of liquid. If the dewar is warm (77 K), a slow initial transfer reduces the liquid consumption. If liquid is already in the dewar, the risk of inflating a warm stream of He gas must be avoided. The simplest transfer tubes are U shaped. There are flexible and demountable transfer tubes.

A junction in a transfer tube is useful when two parts of the tube are fixed to the respective dewars. An example of a transfer tube is shown in Fig. 5.8. Figure 5.9 shows a sleeve coupling. Examples of transfer lines are in ref. [27,28].

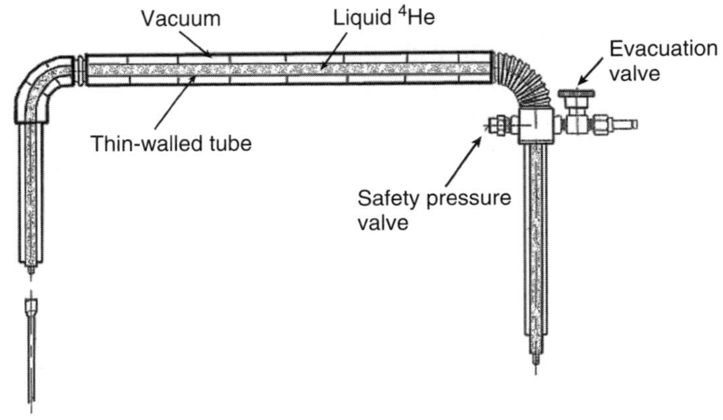

Fig. 5.8. Example of liquid ^{4}He transfer line.

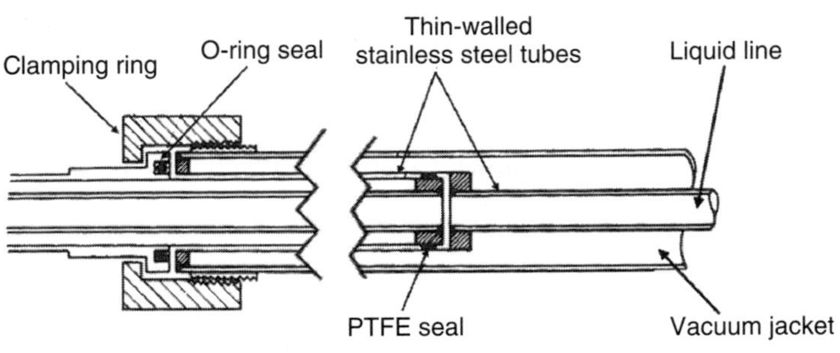

Fig. 5.9. An example of coupling for liquid helium transfer line.

5.6.3 Liquid-level detectors

In containers for cryogenic liquids, we must be able to determine the level of the liquid. Since most dewars are metallic, we cannot detect the level optically.

Many devices for determining the level of a cryoliquid have been proposed: there are acoustic, resistive and capacitive detectors.

Some of these are suitable only for a specific liquid.

1. For locating the liquid surface, the simplest detector is the diaphragm dipstick [29] depicted in Fig. 5.10. When the thin stainless steel tube (2–4 mm diameter) is lowered into the cryoliquid, the pressure in the vapour space increases and the liquid level oscillates. The oscillations are transferred to the diaphragm on top of the tube. The frequency and amplitude of oscillations change when the tube's lower end cross the liquid surface. In practice, the tube is lowered down to the bottom of the dewar and a reference is marked in correspondence with the neck flange. Then, the tube is raised searching for a change in the vibration of the membrane. In correspondence, a second mark is put on the tube. The distance between the two marks gives the height of the liquid from the bottom of the dewar. Usually, on transport dewars, the relation between height and volume of the liquid is written on the body of the container. A simplified version of this detector using only a tube and a rubber fit on top is shown in Fig. 5.10.

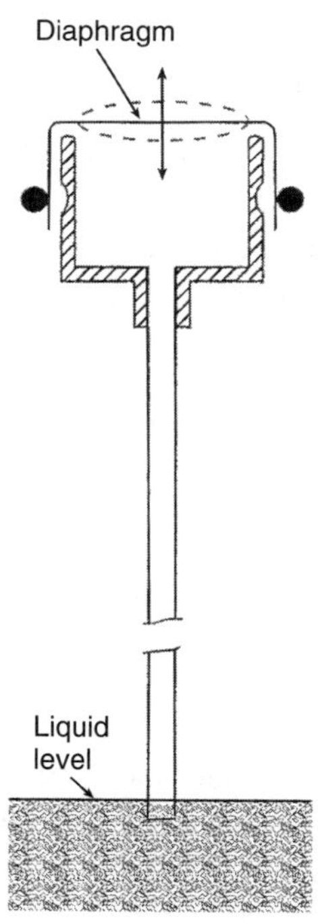

Fig. 5.10. A rubber diaphragm level finder for liquid helium.

2. Another level detector is the capacitance gauge which relies on the difference of dielectric constant of liquid and vapour (1.057 and 1.001 for ^{4}He). In most of these gauges, the capacitor is made of two concentric tubes. A review is reported in ref. [30].
3. In a resistance level detector, a resistive thermometer is powered with 0.1–1 mW. When the thermometer is inside the LHe, its temperature is about 4 K. If it is raised into the vapour phase, its temperature raises because of the weaker thermal contact. The sensing element is usually a NbTi wire (see e.g. ref. [31]).

5.7 Mechanical refrigerators

5.7.1 Introduction

Mechanical cryocoolers are used either to liquefy a gas for use away from the machine or to provide a cold platform for a refrigerator. A cryocooler must be as efficient as possible, whilst taking account of any constraints there may be for particular applications. For this to occur, the maximum possible use must be made of any cold substance that is produced. It is important in a helium liquefier that the fraction of gas which was cooled but did not liquefy is used to precool further incoming gas. This leads us directly to consider the heat exchangers. The combination of a cyclical process with the need for efficient heat exchange led to the idea of a regenerator in which heat may be stored for a short time, so that heat output from one phase of the cooling cycle may be reinserted at some phase.

Since it is the decompression stage which produces cooling, we discuss, for the most part with considerations, how this is achieved.

Several types of mechanical refrigerators have been used in the last decades. Most of them use gas compression–expansion cycle to produce refrigeration: work done on the fluid by compression is rejected to the surrounding, then the expansion of fluid to a lower pressure in a separate chamber achieves cooling.

In this section, we will describe the various types of mechanical coolers with reference to their working cycles. We shall start with a brief description of counterflow heat exchangers. This subject will be further dealt in Chapter 6 when treating dilution refrigerators.

We will then describe the Collins helium liquefier and coolers which use turbo-expanders (Linde He liquefier). At the end of the chapter, we will describe cooling cycles which use regenerators and other cycles used in cryocoolers (Philips-Stirling, Gifford–McMahon (GM), Klimenko cycles and pulse tube refrigerators (PTRs)).

A special attention will be devoted to the latter type of refrigerators (Section 5.8) which seems nowadays one of the more promising cryocoolers.

Before going into the description of coolers, we wish to remind that the work to reach and to maintain low temperatures increases with the decrease of temperature: for example, to provide 1 W of refrigeration at 80 K requires at least 2.6 W of input power, while 1 W of refrigeration at 4 K needs at least 72 W of input power.

5.7.2 Coolers using counterflow heat exchangers

A counterflow current heat exchanger is a basic part of the Joule–Thomson cooling process, which by isenthalpic expansion produces both liquid and cold gas. It ensures that the gas entering the final expansion stage has been cooled sufficiently by the returning unliquefied cold gas and by any other precooling streams, e.g. from other expansion cycles. The principal aims in heat exchanger design are to optimize the heat transfer from the gas stream to the containing tube and vice versa and to maintain a relatively low pressure drop along the exchanger.

More details of the heat-transfer processes in heat exchangers can be found in ref. [32–34].

In this section, we will describe the so-called 'continuous' heat exchangers which are used down to about 1 K. For lower temperatures, due to the increasing importance of the Kapitza resistance (see Section 4.3) 'step exchangers' are used. They will be described in Chapter 6 in connection with the dilution refrigerator.

5.7.2.1 Pressure drop

For a gas of density ρ, viscosity η, flowing at a mass rate $\dot{m}$ in a pipe of effective diameter D and length L, the pressure drop is:

$$\Delta p = \frac{1}{2}\psi\frac{LG^2}{D\rho} \quad \text{where } G = \frac{4\dot{m}}{\pi D^2} \tag{5.9}$$

G is the mass flow per unit area and the dimensionless factor ψ is given by $\psi = 64\eta/GD$ for laminar flow and $\psi = 0.316(GD/\eta)^{-0.25}$ for turbulent flow.

For maximum heat transfer, the flow should be turbulent rather than laminar. This means that Reynolds number $\mathrm{Re} = GD/\eta \geq 2300$.

For non circular tubes, there is an effective hydrodynamic diameter defined by:

$$D_\mathrm{h} = 4 \times \frac{\text{Cross-sectional area of tube } (A)}{\text{Total perimeter of surfaces exposed to stream}} \tag{5.10}$$

Now $G = \dot{m}/A$ and:

$$\Delta p = \frac{1}{2}\psi\frac{LG^2}{\rho D_h} \tag{5.11}$$

If the diameter is increased, the pressure drop along the stream decreases rapidly as $1/D^{4.25}$, where D depends on the cross-sectional shape of the tube.

5.7.2.2 Heat transfer

The heat transfer coefficient h is the rate of transfer of heat $\dot{Q}$ from the gas stream to the tube wall for a temperature difference ΔT, i.e. $\dot{Q} = h\Delta T$.

With turbulent flow, the major temperature drop occurs across the thin layer of gas at the tube wall. The coefficient h depends on the viscosity η, the thermal conductivity

k_f, the specific heat c_p of the gas and the effective tube diameter D_e. Using Nusselt's number, Nu (see e.g. ref. [32]):

$$\mathrm{Nu} = \frac{hD_e}{k_f} = 0.023\left(\frac{D_e G}{\eta}\right)^{0.8}\left(\frac{c_p \eta}{k_f}\right)^{0.4} \tag{5.12}$$

Hence:

$$h = \frac{0.023}{(\mathrm{Pr})^{0.6}} c_p \frac{G^{0.8}\eta^{0.2}}{D_e^{0.2}} \tag{5.13}$$

where Prandtl's number $\mathrm{Pr} = \eta c_p / k_f$, h is in W/m^{2}K, k_f in W/mK, c_p in J/kgK, D_e in m, G in kg/m^{2}s and η is in Pa s.

Representative values of the physical parameters ρ, k_f, η, and Pr are shown in Table 5.3.

From these data, we can approximate Pr by a value 0.75 and use the weak power dependence of 0.2 for η and D ($\sim$0.01 m) to obtain the further approximation $(\eta/D)^{0.2} \approx 0.25$.

Hence, from eq. (5.13): $h \approx 0.006\, G^{0.8} \cdot c_p$ in W/m^2 K.

5.7.2.3 Efficiency and length

Knowing the Reynolds number (to ensure turbulent flow) and the approximate values of the heat-transfer coefficients for the downstream and upstream, we need to relate the length L to h_1, h_2, the specific heats (c_1, c_2) and mass flow ($\dot{m}_1$, $\dot{m}_2$) of the two streams (see Fig. 5.11).

Neglecting the thermal resistance of metallic tube wall, the energy conservation law gives:

$$L = \frac{\alpha}{\gamma} \ln\left(\frac{T_1^{\mathrm{H}} + \beta/\gamma}{T_1^{\mathrm{C}} + \beta/\gamma}\right), \tag{5.14}$$

where

$$\alpha \approx \dot{m}_1 c_1 \left(\frac{1}{h_1 S_1} + \frac{1}{h_2 S_2}\right), \tag{5.15}$$

$$\beta = \left(\frac{\dot{m}_1 c_1}{\dot{m}_2 c_2} T_1^{\mathrm{C}} - T_2^{\mathrm{C}}\right), \tag{5.16}$$

Table 5.3
Selected values of specific heat, density, viscosity, thermal conductivity and Prandtl's number for some gases.

Gas		N_2	O_2	H_2	^4He
c_p [J/kg K]	at s.t.p.	1040	917	14200	5190
ρ [kg/m^3]	at s.t.p.	1.250	1.428	0.090	0.178
η [μPa s]	at 180 K	11.8	13.3	6.3	13.9
	at 50 K	—	—	2.5	6.3
k_f [mW/Km]	at 180 K	16.5	16.5	116	107
	at 50 K	4.9	4.5	36	46
$\mathrm{Pr} = \eta c_p/k_f$	at 180 K	0.74	0.74	0.77	0.68
	at 50 K	—	—	0.84	0.72

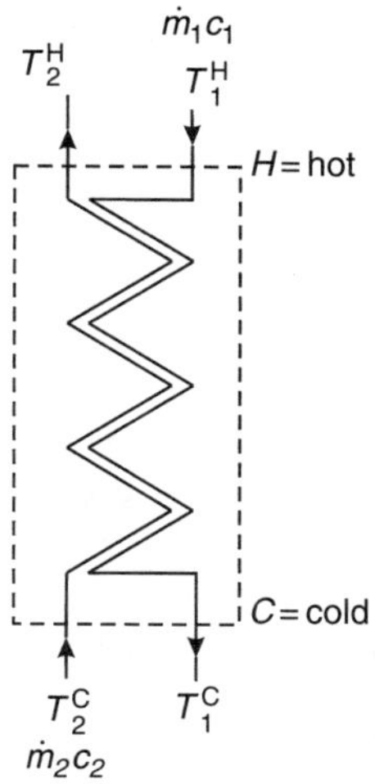

Fig. 5.11. A schematic counter-current heat exchanger. T_1 is the temperature of the input stream, T_2 of the output stream. Superscripts H and C signify hot end and cold end respectively.

$$\gamma = 1 - \frac{\dot{m}_1 c_1}{\dot{m}_2 c_2} \tag{5.17}$$

S_1 and S_2 are the perimeters for heat transfer from the respective streams. The superscripts H and C denote hot and cold ends of the heat exchanger. In the special case where $\dot{m}_1 c_1 \rightarrow \dot{m}_2 c_2$, the equation (5.14) becomes:

$$L = \frac{\alpha}{\beta}\left(T_1^{\mathrm{H}} - T_1^{\mathrm{C}}\right) \tag{5.18}$$

The heat exchanger efficiency ε is the heat transferred from one stream to the other compared with the maximum heat available.

Clearly, if the outgoing stream absorbs all the available heat from the incoming stream, $\Delta T = T_1^{\mathrm{H}} - T_2^{\mathrm{H}} = 0$, and the heat exchanger is 100% efficient.

5.7.2.4 Construction

Linde type

The tube-in-tube or multitube-in-tube heat exchangers are useful in small Linde liquefiers or in the final Joule–Thomson stage of any liquefier. The performance of Linde-type exchangers is easy to calculate, and their realization is simple. In the examples shown in Fig. 5.12 (a)–(c), the tubes are concentric and the outer wall contributes appreciably to the pressure drop in the outer stream without contributing to the heat transfer. Usually, the smaller inner tube is used for the high-pressure stream and the low-pressure stream flows through the outer annular space. The tubes in Fig. 5.12 (d) and (e) are solder bonded while that in (f) is flattened and twisted before insertion into an outer tube.

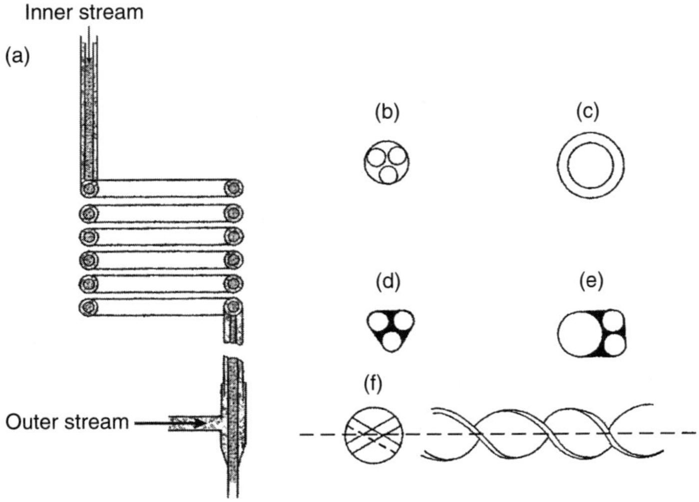

Fig. 5.12. Examples of Linde pattern heat exchangers.

A microversion of the Linde heat exchanger is produced commercially by MMR Technologies [35]. It consists of glass plates which are etched by photolithography to produce fine channels of ~200 μm width and ~30 μm depth. The plates are then bonded together and the gas streams circulate through these fine channels.

Another minicooler using glass tube-in-tube heat exchangers, 10 MPa nitrogen gas input and reaching about 80 K is described in ref. [36].

Hampson type

The calculation for Hampson-type exchangers are more difficult, but in this case the flow resistance (and pressure drop) for the returning low-pressure stream is smaller than in Linde exchangers. Generally, the high-pressure stream is forced to flow around a long spiral path through the shorter, low-pressure stream. They are often made by winding a finned capillary tube on a mandrel and enclosing it in the annular space between two cylinders as in the Collins liquefier.

We wish also to mention the Patel and Brisson's [37] plastic heat exchangers useful down to 0.5 K. Kapton heat exchanger for even lower temperatures have been realized by Frossati [38]. Modelling of heat exchangers is recently treated by Nellis [39].

5.7.2.5 Other liquefier details

Longsworth [40] discusses methods for more rapid cool down by using a mixture of two gases. (for example CF_4 and N_2). He also discusses different designs of expansion valve: a simple fixed orifice for short-period operation or an automatically regulated orifice for longer period operation. Walker and Bingham [41] give a detailed survey of heat exchangers and expansion valves.

For the modelling of cryocooler see ref. [42]. For a low-power cryocooler survey, see for example ref. [43,44]. For vibration problems, see ref. [45–47].

5.7.3 The Collins helium liquefier

This liquefier is a direct descendant of those constructed by Claude and Kapitza [48].

The essential features of the earlier commercial models were that the two piston engines were vertically arranged inside the large finned heat exchanger in a static atmosphere of helium with LHe from the Joule–Thomson expansion collecting in the bottom of the dewar. This liquid could be either used in situ or transferred to all external storage dewar. The energy of expansion was absorbed in a crosshead on top of the dewar assembly.

Problems with the crosshead and linkages to the piston rods resulted in various changes such as the introduction of solid Micarta pistons and an electric crosshead. This together with rearrangement of engines and main exchanger in an evacuated container [49] allowed maintenance-free operation overnight for filling relatively large storage vessels of 500 l or more. The main exchanger is still of the finned tube 'Hampson' construction.

In the later models, the liquid and expanded helium gas was taken through a triple-wall transfer tube to the storage vessel. The returning cold gas was used to shield the transfer siphon before proceeding through the heat exchangers. Storage capacities of over 1000 l and annual production in the region of 10^5 l are commonplace, depending on the model, the number of compressors and whether liquid N_2 precooling is used.

For laboratory production of LHe, other so-called Collins-type liquefiers have been built with one or two stages of GM cooling before the Joule–Thomson expansion (e.g. ref. [50]). The thermodynamic analysis of Collins helium liquefaction cycle can be found in ref. [51].

5.7.4 Klimenko cycle

These coolers are closely related to the Joule–Thomson cycle, but use a gas mixture rather than a single gas. A compressed mixture goes through a counterflow exchanger and expands through a throttle, with the cooled gas returning through the exchanger in a continuous process [52].

The use of mixed gases is designed to 'equalize' the heat capacities of the two counter-flow gas streams and thereby make the exchange process much more efficient. However, the expanded fluid is a mixture of liquid and gas; hence a liquid–vapour separator is needed to allow the expanding liquid to precool the vapour. Figure 5.13 shows the cycle as modified by MMR Technologies [53]

The cycle has also been used in a slight different form by APD Cryogenics to produce the 'Cryotiger' [54].

The fluid mixtures which have been used down to ~80 K include nitrogen with roughly equal parts of methane, ethane and propane. Input pressures are ~10 MPa and overall efficiencies are much greater than single-fluid Joule–Thomson systems and are comparable with the GM and Stirling cycles (see also [55]).

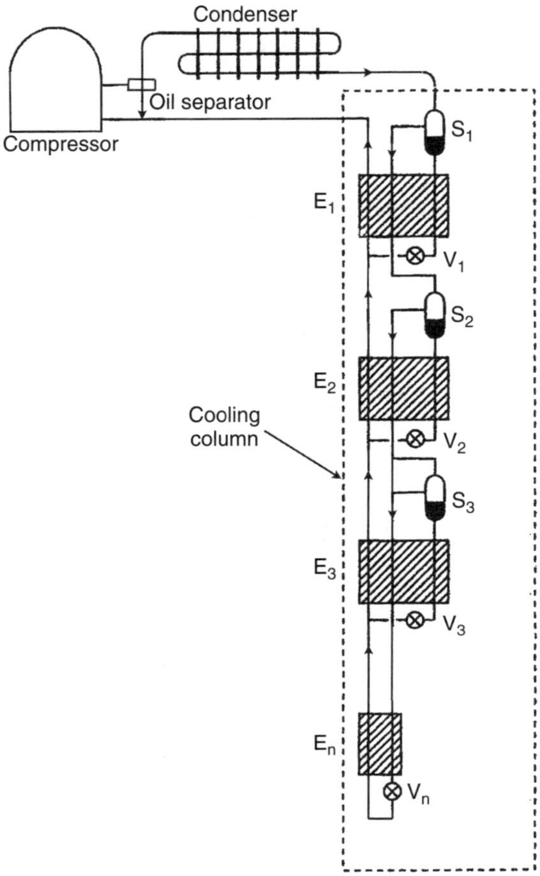

Fig. 5.13. Schematic diagram of a Klimenko cycle cryocooler. $E_{1,2,3,n}$ denote heat exchangers; $V_{1,2,3,n}$ are capillary expansion valves; $S_{1,2,3,n}$ are liquid–vapour separators [53].

5.7.5 Coolers using turbo-expanders

Turbine refrigeration was proposed by Lord Rayleigh in 1898 (see ref. [56]). Gas expands through a turbine (instead of pistons) or, in some cases, the Joule–Thomson throttle. The process is generally more efficient, but the engineering is more sophisticated to implement. Commercial development began in the 1930s. Kapitza did a careful analysis of turbo-refrigeration and developed radial flow turbines for air refrigeration [56,57].

Further developments are due Sixsmith and colleagues in Oxford and Boulder (see ref. [58,59]).

Figure 5.14 shows the flow circuit of the Linde models TCF10, TCF20 and TCF50, which respectively are designed to produce 10–20 l/h, 15–40 l/h and 60–100 l/h, without precooling. The addition of LN_2 precooling (in exchanger E_1) should roughly double the yield.

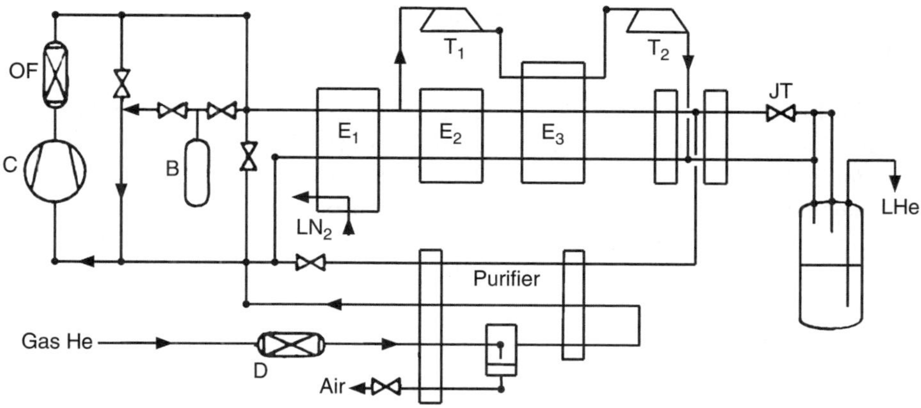

Fig. 5.14. Flow diagram of the Linde TCF series of turbine helium liquefiers. C denotes the screw compressor, OF the oil filter, $E_{1,2,3}$ the heat exchangers, $T_{1,2}$ the turbo-expanders and JT the final Joule–Thomson stage. Cylinder B is for pure He gas storage and Gas He is the normal He gas input to the purifier [60].

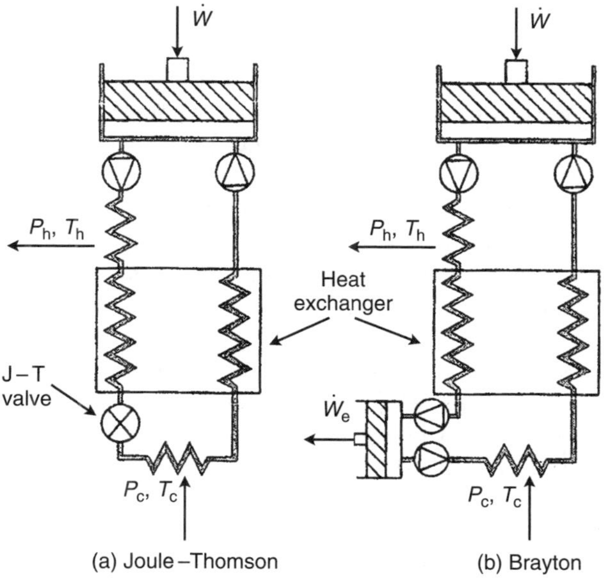

Fig. 5.15. Schematic diagram of Joule–Thomson (a) and Brayton (b) cooling cycles [60].

5.7.6 Brayton cycle

This cycle also uses continuous counterflow heat exchanger and is closely related to the Joule–Thomson and Claude cycles as shown in Fig. 5.15(a) [60]. The cryocooling or reverse Brayton cycle derives from a reciprocating gas engine patented by G. B. Brayton in

USA in 1872. Thermodynamically, the cooling version consists of an adiabatic (isentropic) compression followed by heat transfer to the surroundings, then adiabatic expansion and cooling.

Figure 5.15(b) shows that the final expansion stage occurs in a turbine, rather than in an isenthalpic Joule–Thomson orifice. It has a higher thermodynamic efficiency than that of the Joule–Thomson but is more complex and expensive.

Developments in their use for laboratory scale cryocoolers can be found in reports from cryogenic engineering conferences and cryocooler conferences, e.g. [61–63]. For recent examples of Brayton cooler realization, see [64,65].

5.7.7 Coolers using regenerative heat exchangers

Regenerative heat exchangers differ from the continuous-flow or recuperative exchangers discussed in the last section as they must act as 'cold storage' systems for the discontinuous or pulsed flow processes used in many cooling cycles. They can store the energy from one stream and later transfer the energy to a second stream, for example between the out-of-phase pulses of gas which occur in the Stirling, GM, Vuilleumier and PTRs which are described in Section 5.8.

'Regenerative' exchangers usually consist of a porous matrix of finely divided material in the form of wire mesh, plates or small balls. This form is chosen to maximize surface area and minimize the thermal conduction necessary within the solid in order to promote more efficient heat transfer with the surrounding gas within the duration of a cooling cycle. The exchanger should have a heat capacity (in the operating range of temperature) higher than that of the gas stream. For operation down to 70 or 80 K, stainless steel, phosphor bronze or nickel are used. For cooling stages extending below this (down to 4 K), a combination of lead balls and rare earths is used. For example, ref. [63–66] describe a systematic investigation of a mix of Nd, ErNi, Er_3Ni and $HoCu_2$ with about 25% Pb balls. Here the heat capacities remain large as a result of low Debye temperatures or magnetic ordering effects.

For further information, see ref. [67] or the numerous papers in the proceedings of recent cryogenic engineering conferences which discuss regenerator materials suited to different temperature ranges in GM and pulse tube coolers.

5.7.8 Philips Stirling cycle

In 1954, Kohler and Jonkers developed a small air liquefier using the Stirling engine principle. Later, they produced larger air liquefiers and two-stage systems for lower temperatures [68]. The small portable single-stage version became very popular in the 1980s (e.g. ref. [69]).

Machines of this type were widely used for laboratory air or nitrogen supplies before commercial supplies became more available and convenient. LN_2 was usually produced from air by distillation or sorption, eliminating oxygen for safety reasons.

The basic cycle is illustrated in Fig. 5.16. The working gas (helium) is compressed in one chamber and then transferred through a regenerative heat exchanger to a second chamber where is expanded to cool the 'head' E and later transferred back through the exchanger to the lower chamber. The gas movements are achieved through the approximately quadrature phase shifted motions of two pistons B and C in the cylinder A. Piston C is usually called the displacer.

The distinct stages shown in Fig. 5.16 are:

- Compression in space D by the piston B; the heat of compression is removed by water cooling (or air cooling in the mini-versions).
- Transfer of gas through regenerator G to space E by the displacer; the gas cools through contact with the cold regenerator.
- Expansion and further cooling in space E.
- Transfer of cold gas back through G to space D, thus cooling the regenerator for the next cycle.

Sometimes, the split-cycle cooler, partly shown in Fig. 5.17, is more suitable because of the separation of the 'cold head' from the compressor. The compression is produced

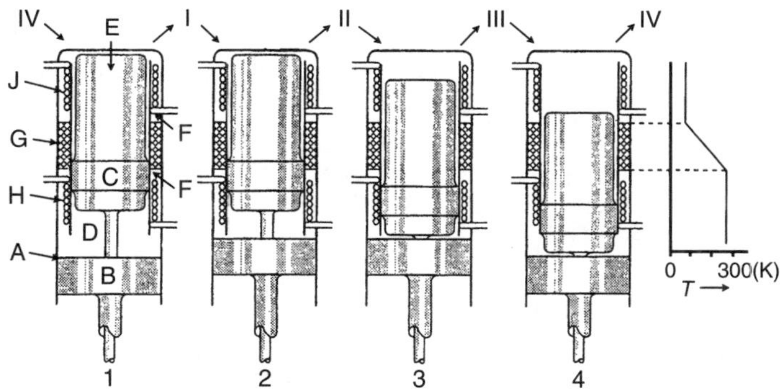

Fig. 5.16. The successive stages of the Philips Stirling cycle. See text for details [68].

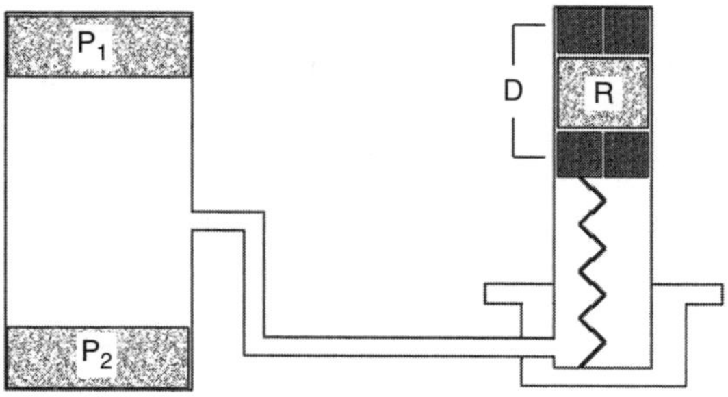

Fig. 5.17. Schematic diagram of a split-cycle Stirling refrigerator.

by two pistons in the left-hand cylinder which move the gas to the bottom of the right-hand cylinder. The combined displacer/regenerator then moves downwards to transfer the compressed gas to the 'head' of the cylinder after which the pistons P_1 and P_2 move apart to allow expansion and cooling. Movement of the displacer/regenerator in the right-hand cylinder is actuated by a spring.

The separation of the expansion chamber (with cold finger) from the compressor by a capillary (~30cm) reduces vibration. Typical small split-cycle single-stage units have a cooling power of ~1 W near 80 K, a cool-down time of 5–10 min and a life of over 5000 h.

Two-stage Stirling refrigerators can achieve temperatures of 4 K and are used for magnet cooling or for precooling in Joule–Thomson liquefiers. In this case, the second-stage regenerators generally use Pb and rare earth mixtures to provide a large heat capacity at low temperatures.

Recent studies and realizations of Stirling refrigerators are reported in ref. [39,44,46, 70–73].

5.7.9 Gifford–McMahon

A single-stage version of the GM cooler is depicted in Fig. 5.18. There is a displacer/regenerator, as in the Stirling cycle, but here the flow from the compressor is

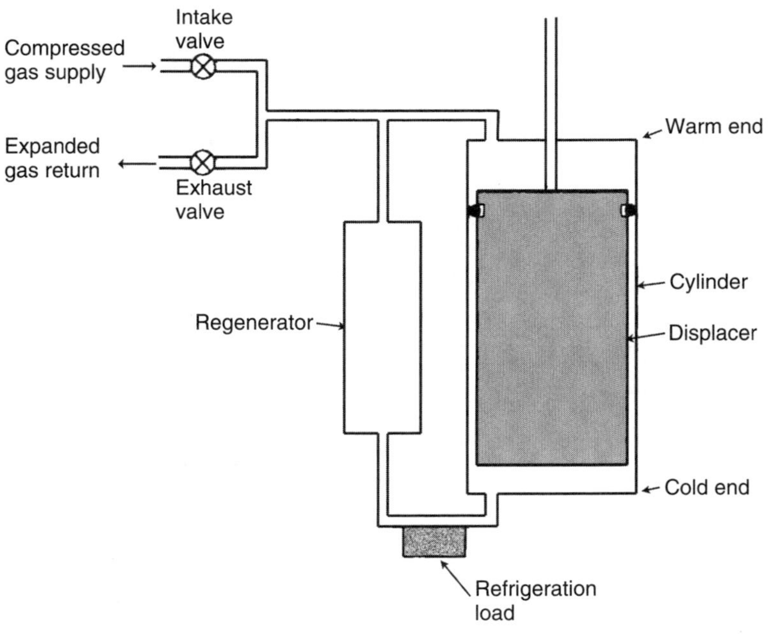

Fig. 5.18. A single-stage version of the Gifford–McMahon refrigerator [74].

controlled by inlet and outlet valves. Also the compressor can be at some meters from the cold head, thereby reducing vibrational problems.

The cycle consists of four phases:

I. The pressurization phase has the displacer piston at the cold end, so that compressed gas flows into the warm volume through the input valve.
II. The input phase, during which the input valve remains open and the displacer moves upwards to enhance the cold volume, with gas from the warm end passing through the regenerator.
III. The expansion phase begins when the input valve is closed and the exhaust valve is opened slowly so that cooling occurs at the cold end.
IV. The exhaust phase occurs when the displacer is moved downwards to push out the remaining cold gas through the regenerator. The exhaust valve is then shut and the cycle is repeated.

Single-stage GM coolers are mainly used for communication electronics and cryopumping, cryogen free magnets and for cooling shields.

Two-stage and three-stage versions have been developed for cooling microwave amplifiers, precooling Joule–Thomson helium liquefiers and in some cryostats.

For the use of GM cryocooler as helium liquefier see ref. [75]; for driving a dilution refrigerator see ref. [76,77].

5.8 Pulse tube refrigerators

5.8.1 Introduction

A drawback of mechanical refrigerators described in the previous section is the presence of moving parts in the cold area.

The PTR is a closed-cycle regenerative mechanical cooler. The main difference of PTR from other mechanical coolers, such as Stirling coolers or GM coolers is that a PTR has no moving parts at low temperatures, and, therefore, has a long lifetime and a low vibration level. The PTR is based on the displacement and the expansion of a gas, usually helium. The working fluid undergoes an oscillating flow and an oscillating pressure. A typical average pressure in a PTR is 10–25 bar, and a typical pressure amplitude is 2–7 bar.

A piston compressor (in case of a Stirling type PTR) or a combination of a compressor and a set of switching valves (GM type PTR) are used to create the pressure oscillations. The regenerator of a PTR stores the heat of the gas during a half-cycle and, therefore, must have a high heat capacity, compared to the heat capacity of the gas.

In 1963, Gifford and Longsworth introduced the basic Pulse tube refrigerator (BPTR) [74]. It is worth noting that the BPTR was quite different from a modern PTR. The cooling principle of the BPTR is the surface heat pumping, which is based on the exchange of heat between the working gas and the pulse tube walls. In modern pulse tube coolers, the heat pumping mechanism is considered a negative effect.

The lowest temperature, reached by Gifford and Longsworth with the BPTR, was 124 K with a single-stage PTR and 79 K with a two-stage PTR (see Fig. 5.19) [78].

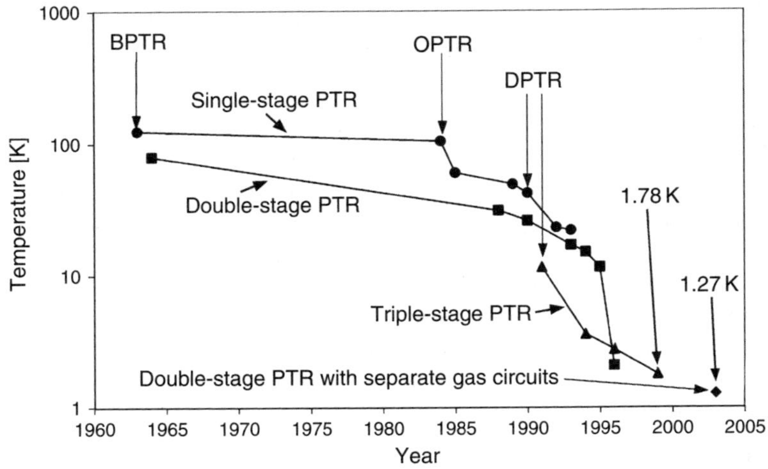

Fig. 5.19. History of temperatures reached with a pulse tube refrigerator.

The main breakthrough came in 1984, when Mikulin and his co-workers introduced the Orifice pulse tube refrigerator (OPTR) [79]. A flow resistance (the orifice) was inserted at the warm end of the pulse tube to allow some gas to pass to a large reservoir. With a single-stage configuration of the OPTR, Mikulin achieved 105K, using air as the working gas. Soon afterwards, Radebaugh reached 60 K with a similar device, using helium [80].

In 1990, Zhu et al. connected the warm end of the pulse tube with the main gas inlet by a tube, containing a second orifice [81]. Thus, a part of the gas could enter the pulse tube from the warm end, bypassing the regenerator.

Because of this effect, such a configuration of the PTR was called the double-inlet pulse tube refrigerator (DPTR). In 1994, Matsubara, using this configuration, reached 3.6 K with a three-stage PTR [82].

In 1999, with a three-stage DPTR, a temperature of 1.78 K was reached [83]. For several years, this temperature remained the lowest temperature reached by a PTR. In 2003, the group of G. Thummes from Giessen University developed a double-circuit ^{3}He/^{4}He PTR that achieved 1.27 K [84].

Nowadays, PTRs provide cooling powers $\sim$1.5 W at 4.2 K and $\sim$40 W at 45 K.

As we said, the material of the regenerator of a PTR must have a high specific heat to provide a good heat storage. Unfortunately, below 20 K, the specific heat of most regenerators rapidly decreases, whereas the heat capacity of helium increases and has a maximum at $\sim$10 K (see Fig. 5.20).

This drawback has been overcome by using rare-earth magnetic materials in the coldest part of the regenerator. These materials have a magnetic phase transitions below 15 K, with an increase in their heat capacities (see Section 3.6). However, to find a material suitable for sub 4 K applications is still a challenge.

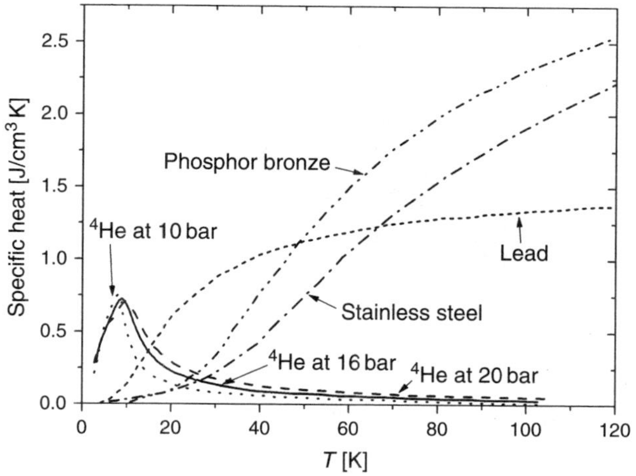

Fig. 5.20. Volumetric heat capacity of ^{4}He and some regenerator materials as functions of temperature.

It is hard to report a complete bibliography about PTR. More than 130 papers about PTRs, for example, appeared in 'Cryogenics' in the last decade. This observation witnesses the interest in this type of mechanical refrigerator. New versions of PTR, with improved characteristics, are continuously proposed and realized.

5.8.2 Two compression methods for the PTR

There are two types of PTRs: a Stirling-type PTR (Fig. 5.21(a)) and a GM-type PTR (Fig. 5.21(b)), depending on the way the pressure oscillations are generated.

In the GM-type PTR (mostly used at present), a compressor produces continuous high and low pressures and uses a rotary valve to generate pressure oscillations in the pulse tube. In the Stirling-type PTR, pressure oscillations are created by the movement of a piston, directly connected to the pulse tube and controlled by a motor.

The main components of the GM-type PTR are shown in Fig. 5.21(b). From the left to the right, the pulse tube system consists of a compressor (CP), a room temperature heat exchanger or an after-cooler (E_1), a rotary valve (RV), a regenerator (RG), a low-temperature heat exchanger (E_2), a pulse tube (PT), another room temperature heat exchanger (E_3), two orifices (O_1 and O_2) and a buffer volume (BF).

The compressed gas enters the regenerator at room temperature T_H. The regenerator consists of a porous material with a high specific heat. The function of the regenerator is to store heat, adsorbed from the gas, when it flows from the compressor to the tube, and to reject this heat back to the gas, when the gas flows from the tube to the compressor. After passing the regenerator, the gas enters the pulse tube, which has the heat exchangers E_2 and E_3 on its two ends. At the low-temperature end T_L, the cold heat exchanger E_2 is used to extract the heat from the cooling object. The hot heat exchanger E_3 rejects this heat to the surroundings. The orifices O_1 and O_2 are flow resistances.

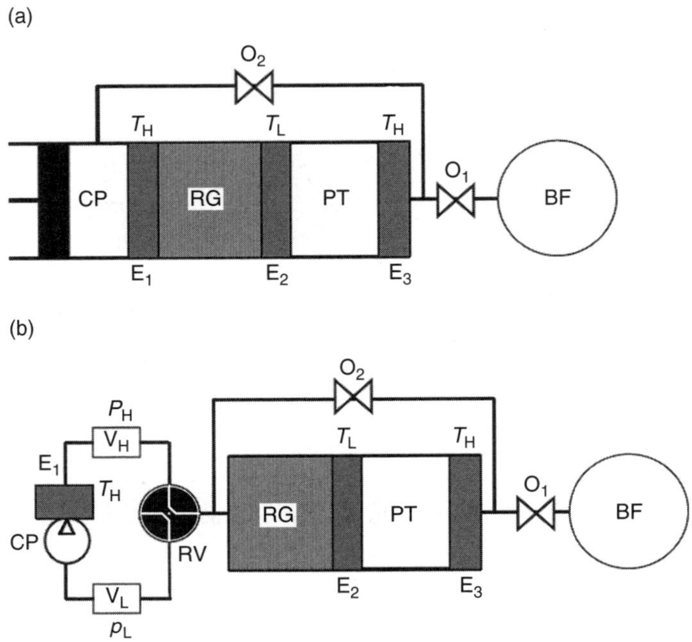

Fig. 5.21. Two types of the PTRs. (a) A Stirling-type PTR. From left to right, it consists of a compressor (CP), an after-cooler (E_1), a regenerator (RG), a cold heat exchanger (E_2), a pulse tube (PT), a hot heat exchanger (E_3), an orifice (O_1) and a buffer volume (BF). Orifice O_2 connects the hot end of the regenerator and the hot end of the pulse tube. (b) A GM-type PTR. Except for the compressor (CP) and rotary valve (RV) combination, the main components of the GM-type PTR are the same as that of the Stirling PTR.

The buffer volume is a reservoir with a volume typically 10 times larger than the volume of the pulse tube. The pressure in the buffer is almost constant and close to the average pressure in the pulse tube. The combination of the orifice O_1 and the buffer provides a phase difference between the flow of the gas in the tube and the pressure oscillation; such phase difference is necessary for the performance of the PTR.

The function of the double-inlet orifice O_2 is to reduce losses. It allows some gas to bypass the regenerator and to enter the pulse tube via E_3.

The choice of the gas compression method depends on the working temperature range. In the case of a single-stage pulse tube cooler for the 20–80 K range, the Stirling-type compressor is more convenient, having a high efficiency, which results in a lower total weight. Typically, the driving frequency of the Stirling-type PTR is from 25 to 50 Hz.

To obtain temperatures below 20 K, the systems are usually operated at lower frequencies (1–5 Hz).

The frequency defines the diffusion depth in the working gas and the regenerator material. The diffusion depth, d, is given by:

$$d = \left(k/\pi\nu C\right)^{1/2} \tag{5.19}$$

where k is the heat conductivity, ν is the frequency and C is the volumetric heat capacity of the regenerator material. From eq. (5.19) we see that, when the frequency is increased, the diffusion depth decreases, and the heat storage in the regenerator degrades. Hence, the regenerator material with a small particle size or fine-mesh screens must be used in the high-frequency PTR. Multistage pulse tubes are required to reach temperatures of LHe range [85].

Large temperature difference over the regenerator requires a rather large length of it. In this case, a high operating frequency leans to a large pressure drop in the regenerator, i.e a poor performance of a system. Therefore, the GM-type PTR is more suitable for temperatures below 20K.

However, there is still an interest in achieving low temperatures with the Stirling-type PTR [86]. This is especially important for space applications, since it is impossible to use a large water-cooled GM-type compressor in space.

5.8.3 Simplified operation principle of PTRs

We are now going to explain the operation principle of PTRs on the example of the OPTR. Let us consider a small gas parcel, which during a cycle travels from the regenerator through E_2 into the pulse tube and back.

Several simplifying assumptions are adopted: (1) the flow in the tube is considered to be one dimensional without turbulence or mixing; (2) the processes in the tube are adiabatic. There is no heat conduction in the tube and no heat exchange between the gas and the tube walls; (3) each heat exchanger is isothermal.

The pressure in the pulse tube is considered to be cyclic with the shape of Fig. 5.22(a). The cycle is divided into four steps. We will assume that the orifice O_1 is closed during steps 1 and 3 and open during steps 2 and 4.

Step 1. The compression step. The system is connected to the high-pressure side of the compressor. The pressure in the pulse tube rises from p_L to p_H. The orifice O_1 is closed.

Step 2. The orifice O_1 is open. As the pressure in the buffer volume is close to the average pressure in the system, the pressure in the pulse tube is higher than the pressure in the buffer. Hence, the gas flows from the tube to the buffer. Since the rotary valve is still open at the high-pressure side of the compressor, the pressure in the tube remains p_H.

Step 3. The expansion step. The system is connected to the low-pressure side of the compressor. The pressure in the pulse tube decreases from p_H to p_L, while the orifice O_1 is closed.

Step 4. The orifice O_1 is open again. Now the pressure in the buffer is higher than the pressure in the pulse tube; hence, the gas flows from the buffer into the tube. Since the rotary valve is still open at the low-pressure side of the compressor, the pressure in the tube remains p_L.

We now follow a gas element, which is inside the regenerator at the beginning of the cycle (point a on Fig. 5.22(a) and Fig. 5.22(b)).

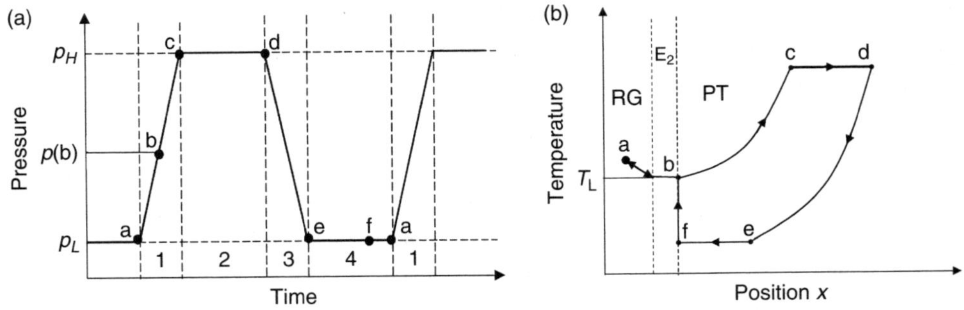

Fig. 5.22. (a) Idealized form of the pressure in a pulse tube as a function of time. The cycle is divided into four steps. The pressure rises from a low pressure, p_L, to a high pressure, p_H, during the compression step 1 (from a to c via b), remains at a value of p_H during step 2 (from c to d), decreases from p_H to p_L during the expansion step 3 (from d to e) and is equal to p_L during step 4 (from e to a via f). (b) Temperature of a gas particle as a function of position in the regenerator E_2 and in the pulse tube PT in the OPTR for a p-time dependence, given in Fig. 5.22 (a).

Step 1. From a to b and c. When the pressure rises, the gas element moves through the regenerator in the direction of the heat exchanger E_2 (from a to b). According to the assumptions, the heat exchange is perfect, and the temperature of the gas element is equal to the temperature of the regenerator. At point b, the gas element leaves E_2 and enters the tube with the temperature T_L. From b to c, the gas element is compressed adiabatically, while it moves towards the orifice. Its temperature rises together with the pressure.

Step 2. From c to d. The gas element moves to the right. The pressure and temperature are constant.

Step 3. From d to e. The orifice is closed and the expansion takes place. Gas moves back in the direction of E_2. Since it is thermally isolated, its temperature and pressure drop. The gas element enters the pulse tube at some intermediate pressure, p(b), after the cycle had already started (see Fig. 5.22(a)). At the end of the expansion step, the pressure of the gas element, p(e), is equal to p_L (Fig. 5.22(a)). As $p(e) < p(b)$, the temperature of the gas element at the end of step 3 is below T_L.

Step 4. From e to f and a. The expansion stops, and the orifice is open again. The gas continues to flow in the direction of E_2. From e to f: the pressure is constant, so the temperature is constant as well. At point f, the gas parcel enters E_2 with $T < T_L$. When passing E_2, the gas warms up to the temperature T_L. The amount of heat, which the gas takes away from the heat exchanger, is the cooling power. In the remaining time of the cycle, the gas element moves inside the regenerator to its original position.

The absence of the gas flow at the hot end means that the PTR operates in the basic mode. This is shown in Fig. 5.23.

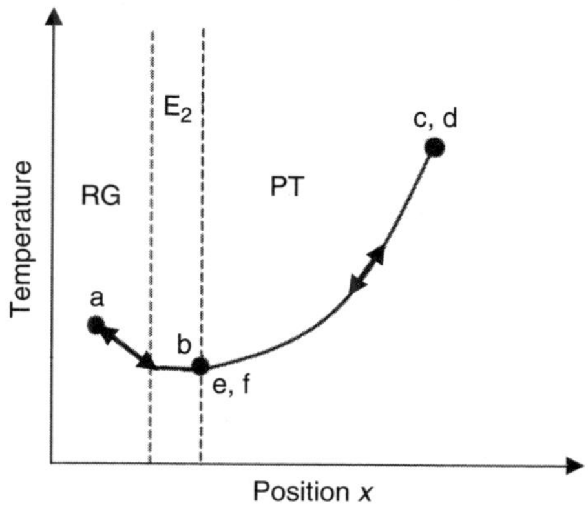

Fig. 5.23. Temperature of a gas particle as a function of position in the regenerator E_2, and in the pulse tube PT in the BPTR for a p-time dependence, given in Fig 5.22(a).

5.8.4 Cooling power

Let us assume a time dependence of the pressure in the pulse tube of the type:

$$p_t = p_0 + p_1 \cos \omega t \tag{5.20}$$

where p_0 is the average pressure, p_1 is the amplitude of pressure oscillation, and ω is the angular frequency, related to the cycle time t_c and the frequency ν by:

$$\omega = 2\pi\nu = 2\pi t_c \tag{5.21}$$

The volume flow of the gas $\dot{V}_H$ at the hot end of the pulse tube is determined by the pressure oscillations δp:

$$\delta p = p_1 \cos \omega t \tag{5.22}$$

and the resistance of the orifice R_1:

$$\dot{V}_H = \frac{\delta p}{R_1} \tag{5.23}$$

The heat, extracted at the cold end of the pulse tube, is rejected to the surroundings by the hot heat exchanger. Therefore, when the gas passes the hot heat exchanger from the pulse tube into the buffer, its temperature T_g is higher than the temperature of the heat exchanger T_H [86]:

$$T_g = T_H + 2T_H \frac{R}{c_p} \frac{\delta p}{p_0} \tag{5.24}$$

where R is the molar ideal gas constant and c_p is the specific heat of the gas.

When the gas returns to the pulse tube, its temperature is equal to the temperature of the heat exchanger: $T_g = T_H$. The heat exchange rate between the gas and the heat exchanger is given by:

$$\dot{Q}_H = \dot{n}_H c_p (T_g - T_H) \tag{5.25}$$

where $\dot{n}_H$ is the molar flow rate at the hot end of the pulse tube, given by:

$$\dot{n}_H = \frac{p_0 \dot{V}_H}{RT_H} \tag{5.26}$$

Integrating $\dot{Q}_H$ in eq. (5.25) over a cycle and dividing by the cycle time gives the average heat exchange rate in the hot heat exchanger:

$$\overline{\dot{Q}_H} = \overline{\dot{V}_H \delta p} = \frac{1}{2} V_1 p_1 \tag{5.27}$$

where V_1 is the volume of the gas passing the orifice during a cycle.

The displacement of the gas at the cold end of the pulse tube is equal to the displacement of the gas at the hot end plus an additional term due to the compressibility of the gas in the tube. Thus, the volume of the gas displaced at the cold end of the tube is:

$$\dot{V}_L = \dot{V}_H + \frac{c_V V_t}{c_p p_0} \frac{dp_t}{dt} \tag{5.28}$$

where V_t is the volume of the tube. Because of the dp_t/dt term, the gas flow is no longer in phase with the pressure in the pulse tube.

From Fig. 5.22(b), we see that the gas enters the pulse tube with the temperature equal to the temperature of the cold heat exchanger, $T_g = T_L$. However, when the gas comes back from the pulse tube its temperature is lower than T_L:

$$T_g = T_L + T_L \frac{R}{c_p} \frac{p_1}{p_0} \frac{2}{1+\beta^2} (\cos \omega t - \beta \cos \omega t) \tag{5.29}$$

where β is the gas piston parameter, expressed as follows:

$$\beta = \frac{c_V}{c_p} R_1 \frac{V_1}{p_0} \omega \tag{5.30}$$

The heat exchange rate between the gas and the cold heat exchanger is:

$$\dot{Q}_L = \dot{n}_L c_p (T_g - T_L) \tag{5.31}$$

where $\dot{n}_L$ is the molar flow rate of the gas at the cold end of the pulse tube, given by:

$$\dot{n}_L = \frac{p_0 \dot{V}_L}{RT_L} \tag{5.32}$$

The average heat exchange rate in the cold heat exchanger (cooling power of the PTR) may be obtained by integration of eq. (5.31) over a cycle and dividing it by the cycle time:

$$\overline{\dot{Q}_L} = \overline{\dot{V}_L \delta p} = \frac{1}{2} V_1 p_1 \tag{5.33}$$

The cooling power given by eq. (5.33) is valid only for an ideal gas. Above 30–40 K, helium can be considered an ideal gas. Below this temperature, this approximation is no longer valid. This means that the energy flows in the PTR are no longer functions of the

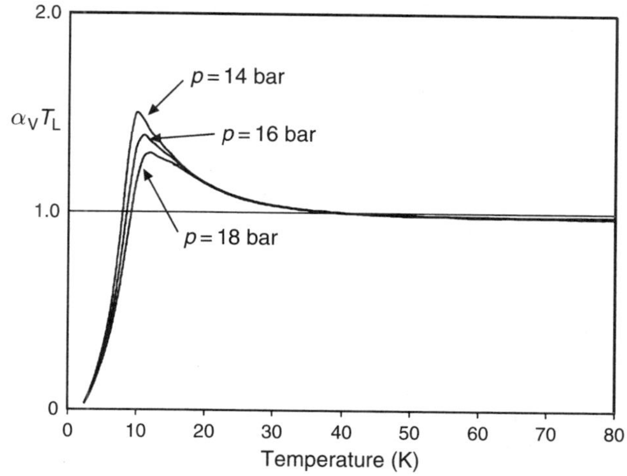

Fig. 5.24. The products of the temperature T_L and volumetric expansion coefficient α_V as functions of temperature at various pressures.

temperature only, but also of the pressure. This results in an additional term $T_L\alpha_V$ in the expression for the cooling power of the PTR:

$$\overline{\dot{Q}_L} = T_L\alpha_V\overline{\dot{V}_L\delta p} \tag{5.34}$$

where α_V is the volumetric thermal expansion coefficient. The product of T_L and α_V is plotted versus temperature in Fig. 5.24: above 30 K, the value of $(T_L\alpha_V)$ is close to 1, which is the value for an ideal gas.

5.8.5 Multistage PTRs

In a multistage PTR, one pulse tube precools the other ones. The hot end of all the pulse tubes is often connected to room temperature. In this configuration, large heat loads on the cold end of the precooling stage, resulting from the heat release in the hot heat exchanger of the following pulse tube, are avoided. The disadvantage of multistage pulse tubes is that the precooling stage reduces the flow to the final stage, thus reducing the cooling power of the final stage.

The subject of the advantages and disadvantages of multistaging has been scarcely treated in the literature [85,86].

References

[1] K. Uhlig, W. Hehn: Cryogenics **33**, 1028 (1993)

[2] G.K. White: *Experimental Techniques in Low Temperature Physics*, 3rd ed., Clarendon, Oxford (1979)

[3] J.E. Jensen, R.B. Stewart, W.A. Tuttle eds.: *Selected Cryogenic Data Notebook*, Brookhaven National Laboratory Report BNL 10200 (1966)
[4] B.H. Billings et al. (eds): *American Institute of Physics Handbook* 3d ed., Mc Graw Hill, New York (1972)
[5] Y.S. Touloukian, C.Y. Ho eds.: *Thermophysical Properties of Matter*, Plenum Press, New York (1973)
[6] U. Smidtchen et al.: Cryogenics **34**, 105 (1994)
[7] R.C. Richardson, E.N. Smith eds.: *Experimental Techniques in Condensed Matter Physics at Low Temperatures*, Addison-Wesley Publishing Company, Inc. Redwood City California (1988)
[8] H. Luck, Ch. Trepp: Cryogenics **32**, 690 (1992)
[9] J.D. Siegwarth, R. Radebaugh: in *Proc. 13th Int. Conf. Low Temp. Phys.*, p. 398, ed. by K.D. Timmerhaus, W.J. O'Sullivan, E.F. Hammel, Plenum Press, New York (1973)
[10] A.C. Rose-Innes: *Low Temperature Laboratory Techniques*, English University Press, London (1973)
[11] D.S. Betts: *Refrigeration and Thermometry Below One Kelvin*, Sussex University Press, Brighton (1976)
[12] K.W. Taconis: in *Prog. In Low Temp. Phys.*, vol. **3**, p.153, ed. by C.J. Goter, North-Holland, Amsterdam (1961)
[13] B.N. Eselson, B.G. Lazarev, A.D. Svets: Cryogenics **3**, 207 (1963)
[14] C.F. Mate et al.: Rev. Sci. Instrum. **36**, 369 (1965)
[15] D. Walton: Rev. Sci. Instrum. **37**, 734 (1966)
[16] A.D. Svets: Cryogenics **6**, 333 (1966)
[17] M. Fruneau, A. Lacaze, L. Weil: Cryogenics **7**, 135 (1967)
[18] W. Wiedemann, E. Smolic: in *Proc. 2nd Int. Cryogenics Eng. Conf.*, p. 559, Brighton (1968)
[19] D. Walton, T. Timusk, A.J. Sievers: Rev. Sci. Instrum. **42**, 1265 (1971)
[20] J.P. Torre, G. Chanin: Rev. Sci. Instrum. **56**, 318 (1985)
[21] O.V. Lounasmaa: *Experimental Principles and Methods Below 1 K*, Academic Press, London (1974)
[22] E.T. Swartz: Rev. Sci. Instrum. **58**, 881 (1987)
[23] H.P. Gush: Opt. Lett. **2**, 22 (1978)
[24] G. Ventura et al.: Infrared Phys. **26**, 347 (1986)
[25] J.P. Torre, G. Chanin: in *Advances in Cryogenic Engineering*, vol. **23**, p.640, Proceedings of the Conference, Boulder, Colo., August 2-5, 1977, Plenum Press, New York (1978)
[26] A. Graziani et al.: Cryogenics 43, 659 (2003)
[27] M.W. Dowley, R.D. Knight: Rev. Sci. Instrum. **34**, 1449 (1963)
[28] J.E. Whitehouse et al.: Rev. Sci. Instrum. **36**, 768 (1965)
[29] J. Gaffney, J.R. Clement: Rev. Sci. Instrum. **26**, 620 (1955)
[30] I.V. Velichkov, V.M. Drobin: Cryogenics **30**, 538 (1990)
[31] K.P. Jüngst, E. Süss: Cryogenics **24**, 429 (1984)
[32] W.H. McAdam: *Heat Transmission*, McGraw Hill, New York (1954)
[33] M. Jakob: *Heat Transfer*, vol. II, Wiley, New York (1957)
[34] A.G. Lenfestey: In C.A. Bailey, *Advanced Cryogenics*, Chapter 6, Plenum Press, New York (1971)
[35] W.A. Little: Rev. Sci. Instrum. **55**, 661 (1984)
[36] H.J. Holland, et al.: *Cryogenics* **38**, 407 (1998)
[37] A.B. Patel, J.G. Brisson: Cryogenics **40**, 91 (2000)
[38] G. Frossati: private communication
[39] G.F. Nellis: Cryogenics **43**, 523 (2003)
[40] R.C. Longsworth: Adv. Cryog. Eng. **35**, 1315 (1990)
[41] G. Walker, E.R. Bingham: *Low-Capacity Cryogenic Refrigeration*, Clarendon Press, Oxford (1994)
[42] J. Fereday et al.: Cryogenics **46**, 183 (2006)
[43] H.J.M. ter Brake, G.F.M. Wiegerinck: Cryogenics **42**, 705 (2002)
[44] Xiaoqin Yang, J.N. Chung: Cryogenics **45**, 537 (2005)
[45] Takayuki Tomaru et al.: Cryogenics **44**, 309 (2004)
[46] A.M. Veprik et al.: Cryogenics **45**, 117 (2005)
[47] C. Minas: Cryogenics, **39**, 617 (1999)
[48] S.C. Collins: Rev. Sci. Instrum. **18**, 157 (1947)
[49] S.C. Collins: Adv. Cryog. Eng. **11**, 11 (1966)
[50] J.M. Poncet, A. Ravex: Adv. Cryog. Eng. **43B**, 1715 (1998)
[51] M.D. Atrey: Cryogenics, 38, 1199 (1998)

[52] A.P. Klimenko: in *Proc.10th Int. Congress. Refrig.*, Copenhagen, Pergamon Press London (1959)
[53] W.A. Little: in *Proc.ICEC 17*, p. 1, Bournemouth, UK 1998, Institute of Physics, Bristol (1998)
[54] M.J. Boiarski et al.: Adv. Cryog. Eng. **43B**, 1701 (1998)
[55] W.A. Little: Adv. Cryog. Eng. **43B**, 1305 (1990)
[56] S.C. Collins, R.L. Cannaday: *Expansion machines for low temperature processes*, Oxford University Press, Oxford (1958)
[57] P.L. Kapitza: Russ. J. Phys. **1**, 7 (1939) (English edn.)
[58] D.B. Mann et al.: Adv. Cryog. Eng. **8**, 221 (1963)
[59] C.A. Bailey: *Advanced Cryogenics*, Plenum Press, New York (1971)
[60] R. Radebaugh: in *Proc.16 ICEC/ICMC, Japan 1996*, p.33, Elsevier Science, Oxford (1997)
[61] J.A. McCormick et al: *Cryocoolers 10*, ed. by E.G. Ross Jr., p.421, Kluwer Academic/Plenum, New York (1999)
[62] G.F. Nellis, J.L. Smith: Adv. Cryog. Eng. **43**, 1767 (1998)
[63] G.F. Nellis et al: *Cryocoolers 10*, ed. by E.G. Ross Jr., p.431, Kluwer Academic/ Plenum, New York (1999)
[64] Mark V. Zagarola, John A. McCormick: Cryogenics **46**, 169 (2006)
[65] Jae Hong Park, Yong Ha Kwon, Young Soo Kim: Cryogenics **45**, 71 (2005)
[66] A. Lang et al.: Adv. Cryog. Eng. **43B**, 1573 (1998)
[67] G. Walker, E.R. Bimgham: *Low-Capacity Cryogenic Refrigeration*, Clarendon Press, Oxford (1994)
[68] J.W.L. Koehler: Prog. Cryogenics **2**, 43 (1960)
[69] G. Walker: *Miniature Refrigerators for Cryogenic Sensors and Cold Electronics* Clarendon Press, Oxford (1989)
[70] Deuk-Yong Koh et al.: Cryogenics **42**, 427 (2002)
[71] Xiaoqin Yang, J.N. Chung: Cryogenics **45**, 537 (2006)
[72] Zhang Cun-quan et al.: Cryogenics **42**, 577 (2002)
[73] S.J. Park et al.: Cryogenics **42**, 419 (2002)
[74] W.E. Gifford, R.C. Longsworth: *Pulse Tube Refrigeration*, Trans. ASME, p. 264 (1964)
[75] P. Schmidt-Wellenburg, O. Zimmer: Cryogenics **46**, 799 (2006)
[76] K. Uhlig, W. Hehn: Cryogenics **37**, 279 (1997)
[77] K. Uhlig: Cryogenics **42**, 569 (2002)
[78] R.C. Longsworth: in *Advances in Cryogenic Engineering* **12**, p. 608, Plenum Press, New York (1967)
[79] E.I. Mikulin, A.A. Tarasov, M.P. Shkrebyonock: in *Advances in Cryogenic Engineering* **29**, p. 629, Plenum Press, New York (1984)
[80] R. Radebaugh et al.: in *Advances in Cryogenic Engineering* **31**, p. 779, Plenum Press, New York (1986)
[81] Sh. Zhu, P. Wu, Zh. Chen: Cryogenics **30**, 514 (1990)
[82] Y. Matsubara, J.L. Gao: Cryogenics **34**, 259 (1994)
[83] M.Y. Xu, A.T.A.M. de Waele, Y.L. Ju: Cryogenics **39**, 865 (1999)
[84] N. Jiang et al.: Cryogenics **44**, 809 (2004)
[85] A.T.A.M. de Waele, I.A. Tanaeva, Y.L. Ju: Cryogenics **40**, 459 (2000)
[86] T. Nast: Cryogenics **46**,164 (2006)

6 Dilution Refrigerators

Contents

6.1 Introduction

In this chapter, we will deal with the most powerful device used in refrigeration. The dilution refrigerator (DR), which uses for refrigeration a mixture of ^{4}He and ^{3}He, not only is capable of reaching temperatures of a few millikelvin, but can maintain such very low temperature for months (theoretically for ever), whereas all the other refrigerators are 'one shot'. Hereafter we will describe four types of DR.

1. the classical DR,
2. the Joule–Thomson (J–T) DR,
3. the so-called 'dry dilution',
4. the 'no-gravity' dilution.

Under the pressure of 1 atm, ^{4}He and ^{3}He boil at 4.2 and 3.3 K respectively. As we have seen in Chapter 5, to obtain lower temperatures using the two cryogens separately, the pressure over the liquids must be decreased. For example, with a pressure of 10^{-2} torr, 0.9 K can be reached with ^{4}He and 0.3 K with ^{3}He. At such a low pressure, to get a throughput necessary to obtain a reasonable cooling power (0.1 mW), the pumping system and lines become cumbersome. For example, in the case of a ^{3}He refrigerator, a 350 m^3/h Root pump, backed by a 60 m^3/h primary pump, circulates 10^{-5} mole/s at 0.3 K with a pressure of 1.4×10^{-3} torr. The theoretical cooling power obtained by this huge system is 0.25 mW. To get the same cooling power with a DR at the same temperature, a ^{3}He circulation of only 35 μmole/s and a 20 m^3/h primary pump working at 3×10^{-2} torr is sufficient.

This new method of refrigeration was proposed in 1962 by H. London, G.R. Clarke and E. Mendoza [1] who developed an idea suggested by the same London eleven years before [2]. A group of the Leiden University built the first prototype of DR in 1965, obtaining a temperature of 0.22 K. Much more efficient refrigerators of this type were realized in the following years. G. Frossati obtained 2 mK in 1977 and 1.85 mK some years later and a record cooling power of 25 μW at 10 mK [3].

Until now, this is the only refrigerating technique capable of producing and maintaining temperature below 300 mK. The DR is the most important commercially available instrument working at temperatures between ~7 mK and 1 K. Moreover, using the DR as first stage, temperatures in the range of microkelvin can be obtained by means of nuclear demagnetization stages (see Section 7.4).

6.2 Properties of ^{3}He–^{4}He liquid mixture

We shall now briefly deal with the properties of the ^{3}He–^{4}He mixture which produces the refrigeration in a DR unit.

The respective concentration of the two helium isotopes are:

$$x = x_3 = \frac{n_3}{n_3 + n_4} \text{ and } x_4 = \frac{n_4}{n_3 + n_4} \tag{6.1}$$

where n represents the number of atoms or moles.

Figure 6.1 shows the x–T phase diagram of the ^{3}He–^{4}He mixture at saturated vapour pressure. Some important characteristics of these mixture are to be noted. In the dilution of ^{4}He with ^{3}He, the temperature of the superfluid phase transition of the former is lowered and the transition disappears for ^{3}He concentration above 67.5%. At such concentration

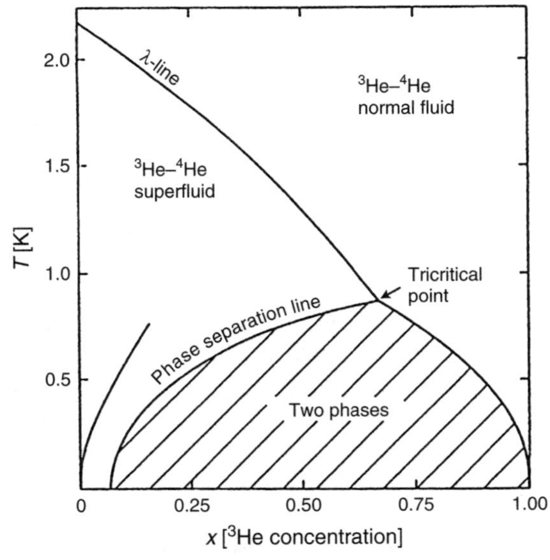

Fig. 6.1. Phase diagram for a mixture of ^{3}He–^{4}He at the saturated vapour pressure [4].

and at the temperature of 0.867 K (tricritical point), the λ-line encounters the phase-separation line. Below this temperature, the two isotopes form a mixture only for certain concentrations which depend on temperature, whereas two separated phases are formed along the phase-separation line. The dashed region in Fig. 6.1 is a non-accessible range of temperatures and concentrations for helium mixtures.

If a He mixture with $x > 6.6\%$ is cooled to temperatures below 0.87 K, the liquid separates into a phase rich in ^{4}He (left branch), the other rich in ^{3}He (right branch). Because of its lower density, the ^{3}He-rich liquid floats on top of ^{4}He-rich liquid. If the temperature is lowered towards 0K, the ^{3}He-rich liquid becomes pure ^{3}He (at $T = 10\,\text{mK}$, $x_4 \sim 10^{-28}$). But, even more surprising, if we observe the left branch of the phase diagram, x_3 does not vanish (for $T \to 0$, $x_3 \to 0.066$). As we shall see, this property is fundamental in the performance of a DR.

The concentrations of the two isotopes on the phase-separation branches at the saturated vapour pressure is given by [5] and [6–8]:

$$x_4 = 0.85 \cdot T^{\frac{3}{2}} \exp\left(-\frac{0.56}{T}\right) \tag{6.2}$$

$$x_3 = 0.066 \times \left(1 + 8.3 \times T^2\right) \tag{6.3}$$

The concentration (at $T = 50\,\text{mK}$) of ^{3}He in ^{4}He as a function of pressure is shown in Fig. 6.2 where it can be observed that, at a pressure of 10 bar, the solubility of ^{3}He in ^{4}He can be increased to 9.5% [6–8].

For $T < 0.5\,\text{K}$, pure ^{4}He is almost in its fundamental state: its entropy, viscosity and specific heat tend to zero. ^{4}He behaves in the mixture as an 'inert solvent' of ^{3}He, since its contribution to the total heat capacity is negligible. ^{3}He atoms in the diluted phase act as a gas: forcing a flux of ^{3}He atoms from the concentrated phase to the diluted one,

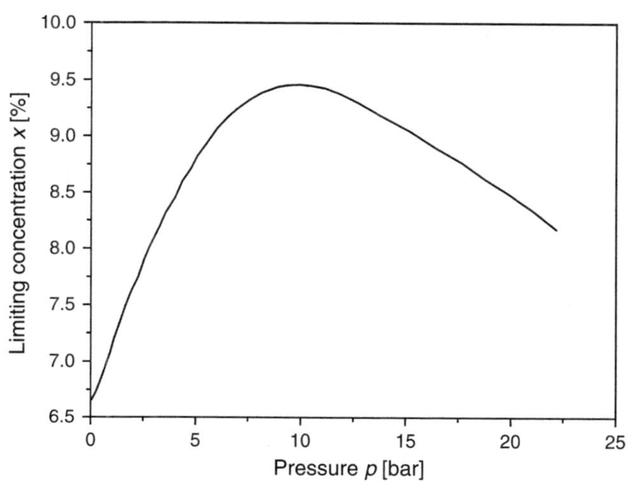

Fig. 6.2. Concentration (at $T = 50\,\text{mK}$) of ^{3}He in ^{4}He as a function of pressure. Data from ref. [7,8].

a process analogous to the evaporation of a liquid is obtained. In fact, the enthalpy of diluted ^{3}He is larger than that of pure ^{3}He.

This is the principle on which a DR works. The advantage of this process over the ^{3}He evaporation is evident: for example, at 0.28 K, the ^{3}He vapour pressure is about 10^{-3} torr, whereas the osmotic pressure from the rich to the diluted phase is about 10 torr for $T \to 0$. Thus, it is always possible to force a ^{3}He flux from the concentrated phase to the diluted one.

Figure 6.3 shows the comparison between the cooling powers supplied by a ^{3}He evaporation refrigerator and a DR. The cooling power produced by the 'evaporation' of ^{3}He from the concentrated to the diluted phase is:

$$\dot{Q} = \dot{n}_3 \left[H_d(T) - H_c(T) \right] \propto x \int \Delta C \mathrm{d}T \tag{6.4}$$

where $\dot{n}_3$ is the number of ^{3}He moles which pass in the unit time from the concentrated phase to the diluted one, H is the molar enthalpy and ΔC is the difference of specific heat of the two phases.

The concentration x of ^{3}He atoms in the diluted phase is practically constant for $T < 0.1$ K (about 6.6%, see eq. (6.3)). Since the ^{3}He specific heat is proportional to T (see Fig. 6.4), the cooling power of the dilution process is:

$$\dot{Q} \propto T^2 \tag{6.5}$$

as shown in Fig. 6.3.

To summarize, the cooling power in the dilution process is $\dot{Q} \propto T^2$, whereas in the ^{3}He evaporation process it is $\dot{Q} \propto e^{-1/T}$.

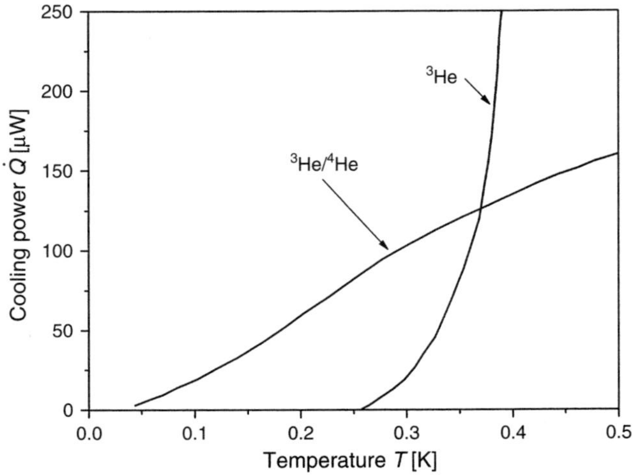

Fig. 6.3. Cooling powers supplied by a ^{3}He evaporation refrigerator and a ^{3}He–^{4}He D.R for the same ^{3}He circulation rate (5 l/s) [9].

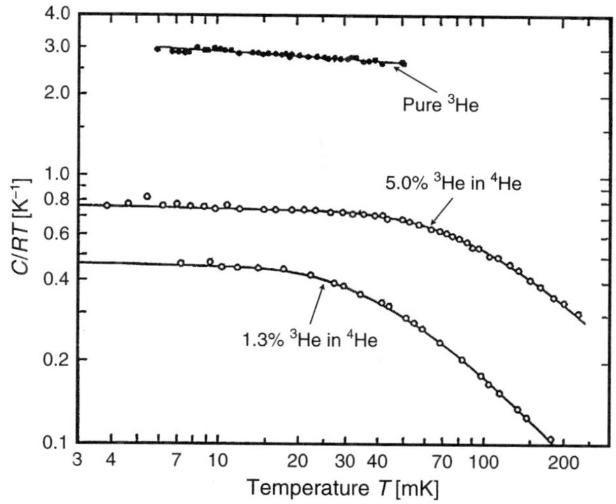

Fig. 6.4. Specific heat C/RT of ^{3}He and of two diluted solution of ^{3}He in ^{4}He at saturated vapour pressure [10,11]. C is the specific heat for the total number of helium moles.

6.3 The classic DR

We shall now describe the so-called 'classic dilution refrigerator'. The main components of a classic DR are shown in Fig. 6.5.

The lowest temperature is reached in the mixing chamber (MC) where the experiments are placed (sometimes inside, see Section 6.5) and where there is the interface between the concentrated and the diluted phase. The MC is in most cases made of Cu. The internal wall of the MC are covered with a sintered metallic powder (Ag or Cu) to reduce the thermal resistance R_k (see Section 4.4) between the liquid mixture of He and the walls.

Let us follow the path of the mixture along the DR starting from the MC. The diluted phase goes up (right branch of Fig. 6.5) and reaches the 'still' chamber through a Linde-type counterflow heat exchangers (see Section 5.7.2.4 and Fig. 6.6). The concentrated 'hot' phase flows in the inner tube towards the MC, whereas the 'cold' diluted phase coming from the MC flows in the outer tube. The cooling of ^{3}He (which comes back to the MC) by means of the outcoming diluted phase is fundamental (see thermal budget of eq. (6.7–6.11)). To optimize the efficiency of a DR, several 'step' heat exchangers (see Fig. 6.10) are put in series between the MC and the counterflow heat exchanger [9,12–26].

We remind (see Section 5.7.2) that a good step exchanger must have the following qualities:

1. small volume and large surface;
2. small flow impedance, to minimize the heating due to viscosity;
3. small thermal resistance between the two fluids to minimize the temperature gradient.

The third point leads to recall what was said about thermal resistances in Section 4.4.

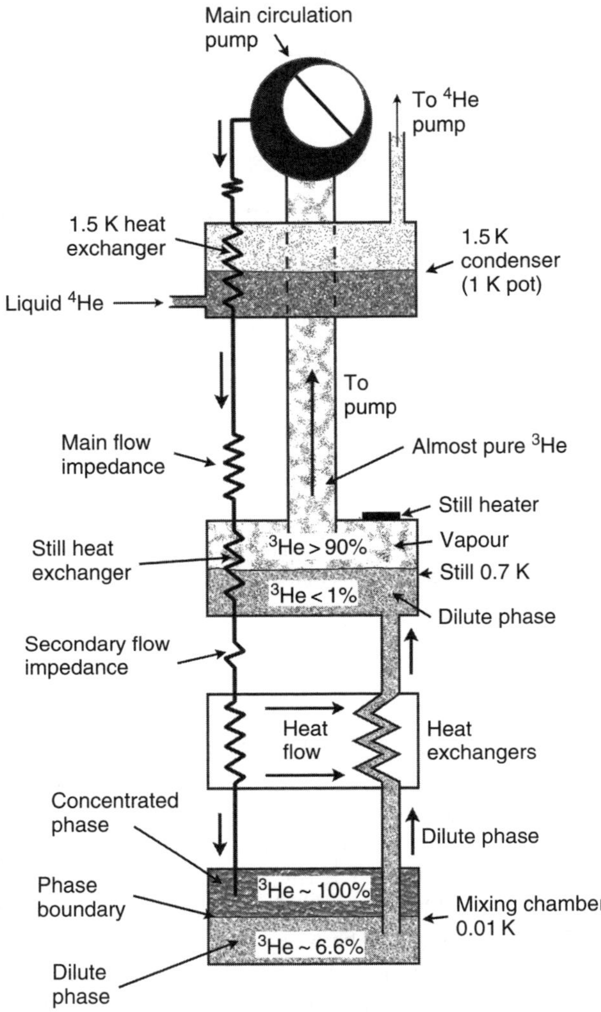

Fig. 6.5. Scheme of a 'classic' dilution refrigerator.

The thermal (Kapitza) contact resistance between helium and other materials is given by (see Section 4.4):

$$R_{\mathrm{K}} = \frac{a}{A} T^{-3} \text{ [K/W]} \tag{6.6}$$

Typical experimental values for a, if A is in [m^2]: $a_c \approx 0.05$ [m^2 K^4/W] (for pure ^{3}He) and $a_d \approx 0.02$ [m^2 K^4/W] (for ^{3}He–^{4}He mixture). The factor a depends also on the material in contact with the fluid. In a DR heat exchanger, the two thermal resistances are in series. Since $a_c > a_d$, we should have $A_c > A_d$. For example, for pure ^{3}He, with $\dot{Q} = 0.1$ mW, $A = 0.1$ m^2 and $T = 100$ mK, we get a $\Delta T \approx 50$ mK; at $T = 20$ mK, and

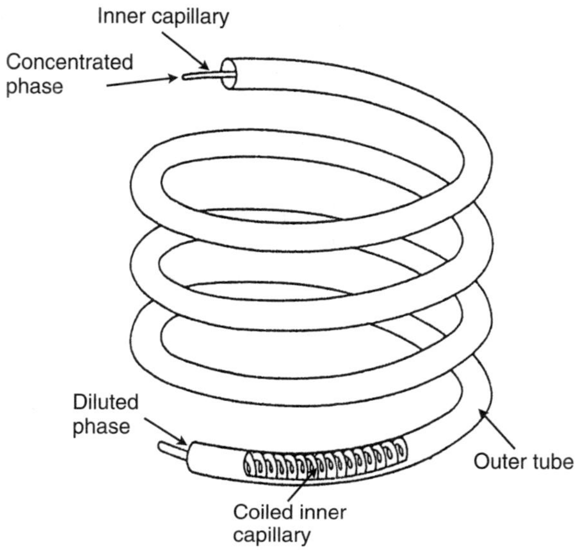

Fig. 6.6. Scheme of a counterflow heat exchanger for dilution refrigerator.

with $\dot{Q} = 0.01\,\text{mW}$, $A = 10\,\text{m}^2$, we get a $\Delta T \approx 6\,\text{mK}$. These temperature differences are too big: it will be necessary either to decrease $\dot{Q}$ or to increase A. Usually, to overcome the problem of Kapitza resistance for $T < 0.1\,\text{K}$, some square meters of surface are needed in the heat exchangers. Moreover, because of the temperature dependence of the Kapitza resistance, the problem is even heavier in the MC where, depending on $\dot{n}_3$ and T_{mc}, tens or even hundreds of square meters of contact surface are necessary [9,12–25].

As mentioned before, for a simple DR with minimum temperature around 30–40 mK, a simple counterflow heat exchanger will be enough (Fig. 6.6). The latter is realized by means of two concentric capillaries (diameters ranging between 0.5 mm and 1 mm, walls 0.1–0.2 mm) with length of a few meters [9,10,12,13,25]. The capillary material must have low thermal conductivity as Cu–Ni alloy, stainless steel or brass. They are wound in coil form for space reasons. In this type of heat exchangers, the temperature continuously changes along the coil: heat is exchanged between the two liquids through the separation wall, while the heat conduction along the capillaries must be negligible. The diluted phase coming from the MC flows in the space between the two capillaries, while the concentrated phase flows in the inner tube. In a more sophisticated configuration, also the inner capillary is wound in a spiral form in order to increase the contact surface, especially in the lower temperature part, as shown in Fig. 6.6 [25,26,28]. With a very good continuous heat exchanger, a DR can achieve temperature as low as 15–20 mK.

To reach lower temperatures, the exchange surface of continuous heat exchanger is not large enough, and step exchangers put in series are added. Step exchangers are usually made from a copper block in which channels for the passage of the two liquids are realized. The latter are filled with sintered Cu (or Ag) powder or with Cu foils to increase the exchange surface.

Step exchangers should have a low flow impedance to minimize heating due to fluid viscosity. As an example, in Fig. 6.7, the Ehnholm and Gylling step exchanger [27] is depicted: there 40 μm diameter sintered Cu powder is used. The packing factor of the powder of about 41% assures a small flux impedance. To get both a large contact surface and a low flow resistance, thin copper 'sponges' of quite large diameter have been built. Thin copper wires were inserted and sintered in the sponge, the other ends have been silver soldered to the exchanger body. Simpler heat exchangers are filled with Cu or Ag powders [19].

The development of step heat exchangers is due especially to Frossati [14–15] who first used Ag powder: an example of Frossati's heat exchanger is shown in Fig. 6.8. This is made up of several rectangular elements filled with sintered Ag powder.

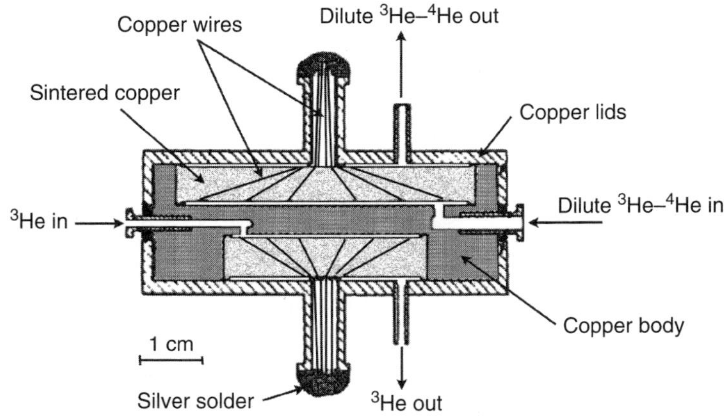

Fig. 6.7. Section of an heat exchanger realized with sintered copper powder.

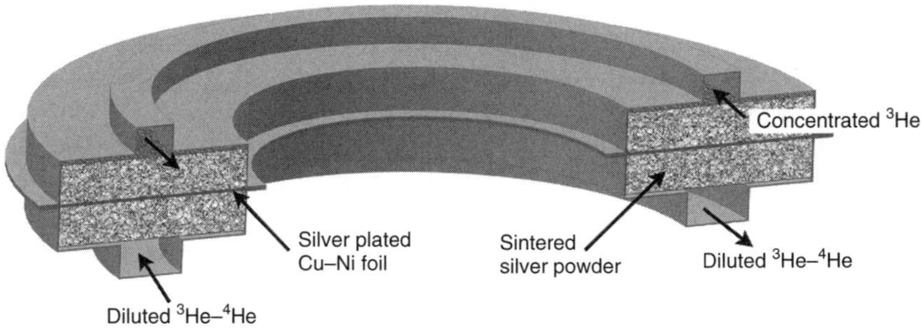

Fig. 6.8. Scheme of a composite heat exchanger realized with sintered silver powder [14–15].

Let us now explain how important the heat exchangers are in the performance of a DR. The need of precooling of ^{3}He before entering the MC is clear if we consider the enthalpy budget:

$$\dot{n}_3\left[H_{3,d}\left(T_{\mathrm{mc}}\right)-H_3\left(T_{\mathrm{mc}}\right)\right]=\dot{n}_3\left[H_3\left(T_{\mathrm{ex}}\right)-H_3\left(T_{\mathrm{mc}}\right)\right]+\dot{Q} \tag{6.7}$$

where $\dot{n}_3$ is the number of moles of ^{3}He crossing the concentrated/diluted interface in the unit time, $\dot{Q}$ is the MC cooling power and T_{ex} is the temperature of ^{3}He coming back to the MC from the exchangers. From ref. [28], the dependence on temperature of the enthalpy H of pure and dilute ^{3}He is:

$$H_3(T)=H_3(0)+11\cdot T^2\ \left[\mathrm{J/mol\,K^2}\right] \tag{6.8}$$

$$H_{3,d}(T)=H_3(0)+95\cdot T^2\ \left[\mathrm{J/mol(^3He)\,K^2}\right] \tag{6.9}$$

From eqs. (6.7), (6.9) and (6.9) we get the cooling power:

$$\dot{Q}=\dot{n}_3\left(95\cdot T_{\mathrm{mc}}^2-11\cdot T_{\mathrm{ex}}^2\right)\quad [\mathrm{W}] \tag{6.10}$$

The maximum power $\dot{Q}_{\max}$ is obtained for $T_{\mathrm{mc}}=T_{\mathrm{ex}}$:

$$\dot{Q}_{\max}=84\cdot\dot{n}_3\cdot T_{\mathrm{mc}}^2 \tag{6.11}$$

The corresponding minimum temperature is:

$$T_{\mathrm{mc,min}}=\left(\dot{Q}/84\cdot\dot{n}_3\right)^{1/2} \tag{6.12}$$

From eq. (6.10), we can calculate the maximum admitted temperature for ^{3}He coming from the last heat exchanger $\left(\dot{Q}=0\right)$:

$$\dot{Q}=\dot{n}_3\left(95\cdot T_{\mathrm{mc}}^2-11\cdot T_{\mathrm{ex}}^2\right)\geq 0\ \Rightarrow\ 3T_{\mathrm{mc}}\geq T_{\mathrm{ex}} \tag{6.13}$$

When the temperature of ^{3}He returning to the MC is higher than the MC temperature for more than about a factor 3, the dilution process does not work. Equation (6.13) clearly shows the importance of having efficient heat exchangers. For example, if $\dot{n}_3=100\,\mu\mathrm{mol/s}$ and the power demanded to the MC is $1\,\mu\mathrm{W}$, we get:

$$T_{\mathrm{ex}}=\begin{cases}18\,\mathrm{mK}\\30\,\mathrm{mK}\end{cases}\Rightarrow T_{\mathrm{mc}}=\begin{cases}12\,\mathrm{mK}\\15\,\mathrm{mK}\end{cases}$$

The diluted phase coming out from the MC, after having passed through the heat exchangers, reaches the still. The latter consists of a chamber (see Fig. 6.9) where the diluted liquid phase is in equilibrium with a vapour that is practically only ^{3}He. In fact, at the typical working temperature of the still (~0.7 K), the ^{4}He vapour concentration is only a few percent (see Section 2.4.1).

The ^{3}He evaporated in the still is pumped by an external circulation pump. The still temperature is kept between 0.6 and 0.8 K by means of a heater. Changing the temperature within this range allows to control the ^{3}He vapour pressure, while keeping a low ^{4}He concentration in the vapour. In fact, the DR efficiency critically depends on the ^{3}He/^{4}He ratio in the circulating mixture.

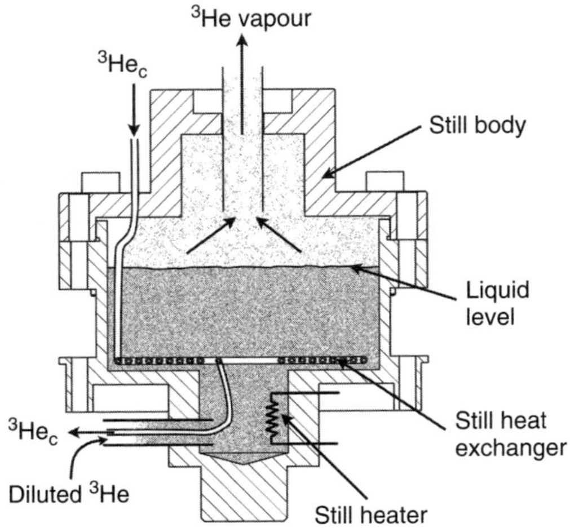

Fig. 6.9. Scheme of the still of a ^{3}He–^{4}He dilution refrigerator.

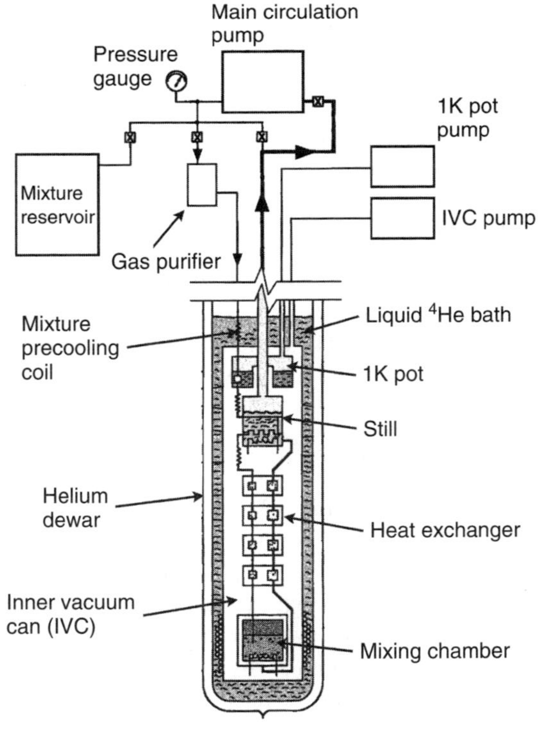

Fig. 6.10. Scheme of a typical ^{3}He–^{4}He refrigerator with four heat exchangers.

The still cooling power $\dot{Q}_{still}$, due to the ^{3}He evaporation process, can be relatively large (up to a few mW in big DR). It is thus possible to thermally connect a thermal shield to the still and also to berth and thermalize capillaries and wires.

The gaseous ^{3}He, removed from the still by the external pumping system at room temperature (see Fig. 6.10), is purified and again cooled at 4.2 K by the ^{4}He bath.

After the cooling at 4.2 K, the mixture undergoes a first condensation at 1.5 K in the '1 K pot'. Such temperature is maintained by the evaporation of an auxiliary ^{4}He bath pumped by an external pump. A flow impedance produces the pressure necessary for condensation and a heat exchanger inside the still provides the recondensation of the fluid partially evaporated in the expansion through the impedance. Finally, the concentrated liquid phase undergoes cooling processes in the heat exchangers before reaching the MC where the cycle is accomplished.

6.4 The J–T DR

Several attempts were made in the past to simplify the construction and operations of the DR [29–31]. A typical drawback which frequently causes problems in the use of a classical DR is the 1K pot. Besides the additional consumption of ^{4}He, the 1K pot may produce vibrations and suffers sometimes from blockages in its filling line.

In 1987, Uhlig [32] proposed and built a DR without a 1K pot. This idea led to the commercial production of a 'dipstick' DR which, thanks to its small diameter, can be immersed directly into the transport dewars, saving time and consumption of ^{4}He in the transfer. More recently, the interest for the J–T DR has strongly increased because of the progress in the field of mechanical refrigerators which allowed the realization of a 'Dry DR' (see Section 6.6).

In the J–T DR (see Fig. 6.11) [32], the condensation of the ^{3}He–^{4}He mixture is produced by a J–T expansion (after Z_1) which takes place after a precooling in a J–T heat exchanger. To achieve the correct decrease of the gas temperature and the right exit pressure, a small compressor is inserted in the circulation line at the output of the circulation pump.

In a J–T multistage heat exchanger, the mixture is cooled by the colder vapours coming out from the still (see Fig. 6.12). This type of exchanger can be seen as a series of isothermal heat exchangers connected in series by low thermal conductivity capillaries.

Typically, in this heat exchanger, the gas mixture temperature is lowered from 4.2 to about 3 K (see Fig. 6.13).

After the J–T heat exchanger, an expansion is realized by flux impedances inserted in the injection line (Z_1 in Fig. 6.11). These impedances can consist of a coiled capillary (smaller than the circulation tube) followed by a much larger (three to four times) tube. Another impedance, placed after the still heat exchanger (Z_2 in Fig. 6.11), is necessary to prevent the pressure in the still exchanger from becoming too low and producing the evaporation of the mixture. Below the still, the mixture undergoes the same processes described in the case of the classic DR, passing through (continuous/step) heat exchangers before reaching the MC and returning to the room temperature.

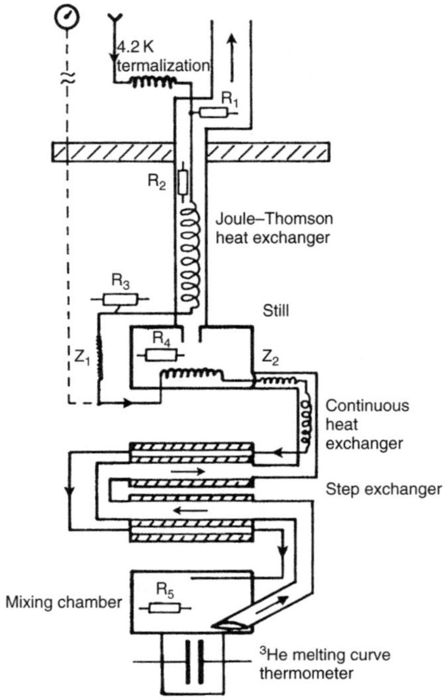

Fig. 6.11. Scheme of a J–T dilution refrigerator made by Uhlig [32]. The R_i represent the thermometers, while Z_i represent the main flux impedances.

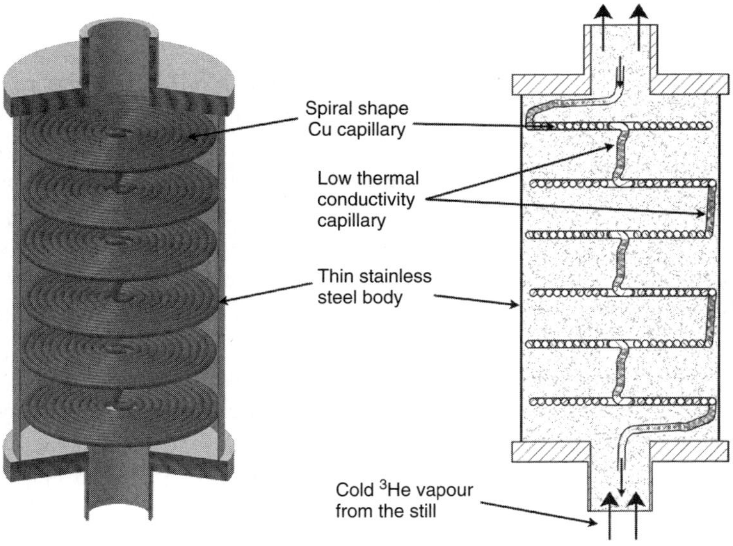

Fig. 6.12. Simplified scheme of a Joule–Thomson heat exchanger for dilution refrigerator.

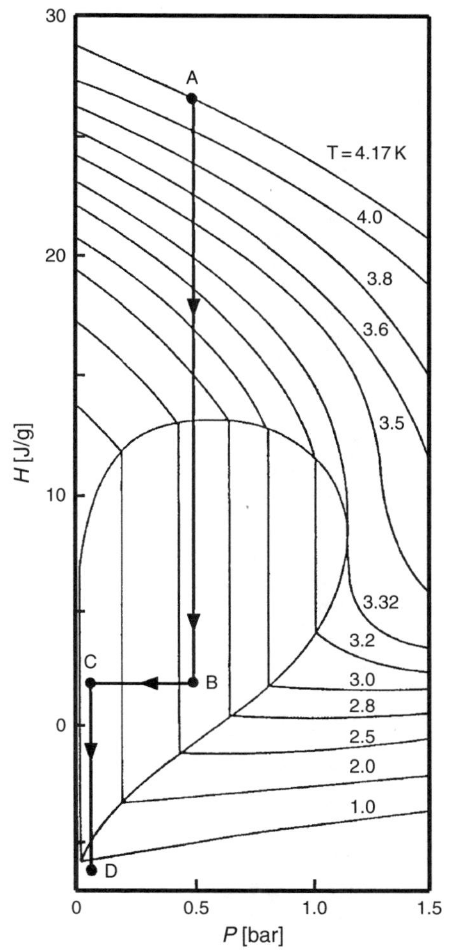

Fig. 6.13. ^{3}He enthalpy as a function of pressure [32].

The cooling process of the mixture can be represented on the enthalpy diagram of Fig. 6.13. Point A represents the starting condition after thermalization at 4.2K. The A–B line corresponds to the decrease of enthalpy in a J–T heat exchanger. In point B, we have a coexistence of liquid and vapour. The percentage of liquid decreases in the isoenthalpic expansion B–C which produces a decrease of p and T. The remaining vapour is condensed in the still exchanger (line C–D).

The thermal charge on the still is balanced by the cooling action of ^{3}He evaporated in the still which (usually) does not necessitate that of a heater [33].

Figure 6.14 shows the photo of a medium power (about 500 μW at 100 mK) J–T DR (MNK-CF-500). Note the 'unusual' shape (coil) of the step heat exchanger.

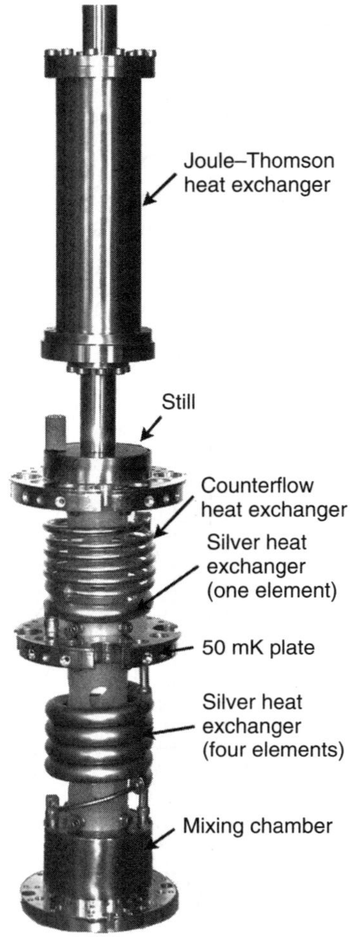

Fig. 6.14. A recent Joule–Thomson dilution refrigerator (courtesy of Leiden Cryogenics).

6.5 Practical operations with a DR

A few naïve suggestions in the practical use of a DR:
At room temperature:

1. Check the degree of vacuum in the OVC (outer vacuum chamber), that is the vacuum in the dewar which will be filled with liquid ^{4}He. In an ambient rich in helium gas, such check is always needed if the dewar neck is made of a plastic material which shows appreciable permeation of ^{4}He at room temperature.
2. Check the ground electrical losses of wiring in the cryostat.
3. Check vacuum in the DR tubing.

Precooling:

4. The precooling (at 77K) of the dewar must be sufficiently long to minimize the liquid ^{4}He consumption. For the same reason, at the end of precooling, no liquid N_2 must remain at the bottom of the dewar.
5. The thermalization of the cold part of a DR is achieved using a helium gas thermal switch (see Section 4.3). The volume of He gas necessary to close the thermal switch in the IVC (inner vacuum chamber) is about 10^{-3} the volume of the IVC.
6. Check for leaks from IVC (usually from indium or Kapton gasket).
7. Check wirings at 77 K.

Cooling to 4.2 K:

8. Check, if possible (with thermometers), that all parts of the DR below the 4 K flange, including the experiment, reach the bath temperature (4.2 K). Parts which remain at temperatures well above 4 K (with high heat capacity) may cause 'mysterious' behaviours of the DR, and most times it is necessary to go back to step 4.
9. Open the thermal switch by pumping away the helium gas from the IVC and start the DR unit. Sometimes one goes on pumping the IVC while temperature decreases. In this case, since also the cold surfaces of DR act as cryogenic pumps, a flux meter (usually present in a leak detector) tells when pumping must be interrupted (at about 1 K). Otherwise, it is safe to close pumping at 4 K.

6.6 DR in high magnetic fields

There exist some situations in which a DR must work in the presence of a high magnetic field, for example when:

1. the magnetic properties of a material are to be studied, or the effect of a magnetic field on a device is to be measured (i.e. magnetoresistance);
2. the DR is used to drive a demagnetization stage (see Section 7.3 and Section 7.4).

The power P_e generated by eddy currents (produced by a change dB/dt) in a cylinder of radius r, length L and electrical resistivity ρ is:

$$P_e = \frac{\pi r^4 L \left(dB/dt\right)^2}{8\rho} \tag{6.14}$$

For a cylinder with $r = 1\,\text{cm}$, $\rho = 20\,\text{n}\Omega\,\text{cm}$ (corresponding to a residual resistivity ratio $=100$ for Cu) and $dB/dt = 1\,\text{T}\cdot\text{h}^{-1}$, we obtain:

$$P_e = 2 \times 10^{-6}\ \text{W/mol Cu} \tag{6.15}$$

A change in dB/dt can be produced either by a demagnetization process or by a vibration in a non-perfectly homogeneous field. To reduce the effect of eddy currents, low-resistivity (plastic) MCs have been built (in which the sample under study is immersed in the mixture, see Fig. 6.15 showing the electrical feedthrough) and even complete plastic DR [34].

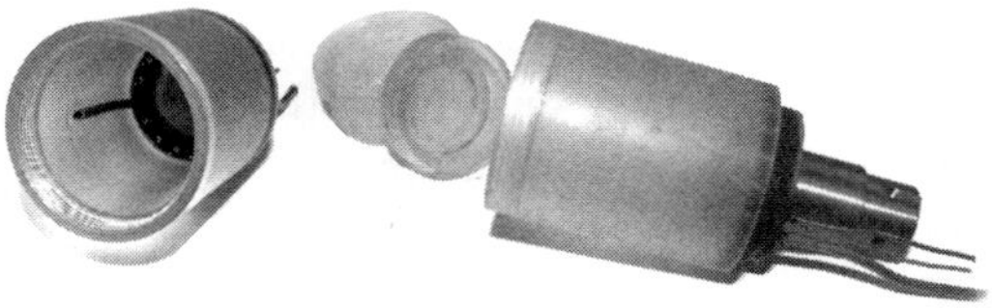

Fig. 6.15. A plastic mixing chamber for a dilution unit.

An interesting example of a DR for optical observations below 1 K in strong magnetic fields is reported in ref. [35]. A multistage nuclear demagnetization refrigerator is described in ref. [36].

For a more detailed discussion of the problems in refrigeration due to a changing B see ref. [4].

6.7 Dry DR

The interest in the J–T DR has strongly increased with the progress in the field of mechanical refrigerators. A 'dry DR', i.e. a DR which does not need cryoliquids in an external dewar, was proposed and realized by Uhlig in 1993 [37]. The two-stage Gifford–McMahon (GM) coolers used to drive the DR led to a relatively complicated construction of the cryostat to prevent the vibrations of the GM cooler from reaching the dilution unit. Instead the pulse tube refrigerator (PTR) (see Section 5.8) has no moving parts except for the rotary valves. Its level of vibration is low compared to a GM refrigerator. The coupling of a DR to a PTR was achieved by Uhlig in 2002 [38]. A cross-cut of the Uhlig's cryostat with PTR, dilution unit and J–T precooling is shown in Fig. 6.16.

Dry DR units, both with and without IVC (for optical experiments), are now commercially available.

6.8 No-gravity dilution

The achievement of temperatures in the millikelvin range in the absence of gravity (on satellites) is nowadays a challenge. Two types of refrigerators have been up to now proposed: the adiabatic demagnetization refrigerator (ADR) (see ref. [39] and Section 7.3) and the space DR. The latter was proposed by A. Benoit and S. Pujol in 1991 [40,41]. The cooling process is made using two independent injection lines: one for pure ^{3}He and one for pure ^{4}He. The gases are precooled, from 300 to 4K by an external cooling source. A 1.6K stage is required to reduce thermal leaks on 0.1K stage. ^{3}He is dissolved in ^{4}He in the 0.1K stage. The result is an endothermic process which extracts heat in the 0.1K

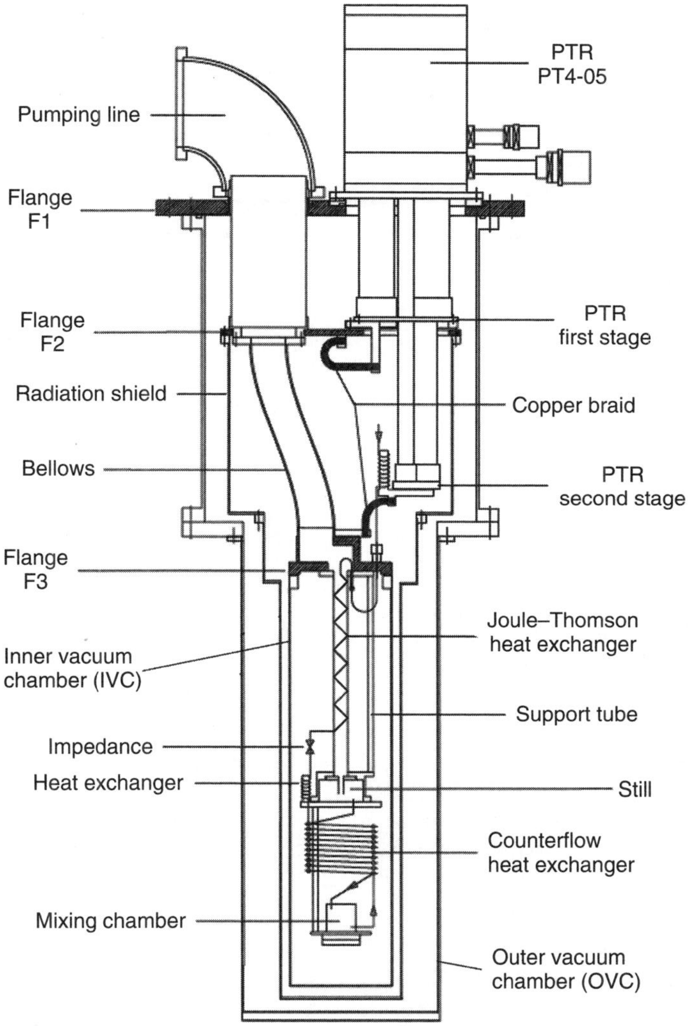

Fig. 6.16. Cross-cut of Uhlig's cryostat with pulse tube refrigerator, dilution unit and Joule–Thomson precooling [38].

stage. Usually the two gases are under pressure. This pressure is used to precool injected gas at 1.6 K by making a J–T expansion in the return line.

The mixture has 20% ^{3}He concentration. Only 6.7% is used in the cooling point, so that the dilution process extracts additional power in the return line until the J–T expansion (see Fig. 6.17).

This additional power is used to precool pure ^{3}He and ^{4}He, electrical wires and mechanical supports.

This refrigerator has been tested in a balloon experiment [42] and will be used in the Plank satellite experiment [43].

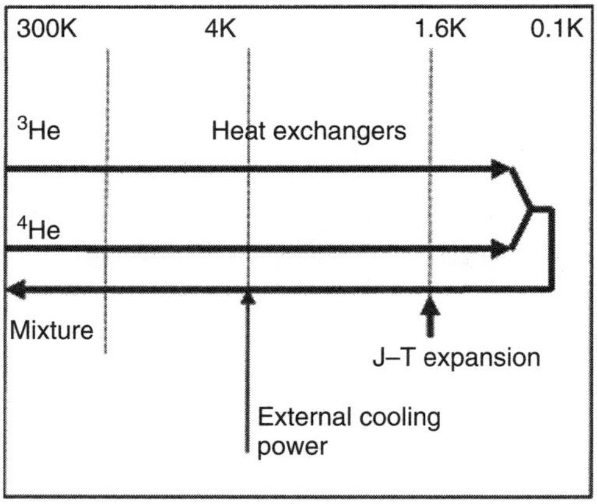

Fig. 6.17. Schematic of the open cycle space dilution refrigerator.

References

[1] H. London, G.R. Clarke, E. Mendoza: Phys. Rev. **128**, 1992 (1962)
[2] H. London: *Proc. Int. Conf. on Low Temp. Phys.*, Clarendon Press Laboratory, Oxford, p. 157 (1951)
[3] G. Frossati: J. Low-Temp. Phys. **87**, 595 (1992)
[4] F. Pobell: *Matter and Methods at Low Temperature*, 2nd ed., Springer-Verlag, New York (1995)
[5] C. Ebner, D.O. Edwards: Phys. Rep. **2C**, 77 (1971)
[6] M. Nakamura et al.: Physica B **165, 166**, 517 (1990)
[7] G.E. Watson, J.D. Reppy, R.C. Richardson: Phys. Rev. **188**, 384 (1969)
[8] S. Yorozu et al.: Phys. Rev. B **45**, 12942 (1992)
[9] O.V. Lounasmaa: *Experimental Methods and Principles Below 1 K*, Academic Press, London (1974)
[10] J.C. Wheatley: Am. J. Phys. **36**, 181 (1968); and in *Progress in Low Temperature Physics*, vol. 6, p. 77, ed. by C.J. Gorter, North-Holland, Amsterdam (1970)
[11] A.C. Anderson et al.: Phys. Rev. Lett. **16**, 263 (1966)
[12] D.S. Betts: *Refrigeration and Thermometry Below One Kelvin*, Sussex University Press, Brighton (1976)
[13] D.S. Betts: *An Introduction to Millikelvin Technology*, Cambridge University Press, Cambridge (1989)
[14] G. Frossati: J. de Physique **39** (C6), 1578 (1978) ; J. Low Temp. Phys. **87**, 595 (1992)
[15] G.A. Vermeulen, G. Frossati: Cryogenics **27**, 139 (1987)
[16] J.C. Wheatley, O.E. Vilches, W.R. Abel: Physics **4**, 1 (1968)
[17] J.C. Wheatley, R.E. Rapp, R.T. Johnson: J. Low Temp. Phys. **4**, 1 (1971)
[18] R. Radebaugh, J.D. Siegwarth: Cryogenics **11**, 368 (1971)
[19] Y. Oda et al.: Cryogenics **23**, 139 (1983)
[20] Y. Oda, G. Fujii, H. Nagano: Cryogenics **18**, 73 (1978)
[21] D.I. Bradley et al.: Cryogenics **22**, 296 (1982)
[22] Yu.M. Bunker et al.: J. Low Temp. Phys. **83**, 257 (1991)
[23] D.J. Bradley et al.: Cryogenics **34**, 549 (1994)
[24] J.D. Siegwarth, R. Radebaugh: Rev. Sci. Instrum. **42**, 1111 (1971); ibid. **43**, 197 (1972)
[25] F.A. Staas, K. Weiss, A.P. Sevarijns: Cryogenics **14**, 253 (1974)
[26] K. Neumaier: in G. Eska, Hyperfine Interact. **22**, 221 (1985)
[27] G.J. Ehnholm, R.G. Gylling: Cryogenics **11**, 39 (1971)
[28] D.S. Greywall: Phys. Rev. B **27**, 2747 (1983); ibid. B **33**, 7520 (1986)
[29] J. Kraus: Cryogenics **17**, 173 (1977)
[30] A.T.A.M. de Waele, A.B. Reekers, H.M. Gijsman: Cryogenics **17**, 175 (1977)
[31] J. van der Mass et al.: Cryogenics **26**, 471 (1986)

[32] K. Uhlig: Cryogenics **27**, 454 (1987)
[33] K. Uhlig: Cryogenics **27**, 2474 (1983)
[34] E. ter Haar, et al.: J. Low Temp. Phys. **99**, 151 (1995)
[35] R. van Rooijen et al.: J. Low. Temp. Phys. **124**, 497 (2001)
[36] W. Yao et al.: J. Low Temp. Phys. **120**, 121 (2000)
[37] K. Uhlig, W. Hehn: Cryogenics **33**, 1028 (1993)
[38] K. Uhlig: Cryogenics **42**, 73 (2002)
[39] P. Shirron et al.: Cryogenics **44**, 581 (2004)
[40] S. Benoit, S. Pujol: Physica B **169**, 457 (1991)
[41] A. Benoit, S. Pujol: Cryogenics **34**, 421 (1994)
[42] A. Benoit et al.: A&A **424**, 571 (2004)
[43] S. Triquenaux et al.: Cryogenics **46**, 288 (2006)

7 Other Refrigerators

Contents

7.1 Introduction

In this chapter, we will describe two refrigeration methods to produce temperatures below 0.3 K. The Pomeranchuck and the adiabatic demagnetization refrigerators (ADRs) are 'one shot' refrigerators. However, multiple ADRs are under study to realize a continuous refrigerator [1]. Both types of refrigerators necessitate a starting temperature in the subkelvin range and hence are usually based on a dilution refrigerator (DR) (see Chapter 6). We will also concisely describe the adiabatic nuclear demagnetization refrigerator which is nowadays the only apparatus capable of temperatures in the microkelvin range.

7.2 Pomeranchuck refrigerator

In 1950, Pomeranchuck (1913–1966) predicted that for ^{3}He on the melting curve below about 0.3 K, the entropy of liquid is smaller than that of solid. It was only after 15 years that Anufriev [2], after Pomeranchuck's suggestion, succeeded in reducing the temperature from 50 to ~18 mK in an experiment based on this ^{3}He property. Four years later, the Pomeranchuck method produced a temperature of 2 mK [3].

The solidification of a liquid takes place usually with heat release according to the equation (see Section 2.4.1):

$$\dot{Q} = \frac{dn}{dt} T \left(S_{sol.} - S_{liq.}\right) < 0 \tag{7.1}$$

where dn/dt is the rate of solidification of the liquid and ($S_{sol.}-S_{liq.}$) is negative since the entropy of the solid is usually less than that of the liquid. However, as shown in Fig. 7.1, below 0.3 K, for ^{3}He, the entropy of the liquid phase is smaller than that of the solid (Pomeranchuck 1950). This property has led to the development (1965–1970) of a

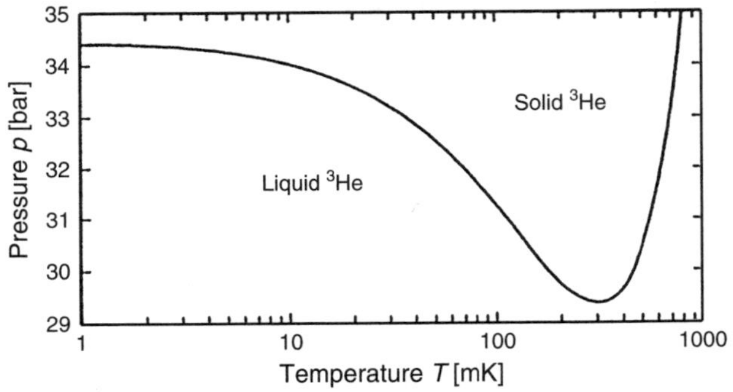

Fig. 7.1. Melting curve of ^{3}He based on data of ref. [8–11].

'one shot' refrigerator capable of temperatures as low as 2–3 mK [3–5]. The main interest in this method is linked to the study of liquid and solid ^{3}He properties [6–7] and in particular to the application to thermometry (see Section 8.4).

7.2.1 The strange behaviour of ^{3}He

The strange behaviour of ^{3}He melting curve below 1 K is shown in Fig. 7.1. The minimum takes place at $T \sim 315\,\text{mK}$ and $p \sim 29$ bar.

If pressure is increased below this minimum, more solid is formed and temperature decreases (Pomeranchuck effect). Along the melting curve, for the Clapeyron law:

$$\frac{\mathrm{d}p}{\mathrm{d}T} = \frac{(S_{\text{liq.}} - S_{\text{sol.}})}{(V_{\text{liq.}} - V_{\text{sol.}})} \tag{7.2}$$

Since the molar volume of the liquid phase is always larger (of $1.31\,\text{cm}^3$/mole per $T < 40\,\text{mK}$ [3–4]) than that of the solid phase, if $S_{\text{liq.}} < S_{\text{sol.}}$, we have:

$$\frac{\mathrm{d}p}{\mathrm{d}T} < 0 \tag{7.3}$$

The negative slope of the melting curve witnesses the anomalous situation of a liquid phase with entropy smaller than the solid phase (see Fig. 7.2).

The cooling power in an isoentropic compression is given by eq. (7.1). It is obviously necessary to start from $T < 0.32\,\text{K}$ and perform an adiabatic solidification of a part of liquid ^{3}He. It can be demonstrated [12] that:

$$\frac{\dot{Q}}{\dot{n}} \approx 5.6 \cdot T - 40 \cdot T^2 \qquad \text{for } 2\,\text{mK} < T < 50\,\text{mK} \tag{7.4}$$

This cooling power is to be compared with the DR cooling power:

$$\frac{\dot{Q}}{\dot{n}} \approx 82 \cdot T^2 \tag{7.5}$$

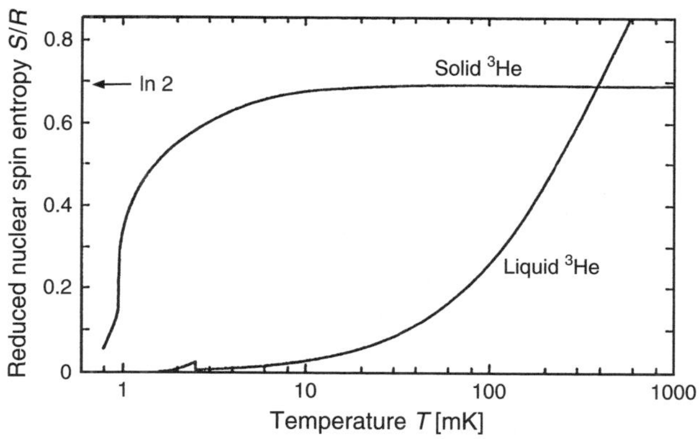

Fig. 7.2. Entropies (divided by the gas constant R) of liquid and solid ^{3}He along the melting curve. The disorder of nuclear spin entropy, corresponding to $S_{sol.}/R = \ln(2I+1) = \ln 2$ is marked. The two curves cross at the minimum of the melting curve at 315 mK and 29 bar [[12] p. 214].

Let us note that below 50 mK, the cooling power of Pomeranchuk process is larger (two orders of magnitude at 2 mK) than that of the dilution process. In Fig. 7.3, the comparison between the two methods is reported.

Around 1 mK, the solid ^{3}He, in zero magnetic field, undergoes a nuclear ordering: its entropy falls by an order of magnitude (see Fig. 7.2) and the cooling power of the process vanishes.

When ^{3}He is compressed, a mechanical work $p\mathrm{d}V$ is done. The ratio between the compression work and the cooling power is shown in Fig. 7.4. If some irreversible process takes place during the compression, the heating may exceed the cooling. In practice, this happens at 0.7–0.8 mK.

If the cell where the Pomeranchuck process takes place is used to cool something else besides ^{3}He, the presence of the solid, which is a very bad conductor, may be a serious drawback. The shape of the ^{3}He melting curve prevents the compression of the gas through a capillary connected to a compressor at room temperature. In fact, a solid block would be formed where the capillary is at $T = 315$ mK. For this reason, a cell with flexible walls, as that shown in Fig. 7.5, is necessary.

The flexible material can be Be-Cu, but also plastic (Kapton). The starting temperature for the refrigeration of a Pomeranchuck cell is usually around 25 mK, produced by a DR. Recent realization of Pomeranchuck refrigerators can be found in ref. [6–7,13]

NOTE

After nearly two decades of controversy, it is now generally accepted that helium crystals are model systems for the general study of crystal surfaces, but also exceptional in having unique quantum properties. Helium crystals exhibit faceting as do ordinary crystals, but no other crystals can grow and melt sufficiently fast to make the propagation of crystallization waves possible at their surfaces.

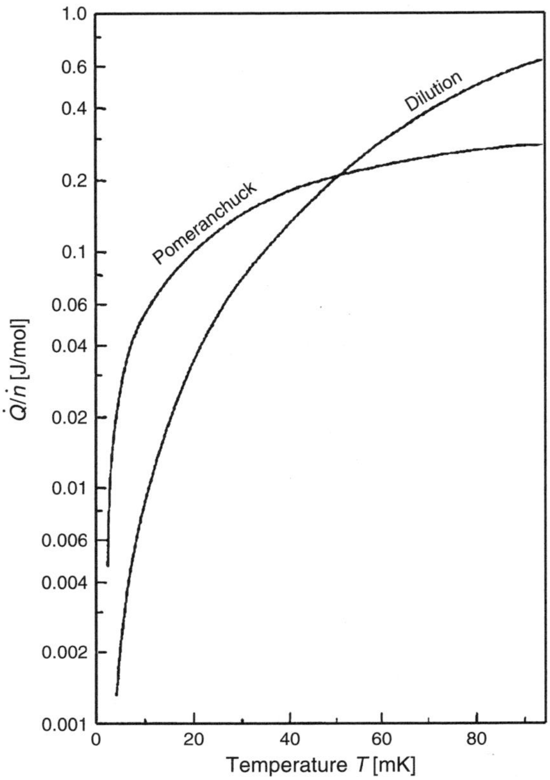

Fig. 7.3. Comparison of the relative cooling powers $\dot{Q}/\dot{n}$ of a Pomeranchuck refrigerator and of a dilution refrigerator. In the former case, the liquid is converted into solid at a rate of 10 μmol/s; in the latter case, ^{3}He is removed from the mixing chamber at the same rate.

The phenomena of crystallization waves and the unusual growth dynamics are linked to the quantum properties of liquid and solid helium at low temperatures. When crystal growth takes place from a superfluid, there are situations in which no dissipation at all takes place at the moving interface; this would be impossible in a classical liquid, where the sticking of individual atoms necessarily dissipates energy because of a momentum exchange at the moving crystal surface. At a temperature of 0.1 K, the growth of ^{3}He crystals is slower than the growth of ^{4}He crystals by 11 orders of magnitude [14]. If one compared classical crystals made of different isotopes of the same element, one would find that they had very similar properties. In contrast, because ^{3}He atoms are Fermi particles while ^{4}He atoms are Bose particles, the growth dynamics of ^{3}He crystals are very different from those of ^{4}He crystals.

A review, which summarizes 25 years of research on the surface of helium crystals, is reported in ref. [15].

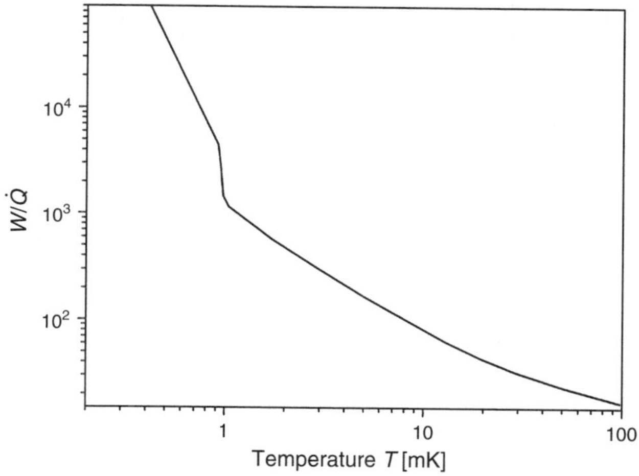

Fig. 7.4. Ratio between the mechanical work $p\mathrm{d}V$ ($p = 34\,\mathrm{bar}$ and $T = 5\,\mathrm{mK}$) and the cooling power for a mole of ^{3}He during the solidification process. Note that at 1 mK, $p\mathrm{d}V$ is ten times the energy absorbed in the solidification of a mole of ^{3}He. The cooling in practice ends at 0.8–0.7 mK. Data from [3–4].

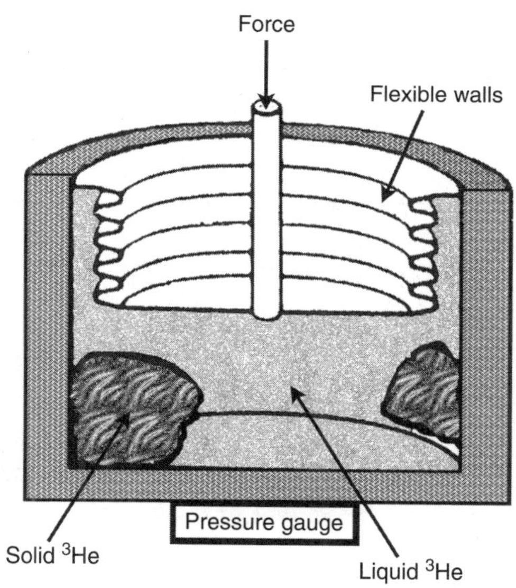

Fig. 7.5. Scheme of a cell with flexible walls: the volume reduction produces a pressure increase.

7.3 Adiabatic demagnetization refrigerator

Adiabatic demagnetization has been historically the first method which allowed to reach temperatures below 1 K. It was proposed in 1926 by P. Debye and W.F. Giauque and converted into a practical realization by W.F. Giauque and D.P. MacDougall in 1933.

Nowadays ADRs can reach temperatures around 2 mK. In most cases, however, the use of a DR is preferred because of it is not "one shot" refrigerator and for the absence of magnetic fields. An exception is when the refrigerator must work in no gravity (space conditions) [16–20].

Let us consider a solid containing paramagnetic ions of electronic magnetic moment μ and suppose that in a certain temperature range, the interaction energy ε_m among the moments and of the moments with an external magnetic field is small compared with $k_B T$. At the temperatures of interest for this type of refrigeration, the magnetic entropy (of the order of a few J/mole) is large in comparison with the other entropies of the system. If we cool the salt, at a certain temperature, the interaction among magnetic moments becomes of the order of $k_B T$, and a spontaneous magnetic alignment takes place.

An applied magnetic field interacts with the magnetic moments, partially orienting them along the field direction. This means that at fixed temperature, the field reduces the entropy.

With reference to Fig. 7.6, the refrigeration process starts from A. The line AB represents an isothermal magnetization. The heat produced along AB is delivered to a heat sink (DR). After thermal isolation, the crystal is adiabatically demagnetized (line BC). Eventually it warms up along the entropy curve at the final magnetic field which is zero in this example. The heat of magnetization is given by the area of the rectangle ABDS$_\infty$, while the cooling power by the shaded area. The limit of the process is the temperature T_c at which a spontaneous magnetic ordering takes place. The performance of a refrigerator based on paramagnetic salts depends on the starting field and temperature and on the type of salt. Typical values for the field are 0.1–1 T and 0.1–1 K for the initial temperature.

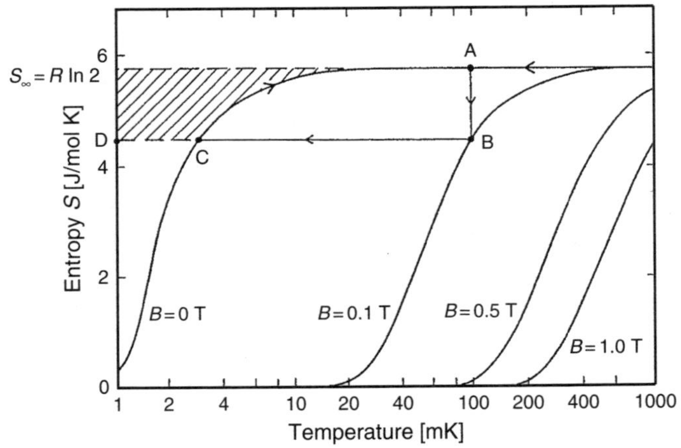

Fig. 7.6. Molar entropy of a single crystal of the paramagnetic salt CMN versus temperature for magnetic fields applied along the crystallographic *a* axis. The refrigeration process starts from A. The line AB represents an isothermal magnetization. The heat produced along AB is delivered to a heat sink (DR). After thermal isolation, the crystal is adiabatically demagnetized (line BC). Eventually it warms up along the entropy curve at the final magnetic field which is zero in this example. The heat of magnetization is given by the area of the rectangle ABDS$_\infty$, the cooling power by the shaded area [21].

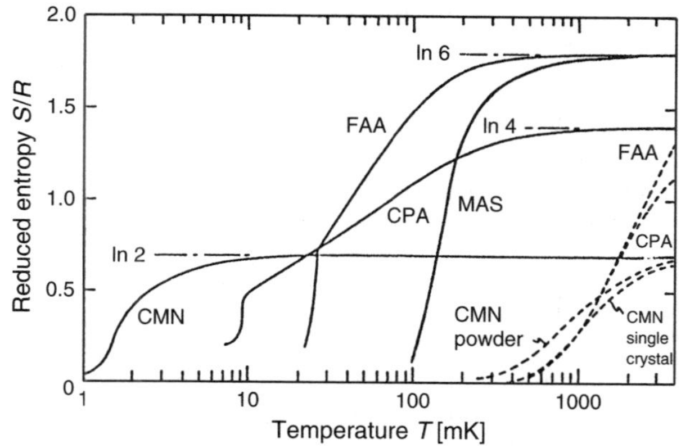

Fig. 7.7. Entropies S (divided by the gas constant R) versus T of four salts used in adiabatic demagnetization in zero field (full lines) and in 2 T (dashed lines) [22].

Table 7.1
Commonly used salt for adiabatic demagnetization refrigerator

Name	Composition	T_c [K]
MAS	$Mn^{2+}\ SO_4 \cdot (NH_4)_2\ SO_4 \cdot 8H_2O$	0.17
FAA	$Fe^{3+}\ (SO_4)_3 \cdot (NH_4)_2SO_4 \cdot 24H_2O$	0.03
CPA	$Cr_2{}^{3+}\ (SO_4)_3 \cdot K_2SO_4 \cdot 24H_2O$	0.01
CMN	$2Ce^{3+}\ (NO_3)_3 \cdot 3Mg(NO_3)_2 \cdot 24\,H_2O$	0.002

The most desirable salt for the refrigeration should have a low T_c and a high magnetic specific heat. The most commonly used salts are (see Fig. 7.7 summarized in, Table 7.1):

Among these salts, because of its low magnetic specific heat, CMN does not represent the optimum choice except when a temperature around 0.01 K is to be reached. For a discussion about the optimal choice of the paramagnetic salt and tricks in the construction of an ADR, see for example ref. [23]. It must be pointed out that all these salts contain crystallization water which guarantees the right distance (10 Å in CMN) among the ions and hence a low T_c. When using these salts in an ADR, the time that the salt spends at room temperature under vacuum must be short to avoid the loss of the crystallization water. Information about paramagnetic salts are reported in ref. [24–30]. Recent realization of ADR refrigerators are reported in ref. [1,31].

7.4 Adiabatic nuclear demagnetization

Many elements and compounds do not present an electronic magnetic moment, but their nuclei possess nuclear magnetic moments. These moments can be used for a magnetic refrigeration as it is done with electronic moments. Also in this case the minimum temperature, now in the microkelvin range, is reached when the spontaneous magnetic ordering takes place. Materials useful for adiabatic nuclear refrigeration are Cu, Ag and $PrNi_5$. Materials are precooled by means of a dilution refrigerator.

The description of these apparatus is beyond the scopes of this book. The reader is referred to the exhaustive treatment of this subject for example in ref. [22,32]. Just to get an idea of the capabilities of this method see ref. [33], where a refrigerator capable of maintaining lattice temperatures below 100 μK for two months is described.

7.5 Electronic refrigeration

Tunneling electric current through the normal metal insulator superconductor junction is accompanied with heat flow out of normal metal when property voltage is biased. The phenomenon enables cooling of electrons and phonons (under special conditions) in the region below 1 K. At lower bath temperatures, two parasitic heat sources decrease refrigerator performance:

(i) Due to tunneling of hot electrons from a normal metal electrode to a superconductive one, an excess quasiparticle density appears in the superconductor which results in tunneling of electrons back to the normal metal and consequently the normal metal electron temperature increases. This phenomenon is called back tunneling.
(ii) The second heat source is the cooper pair formation: the part of energy released by the formation of cooper pairs in a superconductor is dissipated in normal metal.

Both contributions are calculated in ref. [34]. For more details about this method of refrigeration see ref. [[22] p.324, [35–43]].

NOTE

Other more exotic refrigeration methods which are at an early stage of study have been proposed such as the optical refrigeration [44–48].

References

[1] P. Shirron et al.: Cryogenics, **44**, 581 (2004)
[2] Yu.D. Anufriev: Sov. Phys. JEPT Lett. **1**, 155 (1965)
[3] R.T. Johnson et al.: Phys. Rev. Lett. **22**, 449 (1969)
[4] I. Pomeranchuk: Zh. Eksperim. i Teor. Fiz. **20**, 919 (1950)
[5] R.T. Johnson, J.C. Wheatley: J. Low Temp. Phys. **2**, 424 (1970)
[6] R. van Rooijen et al.: J. Low Temp. Phys. **124**, 169 (2001)
[7] R. Blaauwgeers et al.: J. Low Temp. Phys. **134**, 393 (2004)
[8] W.P. Halperin et al.: J. Low Temp. Phys. **31**, 617 (1978)
[9] E.R. Grilly: J. Low Temp. Phys. **4**, 615 (1971); ibid. **11**, 243 (1973)
[10] D.S. Greywall, P.A. Busch: J. Low Temp. Phys. **46**, 451 (1982)
[11] D.S. Greywall: Phys. Rev. B **31**, 2675 (1985)
[12] G.K. White: *Experimental Techniques in Low Temperature Physics*, 3rd ed., Clarendon, Oxford (1979)
[13] E. Terhaar, G. Frossati, W.G. Clark: J. Low Temp. Phys. **94**, 361 (1994)
[14] F. Graner, S. Balibar, E. Rolley: J. Low Temp. Phys. **75**, 69 (1989)
[15] S. Balibar et al.: Rev. Mod. Phys. **77**, 317 (2005)

[16] J. Tuttle et al.: Cryogenics **44**, 383 (2004)
[17] J. Tuttle et al.: Cryogenics **41**, 781 (2001)
[18] S.M. Volz et al.: Cryogenics **36**, 763 (1996)
[19] C. Hagmann, P.L. Richards: Cryogenics **35**, 303 (1995)
[20] C. Hagmann, P.L. Richards: Cryogenics **34**, 221 (1994)
[21] D.S. Betts: *An Introducion to Millikelvin Technology*, Cambridge University Press, Cambridge (1989)
[22] F. Pobell: *Matter and Methods at Low Temperature*, 2nd ed., Springer-Verlag, New York (1995)
[23] R.C. Richardson, E.N. Smith eds.: *Experimental Techniques in Condensed Matter Physics at Low Temperatures*, Addison-Wesley, Redwood City, CA (1988)
[24] O.V. Lounasmaa: *Experimental Principles and Methods Below 1 K*, Academic, London (1974)
[25] D.S. Betts: *Refrigeration and Thermometry Below One Kelvin*, Sussex University Press, Brighton (1976)
[26] R.P. Hudson: *Principles and Application of Magnetic Cooling*, North-Holland, Amsterdam (1972)
[27] B.I. Bleaney, B. Bleaney: *Electricity and Magnetismus*, 3rd ed., Oxford University Press, Oxford (1976)
[28] W.F. Giauque et al.: J. Chem. Phys. **58**, 2621 (1973)
[29] R.A. Fisher et al.: J. Chem. Phys. **58**, 5584 (1973)
[30] O.E. Vilches, J.C. Wheatley: Rev. Sci. Instrum. **37**, 819 (1966); Phys. Rev. **148**, 509 (1966)
[31] M. DiPirro et al.: Cryogenics, **44**,559 (2004)
[32] R.M. Mueller et al.: Cryogenics **20**, 395 (1980)
[33] W. Yao et al.: J. Low Temp. Phys. **120**, 121 (2000)
[34] B. Jug, Z. Trontelj: IEEE Trans. Appl. Superconductivity **11,** 848 (2001)
[35] R.F. Silverberg et al.: NIM A **559**, 630 (2006)
[36] N.A. Miller et al.: NIM A, **559**, 633 (2006)
[37] M. Furlan, E. Kirk, A. Zehnder: NIM A **559**, 636 (2006)
[38] A. Luukanenet al.: in *Low Temperature Detectors: Ninth International Workshop on Low Temperature Detectors* (LTD9), Madison, Wisconsin, 22–27 July 2001, AIP Conference Proceedings **605**, p. 375, ed. by F.S. Porter, D. McCammon, M. Galeazzi, C.K. Stahle, American Institute of Physics, New York (2002)
[39] H.L. Edwards et al.: Phys. Rev. B **52**, 5714 (1995)
[40] P.A. Fisher et al.: J. Low Temp. Phys. **101**, 561 (1995)
[41] M. Nahum et al.: Appl. Phys. Lett. **65**, 3123 (1994)
[42] M.M. Leive et al.: Appl. Phys. Lett. **68**, 1996 (1996)
[43] J.N. Ullom et al.: NIM A **444**, 38 (2000)
[44] P. Pringsheim: Z. Phys. **57**, 739 (1929)
[45] R.I. Epstein et al.: Nature **377**, 500 (1995)
[46] J.L. Clark, G. Rumbles: Phys. Rev. Letters **76**, 2037 (1996)
[47] B.C. Edwards et al.: J. Appl. Phys. **86**, 6489 (1999)
[48] T.R. Gosnell: Opt Lett **24**, 1941 (1999)

PART IV

8 Temperature Scales and Temperature Fixed Points

Contents

8.1 Introduction

In low-temperature experiments, an accurate measurement of the temperature is usually more difficult than reaching that temperature. Knowledge of the temperature in various points of a system is needed in any cryogenic experiment. For example, if we want to measure the thermal conductivity of a material, at least two thermometers measuring a temperature difference are requested. In spite of the fact that temperature is one of the most important parameters in physics and is one of the basic quantity of the SI system, our knowledge of temperature is definitely less accurate than our knowledge of mass, length, time and so on. The starting points of thermometry are the zero-th law and the second law of thermodynamics.

The zero-th law, which justifies the existence of the thermometer, says that two bodies A and B which are in thermal equilibrium with a third body are in thermal equilibrium with each other. There is no heat flow from one to the other, and they are said to be at the same temperature. If A and B are not in thermal equilibrium, A is said to be at a higher temperature if the heat flows from A to B when they are placed in thermal contact. The changes in temperature usually produce changes in physical properties like dimension, electrical resistance and so on. Such property variations can be used to measure the temperature changes.

The second law of thermodynamics says that in a Carnot cycle $Q/T = \text{constant}$. This law allows for the definition of a temperature scale if we arbitrarily assign the value of a reference temperature. If we give the value $T_3 = 273.16\,\text{K}$ to the triple point (see Gibbs law, Section 8.2) of water, the temperature in kelvin units [K] can be expressed as:

$$T = 273.16 \times \frac{Q(T)}{Q_3} \tag{8.1}$$

where $Q(T)$ and Q_3 are the heat exchanged along the two isotherms at T and T_3 respectively of an ideal Carnot cycle. The numerical value of T_3 is based on numerous experimental data and has been chosen in such a way that the minimum asymptotic value in eq. (8.1) is $T = 0$.

The experimental realization of a Carnot cycle to measure the temperature T is unusual. The coincidence of the thermodynamic temperature T with the temperature read by a gas thermometer, for example, allows the use of such thermometer to know T. As we shall see, also other laws of physics relating T with physical parameters other than heat can be used to get an absolute measure of T.

For example the $p(T)$ relation in a ^{3}He melting pressure thermometer (see Section 9.3) can be used to deduce T.

In general, a thermometer is called 'primary' if a theoretical reliable relation exists between a measured quantity (e.g. p in constant volume gas thermometer) and the temperature T. The realization and use of a primary thermometer are extremely difficult tasks reserved to metrological institutes. These difficulties have led to the definition of a practical temperature scale, mainly based on reference fixed points, which mimics, as well as possible, the thermodynamic temperature scale, but is easier to realize and disseminate. The main characteristics of a practical temperature scale are both a good reproducibility and a deviation from the thermodynamic temperature T which can be represented by a 'smooth' function of T. In fact, if the deviation function is not smooth, the use of the practical scale would produce 'steps' in the measured quantities as function of T, using the practical scale. The latter is based on:

(a) a set of fixed points;
(b) one or more interpolating devices;
(c) one or more interpolating equations.

The historical development of the International Temperature Scale (ITS) [1] culminated in the ITS 90 and, for temperatures between 1 and 0.9 K, with the definition of the PLTS 2000.

For cryogenic thermometry, see also ref. [2–5], for the historical development of thermometry see Table 8.1.

We wish to mention the recent proposal for a redefinition of Kelvin in terms of mechanical units through the Boltzmann constant [6–7]: the Kelvin should be defined as the unit of thermodynamic temperature such that the value of the Boltzmann constant is $1.3806505 \times 10^{-23}\,\mathrm{JK^{-1}}$ exactly. Of course, this value of the Boltzmann constant should be consistent with a thermodynamic temperature of the triple point of water of 273.16 K.

8.2 Reference fixed points

For more than half century from the fabrication of the first thermometers (see Table 8.1), the only way of comparing measurements made with different thermometers was that of putting the thermometers in the same 'thermal bath' (water, air and so on).

Table 8.1
Historical development of thermometry

1600	Thermoscope invented by Galileo Galilei	The thermoscope has no graduated scale, therefore temperature evaluation is only qualitative
Mid-1600	Liquid-in-glass graduated thermometers are widely spread	Famous are the alcohol thermometers of the Accademia del Cimento in Florence. Their graduation marks are made of coloured glass bits attached to the stem
1694	Melting ice and boiling water adopted by G. Renaldini as fixed points	Fixed points must be established by reference to natural phenomena, whose temperature is assumed to be intrinsically determined
1724	Modern temperature scale proposed by G. Fahrenheit, defined by a thermometer, a law and three fixed points.	Fahrenheit's thermometer was a mercury-in-glass one. Thermal expansion versus temperature was assumed linear. Three fixed points were defined: 0°F: temperature of a mixture of water, ice and ammonium chloride; 32°F: temperature of melting ice; 96°F: temperature of human body
1742	The temperature interval between melting ice and boiling water subdivided into 100 equal parts by A. Celsius	The degree Celsius was the one eventually adopted
1821	Thermoelectric effect discovered by T.J. Seebeck	The Seebeck effect is the basis for the thermometers designated as thermocouples
1854	The modern absolute temperature definition, suggested by W. Thomson, is based on Carnot cycle	A scale whose definition does not depend on a specific substance is called 'absolute'
1871	First electrical resistance thermometer built by C.W. Siemens.	The electrical resistance thermometer is still today the best practical thermometer
1879	Blackbody radiation law proposed by J. Stefan form the basis for the radiation thermometers, with Planck's law.	The radiation law is strongly nonlinear in temperature, since it depends on T^4
1967	The XIII Conference Générale des Poids et Mesures adopts the Kelvin scale with a single fixed point: 273.16 K.	The unit of temperature is designated 'Kelvin'; its symbol is K.
1990	The XVIII Conference Générale des Poids et Mesures adopts the version of the International Temperature Scale ITS-90	The ITS-90 provides the best to-date practical approximation of the thermodynamic scale and offers a reproducibility that is better than the thermodynamic scale.
2000	PLTS-2000 was formally adopted by the Comité International des Poids et Mesures in October	The PLTS-2000 covers the range of temperature from 1 K down to 0.9 mK and is defined by an equation for the melting pressure of ^{3}He
2005	Consultative Committee for Thermometry: creation of a *mise en pratique* of the definition of the kelvin	The Consultative Committee for Thermometry, considering that the ITS-90 and the PLTS-2000 are internationally accepted practical temperature scales defining temperatures T_{90} and T_{2000} that are good approximations to thermodynamic temperature T

The idea that certain physical states could reproduce always the same temperature rises in the second half of seventeenth century (Hooke, 1664; Renaldini, 1694, see e.g. ref. [8]). Intuitions of this idea can be also found in Aristotele and Galeno. Nowadays, the importance of the control of the thermometric calibration is underestimated and the use of reference fixed points is usually limited to metrological laboratories.

The definition of reference thermometric fixed point is: 'an equilibrium state of a definite substance; the realization of a fixed point must depend only on the composition and on the substance'. Hence boiling points, for example, cannot be considered fixed points, since they depend on pressure. Only triple points fulfil this definition as can be deduced from the Gibbs' rule for pure substances:

$$F = c - f + 2 \tag{8.2}$$

where F is the degree of freedom, c the number of chemical substances and f the number of phases involved. Reference points are of great importance in thermometry, not only because they allow empirical scales (Section 8.3), but because each of them provides a reference temperature for any laboratory and allows the comparison of thermometers without the use of the same thermal bath. Reference points allows also for the control of the stability of a thermometer. The most common combinations of three coexisting phases are:

(a) solid–liquid–vapour;
(b) solid–liquid I–liquid II;
(c) solid I–solid II–liquid;
(d) solid I–solid II–vapour;
(e) liquid I–liquid II–vapour.

Not all of them are equally suitable as fixed points for thermometry. In (b) and (c), the points are characterized by a high equilibrium pressure and may be better considered as pressure fixed points (see, for example, ref. [9]). Types (d) and (e) are not generally referred to as triple points, but as solid–solid and liquid–liquid transitions, respectively. At low temperature, the last solid–liquid–vapour triple point is that of hydrogen (para 99.996%) at 13.8033 K.

The type (e) transition for ^{4}He is very useful as reference point at 2.18 K (normal-to-superfluid lambda transition). Such second orders are represented in Fig. 8.1.

To obtain reference points at lower temperatures, as we shall see (Sections 8.4, 8.5), transitions other than (a)–(e) are necessary.

8.3 The ITS 90

In 1968, an international agreement was reached about the definition of an official (practical) scale of temperature for $T > 14$ K. This temperature scale IPTS-68, corrected in 1975 [11], was defined by reference fixed points given by transitions of pure substances. To extend the low-temperature range of IPTS-68, the EPT 76 [12–13] gave nine reference temperatures defined by phase transition of pure substances; in particular the superconductive transition (between 0.5 and 9 K) of five pure metals was introduced. Moreover,

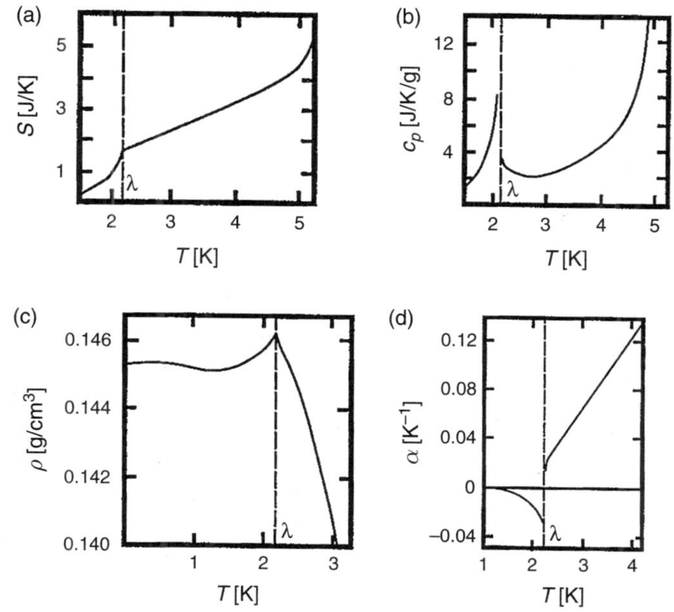

Fig. 8.1. Behaviour of ^{4}He at the λ point: (a) entropy S; (b) specific heat c_p; (c) density ρ; (d) cubic thermal expansion coefficient α [10].

The use of data of ^{4}He and ^{3}He vapour pressure which was accurately reported for $T > 0.5\,\text{K}$ was recommended. Unfortunately it was clear that also for the IPTS-68 errors (order of 10^{-4} K) existed in this temperature scale in comparison with the thermodynamic temperature.

The ITS 90 was adopted by the 'Comité International des Poids et Mesures' in September 1989 [14–16]. The ITS 90 extends from 0.65 K to the highest temperatures, practicably measurable in terms of the Planck radiation law using monochromatic radiation. The defining fixed points of the ITS 90 are mostly phase transition temperatures of pure substances given in Table 8.2.

It has recently been found out that the natural isotopic composition is not a parameter sufficient to describe the substance in an univocal way: for H_2 differences in the T_{triple} exist due to the D_2 content [17]. Also the H_2 vapour pressure and Ne data might be influenced by the isotopic composition.

The scale is divided into four overlapping ranges [14–16]:

(a) 0.65–5 K. Vapour pressure versus temperature of ^{3}He (0.65–3.2 K) and of ^{4}He (1.25–5 K) given by:

$$T = \sum_{i=0}^{9} A_i \left(\frac{\ln p - B}{C} \right)^i \tag{8.3}$$

with T in kelvin and p in Pascal. The constants A_i, B and C are given in Table 8.3. The equation (8.3) is, in fact, valid to 0.5 K.

Table 8.2
Defining fixed points of the temperature scale ITS-90 [14–16]

Material[a]	Equilibrium state[b]	Temperature [K]
He	Vapour pressure point	3÷5
e-H_2	Triple point	13.8033
Ne	Triple point	24.5561
O_2	Triple point	54.3584
Ar	Triple point	83.8058
Hg	Triple point	234.3156
H_2O	Triple point	273.16
Ga	Melting point	302.9146
In	Freezing point	429.7485
Sn	Freezing point	505.078
Zn	Freezing point	692.677
Al	Freezing point	933.473
Ag	Freezing point	1234.93
Au	Freezing point	1337.33
Cu	Freezing point	1357.77

Note that only in the case of gallium, with a deep subcooling, the reference is linked to the melting point instead that to the freezing point.
[a] All substances (except ^{3}He) are of natural isotopic composition. e-H_2 is hydrogen at the equilibrium concentration of the ortho and para molecular forms.
[b] Melting point and freezing point at a pressure of 1.01325 bar.

Table 8.3
Values of the coefficients A_i and of the constants B and C for the ^{3}He and ^{4}He vapour pressure equations and the temperature range for which each equation is valid [14–16]

Coefficient or constant	^{3}He 0.65÷3.2 K	^{4}He 1.25÷2.1768 K	^{4}He 2.1768÷5.0 K
A_0	1.053 477	1.392 408	3.146 631
A_1	0.980 106	0.527 153	1.357 655
A_2	0.676 380	0.166 756	0.413 923
A_3	0.372 692	0.050 988	0.091 159
A_4	0.151 656	0.026 514	0.016 349
A_5	−0.002 263	0.001 975	0.001 826
A_6	0.006 596	−0.017 976	−0.004 325
A_7	0.088 966	0.005 409	−0.004 973
A_8	−0.004 770	0.013 259	0
A_9	−0.054 943	0	0
B	7.3	5.6	10.3
C	4.3	2.9	1.9

For use convenience, ^{3}He and ^{4}He vapour pressures, calculated from eq. (3.44), are given in Tables 8.4 and 8.5.

(b) 3.0–24.5561 K (triple point of Ne). Interpolating constant-volume helium gas thermometer (Section 9.2) calibrated at the ends of the range and at the H_2 triple point, and using the interpolation procedures reported in ref. [14–16].

Table 8.4
^{3}He vapour pressure [kPa] according to the ITS 90 [14–16].

T [K]	0.00	0.01	0.02	0.03	0.04	0.05	0.06	0.07	0.08	0.09
0.6	0.071	0.079	0.087	0.096	0.105	0.116	0.127	0.139	0.152	0.166
0.7	0.180	0.195	0.211	0.229	0.247	0.267	0.287	0.308	0.330	0.353
0.8	0.378	0.404	0.431	0.459	0.489	0.520	0.552	0.586	0.621	0.657
0.9	0.695	0.734	0.775	0.817	0.861	0.907	0.954	1.003	1.054	1.106
1.0	1.160	1.216	1.274	1.333	1.395	1.459	1.523	1.590	1.660	1.731
1.1	1.804	1.880	1.957	2.037	2.118	2.202	2.288	2.376	2.466	2.559
1.2	2.654	2.752	2.851	2.954	3.059	3.165	3.275	3.387	3.501	3.618
1.3	3.738	3.860	3.985	4.112	4.242	4.375	4.511	4.649	4.790	4.934
1.4	5.081	5.231	5.383	5.538	5.697	5.858	6.022	6.189	6.360	6.533
1.5	6.709	6.889	7.071	7.257	7.446	7.638	7.834	8.033	8.235	8.440
1.6	8.649	8.861	9.076	9.295	9.517	9.742	9.972	10.20	10.44	10.68
1.7	10.92	11.17	11.42	11.68	11.93	12.19	12.46	12.73	13.00	13.28
1.8	13.56	13.84	14.13	14.42	14.72	15.02	15.32	15.63	15.94	16.26
1.9	16.58	16.99	17.23	17.56	17.90	18.24	18.58	18.93	19.28	19.64
2.0	20.00	20.37	20.74	21.11	21.49	21.87	22.26	22.65	23.05	23.45
2.1	23.85	24.26	24.68	25.10	25.52	25.95	26.38	26.82	27.26	27.71
2.2	28.16	28.61	29.08	29.54	30.01	30.49	30.97	31.45	31.94	32.44
2.3	32.94	33.44	33.95	34.47	34.99	35.51	36.04	36.58	37.12	37.67
2.4	38.22	38.77	39.33	39.90	40.47	41.05	41.63	42.22	42.82	43.41
2.5	44.02	44.63	45.24	45.87	46.49	47.12	47.76	48.40	49.05	49.71
2.6	50.37	51.04	51.71	52.38	53.07	53.76	54.45	55.15	55.86	56.58
2.7	57.29	58.02	58.75	59.49	60.23	60.98	61.74	62.50	63.27	64.04
2.8	64.82	65.61	66.41	67.21	68.01	68.83	69.65	70.47	71.30	72.14
2.9	72.99	73.84	74.70	75.57	76.44	77.32	78.21	79.10	80.00	80.91
3.0	81.83	82.75	83.68	84.61	85.96	86.51	87.46	88.43	89.40	90.38
3.1	91.37	92.36	93.37	94.38	95.39	96.42	97.45	98.49	99.54	100.60
3.2	101.66	102.73	103.82	104.90	106.00	107.10	108.22	109.34	110.47	111.61

(c) 13.8033 K (triple point of H_2)–1234.93 K (freezing point of Ag). Several Pt thermometers (Section 9.5); for details see ref. [9, pp. 449–453].
(d) Above 1234.93 K. Planck's radiation law.

The ITS 90 contains detailed instructions about how to calibrate a thermometer in the various temperature ranges.

8.4 The provisional Low-Temperature Scale 2000 (PLTS)

The PLTS 2000 extends the range of the ITS 90 to temperature below 650 mK, defining the temperatures between 0.9 mK and 1 K by means of the ^{3}He melting curve pressure (see Section 9.3 and Fig. 8.2).

The choice of a primary thermometer such as the ^{3}He melting pressure thermometer to define the PLTS 2000 witnesses the great difficulties encountered in the measurement of very low temperatures. For example, at the beginning of 1980s, it was realized that differences up to 40% existed in the data of ^{3}He specific heat obtained

Table 8.5
^{4}He vapour pressure [kPa] according to the ITS 90 [14–16].

T [K]	0.00	0.01	0.02	0.03	0.04	0.05	0.06	0.07	0.08	0.09
1.2	0.087	0.087	0.093	0.100	0.107	0.115	0.123	0.131	0.139	0.148
1.3	0.158	0.168	0.178	0.189	0.201	0.213	0.226	0.239	0.252	0.267
1.4	0.282	0.298	0.314	0.331	0.348	0.367	0.387	0.407	0.428	0.449
1.5	0.472	0.495	0.519	0.544	0.570	0.597	0.625	0.654	0.684	0.715
1.6	0.747	0.780	0.814	0.849	0.885	0.922	0.961	1.001	1.042	1.084
1.7	1.128	1.173	1.219	1.266	1.315	1.365	1.417	1.470	1.525	1.581
1.8	1.638	1.697	1.758	1.820	1.883	1.948	2.015	2.084	2.154	2.226
1.9	2.299	2.374	2.451	2.530	2.610	2.692	2.776	2.862	2.949	3.039
2.0	3.130	3.223	3.317	3.414	3.512	3.613	3.715	3.818	3.925	4.032
2.1	4.141	4.253	4.366	4.481	4.597	4.716	4.836	4.958	5.082	5.207
2.2	5.335	5.465	5.597	5.731	5.867	6.005	6.146	6.288	6.433	6.580
2.3	6.730	6.882	7.036	7.192	7.351	7.512	7.675	7.841	8.009	8.180
2.4	8.354	8.529	8.708	8.889	9.072	9.258	9.447	9.638	9.832	10.03
2.5	10.23	10.43	10.64	10.84	11.05	11.27	11.48	11.70	11.92	12.15
2.6	12.37	12.60	12.84	13.07	13.31	13.55	13.80	14.05	14.30	14.55
2.7	14.81	15.07	15.33	15.60	15.87	16.14	16.42	16.70	16.98	17.26
2.8	17.55	17.84	18.14	18.44	18.74	19.05	19.36	19.67	19.98	20.30
2.9	20.63	20.95	21.28	21.61	21.95	22.29	22.64	22.98	23.33	23.69
3.0	24.05	24.41	24.77	25.14	25.52	25.89	26.27	26.66	27.05	27.44
3.1	27.84	28.24	28.64	29.05	29.46	29.87	30.29	30.72	31.14	31.58
3.2	32.01	32.45	32.89	33.34	33.79	34.25	34.71	35.17	35.64	36.11
3.3	36.59	37.07	37.56	38.05	38.54	39.04	39.54	40.05	40.56	41.08
3.4	41.60	42.12	42.65	43.18	43.72	44.26	44.81	45.36	45.92	46.48
3.5	47.05	47.62	48.19	48.77	49.35	49.94	50.54	51.13	51.74	52.35
3.6	52.96	53.57	54.20	54.82	55.46	56.09	56.73	57.38	58.03	58.69
3.7	59.35	60.02	60.69	61.37	62.05	62.73	63.43	64.12	64.83	65.53
3.8	66.25	66.96	67.69	68.41	69.15	69.89	70.63	71.38	72.14	72.90
3.9	73.66	74.43	75.21	75.99	76.78	77.57	78.37	79.17	78.98	80.80
4.0	81.62	82.44	83.27	84.11	84.95	85.80	86.66	87.52	88.38	89.26
4.1	90.13	91.02	91.91	92.80	93.70	94.61	95.52	96.44	97.37	98.30
4.2	99.23	100.18	101.13	102.08	103.04	104.01	104.98	105.96	106.95	107.94
4.3	108.94	109.94	110.95	111.97	113.00	114.03	115.06	116.11	117.15	118.21
4.4	119.27	120.34	121.42	122.50	123.51	124.68	125.79	126.89	128.01	129.13
4.5	130.26	131.40	132.54	133.69	134.84	136.01	137.18	138.36	139.54	140.73
4.6	141.93	143.13	144.35	145.57	146.79	148.03	149.27	150.52	151.77	153.04
4.7	154.31	155.58	156.87	158.16	159.46	160.77	162.09	163.41	164.74	166.08
4.8	167.42	168.78	170.14	171.51	172.89	174.27	175.66	177.07	178.47	179.89
4.9	181.32	182.75	184.19	185.64	187.10	188.56	190.04	191.52	193.01	194.51
5.0	196.08	197.53	199.06	200.59	202.13	203.68	205.24	206.81	208.39	209.97

from different laboratories [18]. Still in 1996, differences of the order of 0.3% around 500 mK and of 6% at 0.9 mK were found among the measurements carried out in different laboratories [18–23]. In the range of overlapping with the ITS 90 (0.65–1 K), the PLTS 2000 is considered more accurate with an error of 0.3 mK at 1 K [24].

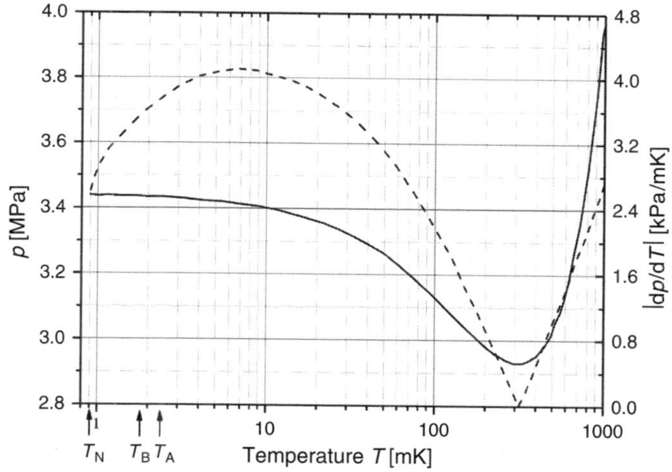

Fig. 8.2. Melting pressure p_m of ^{3}He (full line) and absolute value of the derivative $\mathrm{d}p_m/\mathrm{d}T$ (dashed line) as a function of temperature. T_N, T_B, T_A are the temperatures of the three phase transitions in solid and liquid ^{3}He.

Table 8.6
Coefficients of eq. (8.4) of the PLTS 2000 [24]

a_{-3}	$-1.385\,544\,2 \times 10^{-12}$
a_{-2}	$4.555\,702\,6 \times 10^{-9}$
a_{-1}	$-6.443\,086\,9 \times 10^{-6}$
a_0	$3.446\,743\,4 \times 10^{0}$
a_1	$-4.417\,643\,8 \times 10^{0}$
a_2	$1.541\,743\,7 \times 10^{1}$
a_3	$-3.578\,985\,3 \times 10^{1}$
a_4	$7.149\,912\,5 \times 10^{1}$
a_5	$-1.041\,437\,9 \times 10^{2}$
a_6	$1.051\,853\,8 \times 10^{2}$
a_7	$-6.944\,376\,7 \times 10^{1}$
a_8	$2.683\,308\,7 \times 10^{1}$
a_9	$-4.587\,570\,9 \times 10^{0}$

The PLTS 2000 is defined by the following relation between the melting pressure p and the temperature T_{2000} (see Table 8.6):

$$p/\mathrm{MPa} = \sum_{i=-3}^{+9} a_i \cdot (T_{2000}/\mathrm{K})^i \tag{8.4}$$

The ^{3}He melting pressure thermometer has been chosen to extend the ITS 90 for several reasons, such as the good sensitivity over three temperature decades, except around the pressure minimum at 315.24 mK. On the other hand, such a minimum is a reference point in the calibration of the pressure transducer; in fact, the pressure must be measured in situ since, below 315.24 mK, the entrance of the measurement cell is blocked by solid ^{3}He.

Table 8.7
Notable temperature references in liquid and solid ^3He [24]

Minimum of melting curve	315.24 mK	2.93113 MPa
Superfluid A transition at p_{melt}	2.444 mK	3.43407 MPa
Superfluid B transition at p_{melt}	1.896 mK	3.43609 MPa
Superfluid B transition at Saturated Vapour Pressure (SVP)	0.94 mK	0
Nuclear magnetic Transition of solid ^3He at p_{melt}	0.902 mK	3.43934 MPa

Besides the pressure minimum, the ^3He melting curve presents a few characteristics which can be used as temperature and pressure reference points: the superfluid A transition, the A–B transition in the superfluid and Nèel transition in the solid. The reference values are reported in Table 8.7 (see also Section 2.2.4).

The standard uncertainty of the scale in thermodynamic terms is estimated to be 0.5 mK down to 500 mK, decreasing linearly to 0.2 mK at 100 mK. It decreases further with falling temperature, but in percentage terms, it increases to about 0.3% at 25 mK and 2% at 0.9 mK. The standard uncertainties in the absolute pressures are estimated to be about 60 Pa (around 3 MPa).

Taking up a 2005 recommendation of the Consultative Committee for Thermometry (CCT) [25], the Comité International des Poids et Mesures (CIPM) created a 'mise en practique' of the definition of the kelvin containing clarifications and recommendations of the CCT concerning the realization of the definition of the kelvin and the implementation of practical temperature scales.

Both the ITS 90 and the PLTS 2000 appear in the 'mise en practique' with the same level of formality as 'internationally accepted practical temperature scales defining temperatures T90 and T2000 that are good approximations to thermodynamic temperature T'.

8.5 NBS-SRM 767a, 768 and SRD 1000 fixed point devices

In 1972, the National Bureau of Standard (now NIST, National Institute of Standard and Technology) started producing and selling two fixed point devices based on the normal-superconducting transition of some metals:

1. the SRM-767a with six elements with superconducting transitions in the 0.5–9.3 K range (see Table 8.8).

Table 8.8
Superconducting transition temperatures of the six elements in the SRM-767a

Element	T_C [K]
Nb	9.3
Pb	7.200
In	3.415
Al	1.180
Zn	0.851
Cd	0.519

Table 8.9
Properties of the materials used in the SRM-768

Material	T_C [mK]	B_C [mT]	RRR
W	$15.5 \div 15.6$	0.12	10^3
Be	$22.6 \div 22.8$	0.114	79
$Ir_{0.8}Ru_{0.2}$	$99 \div 100$	type II	2.5
$AuAl_2$	$159 \div 161$	1.21	50
$AuIn_2$	$203 \div 206$	1.45	50

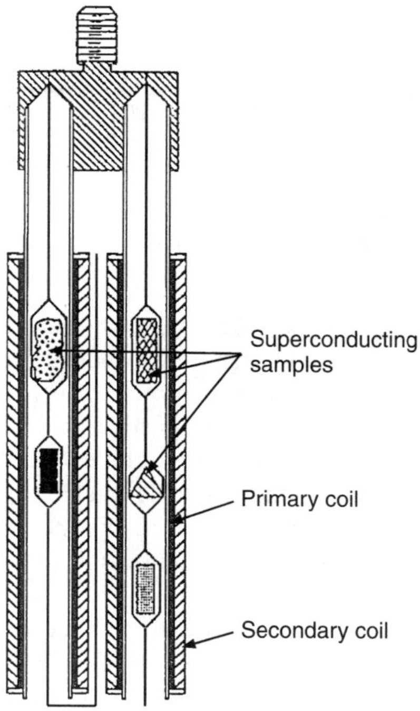

Fig. 8.3. Schematic of the SRM-768 device containing five superconducting samples, as well as the transformer coils for measuring the transition temperatures of the samples [27–28].

2. SRM-768 using five materials with transition between 15 and 200 mK (see Table 8.9 and Fig. 8.3) [26,27].

It is worth remarking that, in fact, such devices do not supply true reference fixed points as explained in Section 8.2, since these devices were calibrated one by one against primary thermometers. Their accuracy is 0.1–0.2 mK.

One device is made up of samples of superconducting materials put inside a transformer: a low-frequency small current in the primary creates a weak magnetic field.

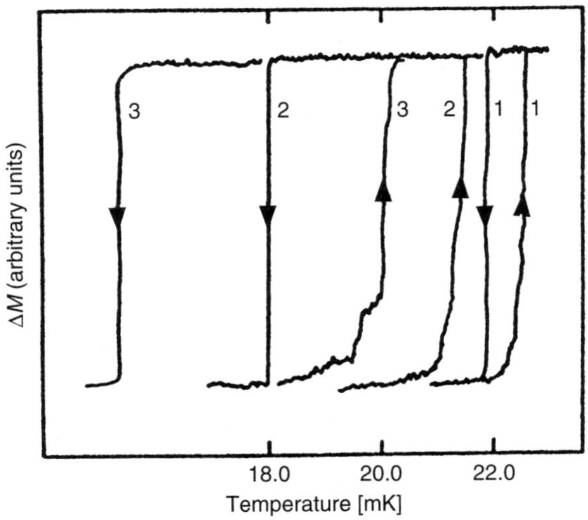

Fig. 8.4. Superconductive transition of pure Be showing the dependence on the magnetic field. Data in the cooling and warming correspond to fields of 0.5 μT (1), 9.5 μT (2) and 19 μT (3) [27–28].

When one of the samples enters the superconducting state, the sample expels the field (Meissner effect), thus producing a change of the mutual inductance between primary and secondary coils which is detected as a change of the voltage at the secondary output. The two devices were supplied neither with magnetic shield nor with electronics. Since a magnetic field modifies both the transition temperature T_C and the transition width (see Figs 8.4 and 8.6), the user had to realize a suitable shield around the measuring head. In a magnetic field, the superconducting transition becomes a first-order transition. Hence, supercooling effects may take place in pure elements as shown in Fig. 8.4 for pure Be used in the SRM-768.

In the SRM-768, this effect is reduced by spot welding small pieces of Al to the W and Be samples. The Al, with its $T_C = 1.18$ K, serves as a nucleation centre to induce superconductivity in the samples by proximity effect (see Section 15.2.2).

External fields must be shielded to about 10^{-7} T, i.e. well below earth magnetic field (~50 μT). The maximum recommended field is 10^{-6} T and the dT_C/dH (about 100 K/Tesla) is supplied around $H = 0$. The maximum power which may be supplied to the coils is between 10^{-9} and 10^{-10} W. The sweep rate used to trace out the superconducting transitions should be less than 0.1 mK/min (for W and Be) to avoid hysteresis effects. Unfortunately, nowadays SRM-767a and SRM-768 are no longer produced.

Recently, a superconductive reference device, SRD1000 (Hightech Development Leiden and Kamerling Onnes Laboratory) [29], with ten reference materials (15 mK–1K, see Table 8.10) has been developed to enable simplified in situ calibration of interpolating thermometers. An evaluation of SRD1000 prototypes at various laboratories is reported in ref. [30].

A very attractive feature of the SRD1000 device is the presence of the electronics to trace out the superconducting transitions of the samples, of the magnetic shields and of

Table 8.10
Properties of superconducting materials in SRD1000 device.

Proposed reference temperature [mK]	Selected metal	Observed T_C and W_C [mK]	Accuracy determination of T_C
1180	Al	1182 (4.0)	0.007
850	Zn	850 (11)	0.004
520	Cd	519.5 (4.5)	0.01
208	$AuIn_2$	208.0 (0.4)	0.0005
160	$AuAl_2$	161.3 (0.3)	0.001
100	Ir	99.8 (0.5)	0.002
60	$Ir_{0.92}Rh_{0.08}$	61.3 (0.5)	0.002
35	$Ir_{0.80}Rh_{0.20}$	34.0 (0.9)	0.008
22	$Ir_{0.73}R_{0.27}$	20.5 (1.5)	0.03
15	W	15.0 (0.3)	0.07

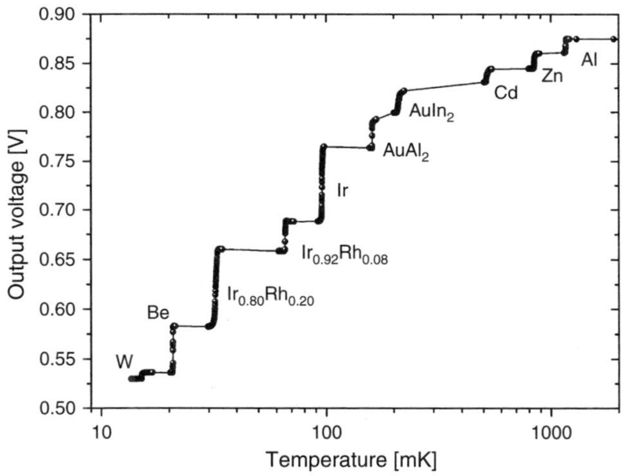

Fig. 8.5. Temperature dependence of SRD1000 mutual inductance measured as output voltage.

two Helmholtz coils to compensate for the residual field. Figure 8.5 shows temperature dependence of SRD1000 mutual inductance measured as output voltage.

8.6 APPENDIX: Superconductive transitions and influence of purity and magnetic fields

In Table 8.11, the approximate transition temperature of some elements in zero magnetic field and critical fields are reported [31].

Table 8.11
Superconductivity parameters for elements.

Element	T_C [K]	Critical field [gauss]
Be	0.026	—
Al	1.14	105
P	5	—
Ti	0.39	100
V	5.38	1420
Cr*		—
Zn	0.875	53
Ga	1.091	51
Ge*	5	—
As*	0.5	—
Se*	7	—
Zr	0.546	47
Nb	9.50	1980
Mo	0.90	95
Tc	7.77	1410
Ru	0.51	70
Rh	0.0003	0.049
Cd	0.56	30
In	3.4035	293
Sn(w)	3.722	309
Sb*	3.5	—
Te*	4	—
Cs*	1.5	—
Ba*	5	—
La(fcc)	6.00	1100
Hf	0.12	—
Ta	4.483	830
W	0.012	1.07
Re	1.4	198
Os	0.655	65
Ir	0.14	19
Hg	4.153	412
Tl	2.39	171
Pb	7.193	803
Bi*	8	—
Ce*	2	—
Lu	0.1	—
Th	1.368	1.62
Pa	1.4	—
U*	2	—

Transition temperature in Kelvin, critical magnetic field in gauss (10^{-4} T).

* Superconducting only in thin films or under high pressure in a crystal modification not normally stable. Critical temperatures for those elements from [32, Chapter 12].

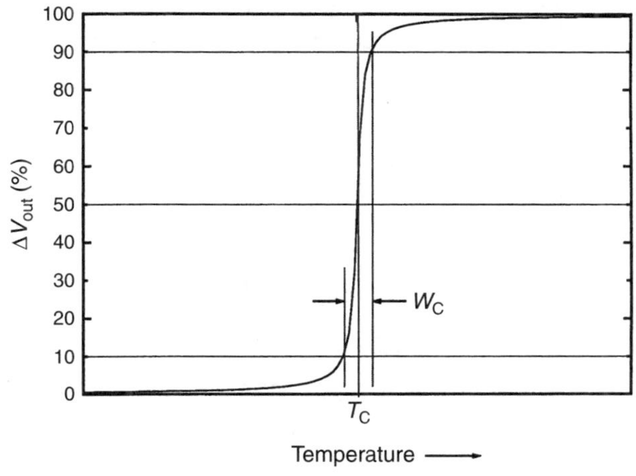

Fig. 8.6. Definition of transition temperature T_C and transition width W_C.

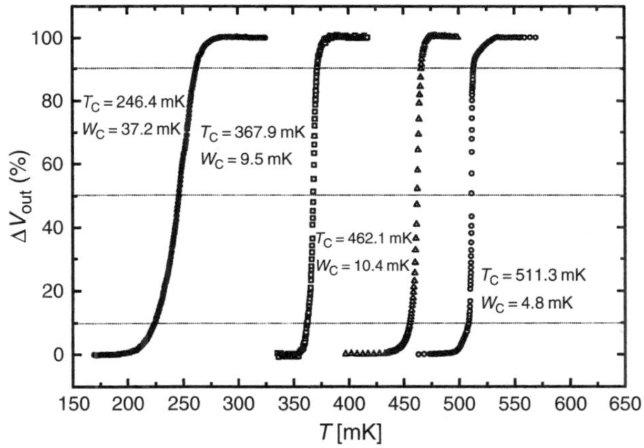

Fig. 8.7. Superconductive transitions of four Ti samples of different purity: • sample A_1, $T_C = 246.4\,\text{mK}$, $W_C = 37.2\,\text{mK}$; □ sample B_1, $T_C = 367.9\,\text{mK}$, $W_C = 9.5\,\text{mK}$; Δ sample C_1, $T_C = 462.1\,\text{mK}$, $W_C = 10.4\,\text{mK}$; o sample D_1, $T_C = 511.3\,\text{mK}$, $W_C = 4.8\,\text{mK}$.

In superconducting transitions, the purity of materials and the shielding from magnetic fields are fundamental to obtain steep transitions (see Figs 8.6–8.8).

Figure 8.6 shows the definition of transition temperature T_C and transition width W_C. To remark how the purity of the material and the magnetic field influence the two parameters, we will report hereafter the measurements of the superconducting transition of titanium samples of different purity and in the presence of different magnetic fields. The transitions of four Ti samples, whose characteristics are reported in Table 8.12, are shown in Fig. 8.7 where a change of only 0.2% in purity dramatically moves both the transition

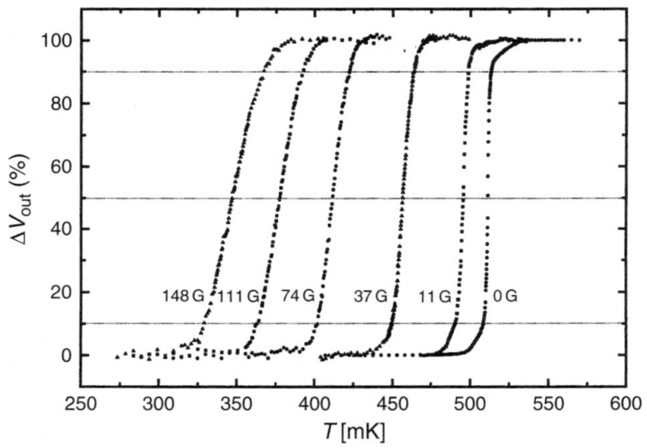

Fig. 8.8. Superconductive transitions of the Ti sample D of Table 8.12 in a magnetic field (0–148 G).

Table 8.12
Characteristics of Ti samples of Figs 8.6 and 8.7.

Sample	A	B	C	D
Structure	Polycrystal	Polycrystal	Polycrystal	Polycrystal
Purity (% of weight)	99.8%	99.975%	99.99%	99.999%
Treatment	Annealed	Annealed	As drawn	As rolled
Geometry	wire	wire	wire	bar
Dimension	Ø=1mm	Ø=0.5mm	Ø=0.5mm	$2.6 \times 1\ mm^2$
Impurity (ppm in weight)	N 30 C 30 H 11 Fe 250 O 540	Al, Cr, Mg, Mo<10 Ca, Cu<5 Fe<160 K, Li, Si, Na<1	Al 0.63, O 211, C 18, Fe 3.15, Ca, Nb<0.2 Cd, Mg<0.1 Cr<0.945 H 3.0, N 9, Ni<0.055, S 1.0, Si 0.24, Rb<5, Ta<1.0, V 0.825, Zr 0.38	Fe 5.5 V 1.3
Distributor	Advent	Goodfellow	Goodfellow	Goodfellow
Product code	Ti555311	TI005127	TI005135	TI000433

temperature T_C and the transition width W_C. Moreover, the presence of a magnetic field also increases the transition width and moves the transition toward a lower temperature (see Fig. 8.8).

References

[1] T. Quinn: *Temperature*, 2nd edn., Academic Press, London (1990)
[2] L.G. Rubin: Cryogenics **10**, 14 (1970)

[3] L.G. Rubin, B.L. Brant, H.H. Sample: Cryogenics **22**, 491 (1982)
[4] L.G. Rubin Cryogenic **37**, 341 (1997)
[5] G. Schuster et al.: Physica B **165** & **166**, 31 (1994)
[6] I.M. Mills et al.: Metrologia **43**, 227 (2006)
[7] J. Fischer, B. Fellmuth: Reports on Progress in Physics **68**, 1043 (2005)
[8] M. Arnold, R. Miller: Science Education **80**, p. 249 (1996)
[9] F. Pavese, G. Molinar: *Modern gas-based temperature and pressure measurements*, Plenum Press, New York (1992)
[10] S.W. Van Sciver: *Helium cryogenics*, Plenum Press, New York (1986)
[11] IPTS-68: Metrologia **12**, 1 (1976)
[12] EPT-76: Metrologia **15**, 65 (1979)
[13] M. Durieux et al.: Metrologia **15**, 57 (1979)
[14] H. Preston-Thomas: Metrologia **27**, 3 (1990)
[15] B.W. Magnum: J. Res. Nat. Inst. Stand. Technol. **95**, 69 (1990)
[16] B.W. Magnum, G.T. Furukawa: Nat. Inst. Stand. Technol. Technical Note 1265 (1990)
[17] B. Fellmuth et al.: Metrologia **42**, 171 (2005)
[18] D.S. Greywall: Phys. Rev. B **27**, 2747 (1983); ibid. B **33**, 7520 (1986)
[19] D.S. Greywall, P.A. Busch: J. Low Temp. Phys. **46**, 451 (1982)
[20] D.S. Greywall: Phys. Rev. B **31**, 2675 (1985)
[21] W.P. Halperin et al.: J. Low Temp. Phys. **31**, 617 (1978)
[22] E.R. Grilly: J. Low Temp. Phys. **4**, 615 (1971); ibid. **11**, 243 (1973)
[23] W. Ni et al.: J. Low Temp. Phys. **99**, 167 (1995)
[24] R.L. Rusby et al.: J. Low Temp. Phys. **126**, 633 (2002)
[25] CCT document CCT/05-Rec-T3, see http://www.bipm.org/cc/CCT/Allowed/23/CCT_05_32.pdf
[26] R.J. Soulen, R.B. Dove: N.B.S. Special Publication 260–262 (1979)
[27] J.H. Colwell, W.E. Fogle, R.J. Soulen: *Proc. 17th Int. Conf. On Low Temp. Phys.* p. 395, North Holland (1984)
[28] R.J. Soulen, R.B. Dove: Standard Reference Materials: SRM 768: *Temperature reference standard for use below 0.5 K*. National Bureau of Standards, US Department of Commerce, Special Publication 260-62 (1979)
[29] W.A. Bosch et al.: in *Proc. TEMP-MEKO 2001*, p. 397, ed. by B. Fellmuth et al., VDE Verlag, Berlin (2001)
[30] S. Schottl et al.: J. Low Temp. Phys **138** , 941 (2005)
[31] C. Kittel: *Introduction to Solid State Physics*, 7th ed., Wiley (1996).
[32] H.P. Myers: *Introductory Solid State Physics*, 2nd Ed., Taylor & Francis (1997)

9 Low-Temperature Thermometry

Contents

9.1 Introduction

A thermometer is a device by which we can measure a property of matter function of temperature. If a relation, based on fundamental laws of physics, between such property and the thermodynamic temperature is considered reliable, the thermometer does not need a calibration and is called *primary thermometer*. In the other cases, the thermometer needs a calibration and is called *secondary*. Examples of primary thermometers are gas thermometers and noise thermometers.

Primary thermometry is usually quite difficult and reserved to metrological laboratories.

Secondary thermometers must be calibrated by means of a primary thermometer or at fixed points as discussed in Chapter 8. Secondary thermometers are often quite easy to use and more sensitive than primary thermometers. A typical example of secondary thermometer is the electrical resistance thermometer (see Section 9.6).

Table 9.1
Common low-temperature thermometry

1	Gas thermometry
2	Vapour pressure thermometry
3	^{3}He melting pressure thermometry
4	Thermoelectric power thermometry
5	Resistance thermometry
6	Noise thermometry
7	Dielectric constant thermometry
8	Electronic paramagnet thermometry
9	Nuclear paramagnet thermometry (range 1–100 μK)
10	Nuclear orientation thermometry
11	Coulomb blockade thermometry

In most cases, one thermometer is useful only in a limited temperature range. Usually, out of the useful range, it simply loses sensitivity: for example a Pt thermometer (see Section 9.6.1) shows an electrical resistance which decreases of ~0.13%/K down to ~20 K, but only of ~0.01%/K at 10 mK.

A good low-temperature thermometer should possess several qualities:

- reproducibility and high sensitivity;
- wide operating range and low sensitivity to environmental changes such as magnetic or stray fields;
- short time to reach equilibrium and fast reading system. A low heat capacity of the sensor is not sufficient to obtain fast readings of temperature (see Section 9.11);
- at very low temperature, the heat introduced by the measurement must be as low as possible to avoid overheating of the sensor (see Section 9.5).

Since most of the properties of materials depend on temperature, there are a lot of possible choices for a thermometer. Some thermometric properties like Mössbauer effect or osmotic pressure, of historical interest, but no longer in use, are reported in ref. [[1], pp. 200–206]. Hereafter, some thermometric properties useful at low temperature are described (see Table 9.1). Due to the enormous amount of papers on the subject, the bibliography cannot be complete. References before 1980 are reported in ref. [2].

Useful source of information about thermometry are ref. [3–8].

9.2 Gas thermometry

The three modern types of gas thermometry – constant volume gas thermometry (CVGT), acoustic gas thermometry (AGT) and dielectric constant gas thermometry (DCGT) – are presently considered ‘primary’. They are based on simple relations between the properties of an ideal gas and temperature T. However, the departure from the ideal behaviour must be carefully considered in view of the desired level of accuracy. This is done by measuring the thermometric property as a function of density. Then the ideal behaviour is deduced by fitting an appropriate expansion to the measured isotherm and extrapolating to zero density. We wish to remark the fact that AGT and DCGT are based on the variation with T of an intensive property of the gas (speed of sound and dielectric constant, respectively), whereas primary CVGT requires the knowledge of the number of moles of gas present in the gas bulb.

9.2.1 Constant volume gas thermometry (CVGT)

In the CVGT, the pressure in a fixed amount of gas is related to the temperature using an equation of state for the gas in the form:

$$pV = nRT \cdot \left[1 + B(T) \cdot n/V + C(T) \cdot (n/V)^2\right] \tag{9.1}$$

where $n =$ moles of gas in the volume V.

At low temperature, in eq. (9.1) only the coefficient B is significant [3], whereas at very low temperature also C must be taken into account. Values of the coefficients for ^{3}He and ^{4}He are reported in ref. [9–14].

The filling gas pressure of a CVGT must be carefully chosen: it must be high enough to get a good sensitivity and low enough to approximate the ideal gas behaviour.

The accuracy in the measurement of pressure p depends on the lowest temperature down to which the CVGT is used. For example, with an allowed pressure variation $\Delta p = 0.133\,\mathrm{Pa}$, the minimum allowed temperature is $\approx 2\,\mathrm{K}$ for ^{4}He and $\approx 1\,\mathrm{K}$ for ^{3}He respectively. With a fixed number of moles N, the CVGT gives T carrying out two pressure measurements: at the temperature T to be determined and at a known temperature $T_{\mathrm{ref.}}$. The requirements for the constancy (practically the knowledge) of N in a CVGT are the following:

(a) Dead volume (manometers, valves and capillary tube connecting the pressure gage) should be small and possibly constant (e.g. the amount of gas in the dead volume, such as the capillary, depends also on the temperature distribution and hence on the reproducibility of the gradient along the capillary), since corrections for their contribution are to be taken into account;
(b) Thermal and elastic volume changes (as a function of T) of various components (e.g. the gas bulb) of the CVGT should be known;
(c) Adsorption and desorption of gas from the walls should be evaluated. To minimize these phenomena, helium is the most favourable gas, since its chemical adsorption is negligible and its adsorption energy is the lowest among noble gases. Nevertheless, it is worth reminding that desorption is an irreproducible phenomenon because it depends not only on temperature but on the previous 'thermal history' of the device;
(d) Impurities (usually H_2 in ^{4}He and ^{4}He in ^{3}He) in the gas change the value of B in eq. (9.1);
(e) The 'aerostatic head' must be considered (see ref. [15]): it is defined as the pressure difference Δp due to a gas column of height h and density $\rho(T)$. Δp is proportional to $1/T$. The effect is mainly due to the low-temperature region, i.e. the bulb itself. Hence the aerostatic head correction depends on the bulb temperature T;

(f) Thermomolecular pressure difference along the capillary connecting the bulb to the room temperature gage (see e.g. ref. [16]);
(g) A low-temperature pressure gage, of course, eliminates or minimizes many of the aforementioned problems (see e.g. ref. [17]). Note that it must stand the room temperature gas pressure in the bulb.

The realization of a CVGT with a high level of accuracy requires complex techniques and skill. It is mostly used by national laboratories for calibration purposes. For a detailed description of CVGT, see e.g. ref. [18].

Over the years, a simpler instrument called 'interpolating constant volume gas thermometer' (ICVGT) has been developed. It differs from traditional CVGT because it is calibrated at a number of temperatures known from independent measurements.

In the ITS-90, the ICVGT has been introduced instead of the CVGT for its simplicity and better reproducibility. By means of the ICVGT, temperatures are defined in the range from 3 K to the triple point of Ne (24.5561 K) by interpolating curves (established according to an international agreement) referring to ^{3}He or ^{4}He gas pressure [19–21]. Suppose that everything is in the same situation both during calibration and measurement (e.g. the same thermal distribution in the capillary), by means of the calibration at fixed points of the ICVGT, the corrections listed before ((a) through (f)) are not needed. In particular the p used in the interpolating formula is the pressure measured by the manometer and not the pressure inside the bulb.

In the ICVGT, corrections (a), (e) and (f) are not taken into account. When using the established quadratic interpolation curve, these corrections (within 0.3 mK) are taken into account through the factor B of eq. (9.1). Details of construction and operation for ^{3}He and ^{4}He ICVGT are reported in ref. [22,23] for ^{3}He and [24–26] for ^{4}He.

The problems encountered in performing accurate measurements of T with a primary CVGT can be overcome by the use of the dielectric constant gas thermometer [27–31] or by the acoustic gas thermometer [3,32,33].

9.2.2 Acoustic gas thermometry (AGT)

Sound waves are propagated in a gas as longitudinal waves of speed v_s:

$$v_s = (B_S/\rho)^{1/2} \tag{9.2}$$

where B_S is the adiabatic bulk modulus and ρ the gas density.

For an ideal gas:

$$v_s = (\gamma \cdot RT/M)^{1/2} \tag{9.3}$$

with M the molar mass and $\gamma = C_P/C_V$ ratio. Corrections are small except at very low temperatures. Difficulties or errors have been nevertheless encountered because of the propagation of higher frequency modes or boundary effects. Descriptions of the NIST and NPL acoustic thermometers can be found in ref. [3]. Details of a spherical interferometer giving uncertainties below 0.3 mK can be found in ref. [32–34].

9.2.3 Dielectric constant gas thermometry (DCGT)

The basic idea of DCGT is to replace the density in the state equation of a gas by the dielectric constant ε.

For an ideal gas, this yields the simple relation between the pressure p and ε:

$$p = k_B T \cdot (\varepsilon - \varepsilon_0)/a_0 \tag{9.4}$$

where ε_0 is the exactly known vacuum dielectric constant and a_0 the static electric dipole polarizability of a gas atom (see Fig. 9.1). The measurement of DCGT isotherms yields A_ε/R and thus:

$$k_B = (R/A_\varepsilon) \cdot a_0/(3\varepsilon_0) \tag{9.5}$$

where A_ε is the molar polarizability.

As AGT, DCGT avoids the troublesome density determination of the conventional gas thermometry. In addition, the pressure sensing tubes can be of any convenient size, and the thermometric gas can be moved in or out the bulb without the need to allow for the amount of the gas involved.

Absolute DCGT requires knowing a_0 with the necessary accuracy. Nowadays this condition is fulfilled for helium. Recent progress has decreased the uncertainty of the value of a_0 well below 10^{-6} [35,36]. Up to now, DCGT using cylindrical capacitors for

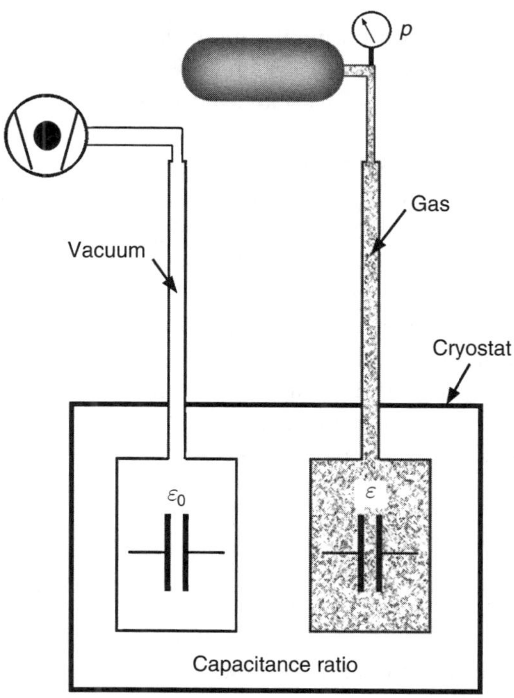

Fig. 9.1. Schematics of dielectric constant gas thermometry.

determining ε has been performed by two groups in the range from about 3 to 27 K [37,38]. The obtained relative uncertainties are of order of 10^{-5}, but it seems to be possible to decrease the uncertainty at least by an order of magnitude by optimizing the capacitor design and with in situ measurements [39].

Further progress depends essentially on the developments in p and C measurements. A new proposal is to apply quasi-spherical resonators for DCGT [40].

9.3 Vapour pressure thermometry

The vapour pressure of a liquid depends on the kind of liquid and changes rapidly with temperature (see Section 2.4). Hence, there is the possibility of realizing convenient secondary thermometers, which were widely used before the advent of commercially calibrated resistance thermometers.

From eq. (2.3) of Section 2.2.4.1, the vapour pressure over a liquid (with latent heat of evaporation $L = \text{constant}$) can be written as:

$$\ln(p) = \frac{A}{T} + B \tag{9.6}$$

However, in reality, $L = L(T)$; hence an empirical equation is commonly used:

$$\ln\left(\frac{p}{p_0}\right) = a + \frac{b}{T} + c \cdot T + \ldots \tag{9.7}$$

A compilation of the vapour pressure of various liquids can be found in ref. [41].

The most interesting liquids for low-temperature thermometry are ^{3}He and ^{4}He, especially for the calibration of resistance thermometers in the range from 0.5 to 4.2 K. Vapour pressure of H_2 is also interesting to realize vapour pressure-fixed points included in ITS-90. The measure of He vapour pressure has been carried out with great accuracy [42,43] to establish the ITS-90 (see Section 8.3). There are several experimental precautions to be observed in order to obtain reliable measurements [2].

Also in the vapour pressure thermometer, the presence of impurities, isotopes or different spin states must be evidenced and minimized.

The main problem encountered in the realization of a vapour pressure thermometer is the correct measurement of the pressure p. In this case, however, the dead space is not a problem and larger tubes may be used (instead of capillaries). Since the pressure gage is usually at room temperature, also in this case, a gradient in temperature between the cold and the warm end of the sampling tube produces a thermomolecular pressure difference [44,45, p. 48]:

$$\frac{\mathrm{d}p}{\mathrm{d}T} = \frac{p}{2T} \cdot f\left(r/\lambda\right) \tag{9.8}$$

where r is the radius of the tube, λ is the mean free path of the gas particles and f is a function of the ratio r/λ.

Thermomolecular pressure difference is present in vapour pressure with any gas. In the case of ^{4}He, additional problems occur: above the lambda point (see Section 2.2.4.1), the result is that the temperature above the surface may be a few millikelvin lower than that of the liquid. These effects can be reduced by containing the liquid in a high-conductivity chamber. Below lambda point, He II is a very good heat conductor so that temperature gradients do not exist in the liquid. Anyway, the flow of a superfluid He film creeping to the warmer part of the tube wall may be a problem: this steady heat flow into the container gives raise to a temperature difference between the liquid and the walls of the container, accentuated by the presence of a Kapitza resistance at the solid–liquid interface. Temperature gradients can only occur in the case of liquid ^{3}He which is a poor heat conductor.

A detailed discussion of the problems encountered in vapour pressure measurements at low temperature is given in ref. [46,47], where also the use of an 'in situ' manometer is described. Vapour pressure gas thermometry with other liquids besides He is discussed in ref. [43, p. 49].

9.4 ^{3}He melting curve thermometry

As we saw in Chapter 7, the ^{3}He melting curve, shown in Fig. 7.1, is described by the Clausius–Clapeyron equation:

$$\left(\frac{\mathrm{d}p_f}{\mathrm{d}T}\right)_{\text{vap.}} = \frac{S_{\text{liq.}} - S_{\text{sol.}}}{V_{\text{m,liq.}} - V_{\text{m,sol.}}} \tag{9.9}$$

where f refers to fusion.

Since the liquid molar volume is always larger than that of solid [48,49], $\mathrm{d}p/\mathrm{d}T < 0$ if $S_{\text{liq.}} < S_{\text{sol.}}$.

For temperatures above 0.32 K, the entropy of liquid is higher than that of the solid as happens in all other materials (see Section 7.2.1); below 0.32 K, the situation is reversed. We remind that this property allows to cool the liquid by isoentropic compression (see Section 7.1).

To use the ^{3}He melting pressure as thermometric property, pressure is measured in situ by means of capacitance pressure gage [50]. With a pressure resolution of 10 microbar, the precision in the measurement of temperature is 3×10^{-4} at 1 mK, 3×10^{-5} at 10 mK and 5.10^{-6} at 100 mK [51].

The confidence in the theoretical relationship between pressure and temperature along the melting curve led to state as primary this type of thermometer and put it as the base of PLTS 2000 down to 0.9 mK (see Section 8.5).

For the realization of a primary melting curve thermometer (MCT), three different methods have been suggested and tested:

1. From the knowledge of the quantities (known by independent measurements) in Clausius–Clapeyron equation, the slope of the fusion curve can be evaluated and integrated to get $p_f(T)$. For example, in the 5–20 mK range, the Clausius–Clapeyron equation gave temperature values with an 1% accuracy [52,53];
2. From the temperature T defined as:

$$T = \frac{\mathrm{d}Q}{\mathrm{d}S} = \left(\frac{\mathrm{d}Q}{\mathrm{d}p_f}\right) \cdot \left(\frac{\mathrm{d}p_f}{\mathrm{d}S}\right) \tag{9.10}$$

where $\mathrm{d}Q/\mathrm{d}p_f$ is the measured $\mathrm{d}Q$ needed to produce a pressure change $\mathrm{d}p_f$. Instead, $\mathrm{d}p_f/\mathrm{d}S$ can be evaluated from a series of thermodynamic transformation [52,54];

3. The third method is a variation of the second: a heat pulse $\mathrm{d}Q$ is supplied, keeping both the pressure and temperature constant by means of a feedback system. In this process, some moles of liquid ^{3}He are solidified: the latent heat of solidification is the measured $\mathrm{d}Q$. Since from Clapeyron equation:

$$\frac{\mathrm{d}p_f}{\mathrm{d}T} = \frac{\mathrm{d}S}{\mathrm{d}V} \tag{9.11}$$

from $\mathrm{d}Q = T\mathrm{d}S$, we get:

$$\frac{\mathrm{d}Q}{\mathrm{d}V} = T \cdot \frac{\mathrm{d}p_f}{\mathrm{d}T} \tag{9.12}$$

By measuring $\mathrm{d}Q/\mathrm{d}V$ for some pressure values along the melting curve, by integration, we get:

$$T = \overline{T} \times \exp\left(\int_0^{p-\overline{p}} (\mathrm{d}V/\mathrm{d}Q)\,\mathrm{d}p'\right) \tag{9.13}$$

where p' is the pressure difference measured along the melting curve with a fixed point $\overline{p}$. Integration of eq. (9.13) gives the temperature from the knowledge of one fixed point $\overline{T}$.

The method, developed at Cornell for the 1–25 mK, allows one to get T without the knowledge of $\overline{T}$. In fact, all the quantities which describe the system can be evaluated as functions of the ratio $T/\overline{T}$. On the other hand, $\overline{T}$ can be evaluated at the end as a fit parameter [55].

Even if nowadays, the MCT may be considered a primary thermometer only on a narrow temperature range, it is considered the best dissemination standard in the millikelvin range [56–59]. In fact, the ^{3}He melting pressure is a good thermometric property because of its sensitivity over three decades of temperature with a resolution $\Delta T/T$ up to 10^{-5} [56]. The good repeatability, the insensitivity to magnetic fields up to 0.5 T [60] and the presence of temperature-fixed points allow for the control of possible shifts in the calibration curve of the pressure transducer. The usefulness of these fixed points is evident, considering that the ITS-90 is based just on the definition of fixed points.

The precautions in a thermometric measurement with the ^{3}He melting curve thermometer are the control of the impurity from ^{4}He [61] and of presence of high magnetic fields [62,63].

The advantages of this thermometer are the high resolution (about 1 μK), repeatability (a few ppm), the zero power dissipation and the insensitivity to radio frequency (RF) fields. The experimental realization is however quite complicated.

9.5 Thermocouples

The use of the thermoelectric power of metallic junctions presents the following advantages:

- it is a local measurement with point sensors;
- the apparatus is simple and the measurements are quite easy;

- the thermometer is usually insensitive to magnetic fields;
- the sensors have low heat capacity;
- no power is involved in the measurement;
- measurements are reproducible.

Unfortunately, the thermoelectric power vanishes when temperature tends to zero: pairs as the classic Cu/constantan thermocouples show very low sensitivity below 10 K (see Fig. 9.2).

However, new materials have been found which allow the extension of the measurements to the millikelvin region. These materials contain a small amount of magnetic dopants as Au/Fe or Pd/Fe, see Fig. 9.3 [64].

Obviously, the small thermoelectric powers produced by metal pairs at low temperatures cannot be measured using a reference temperature of 0°C, where the thermoelectric power is high; a 4.2 K bath is a typical reference.

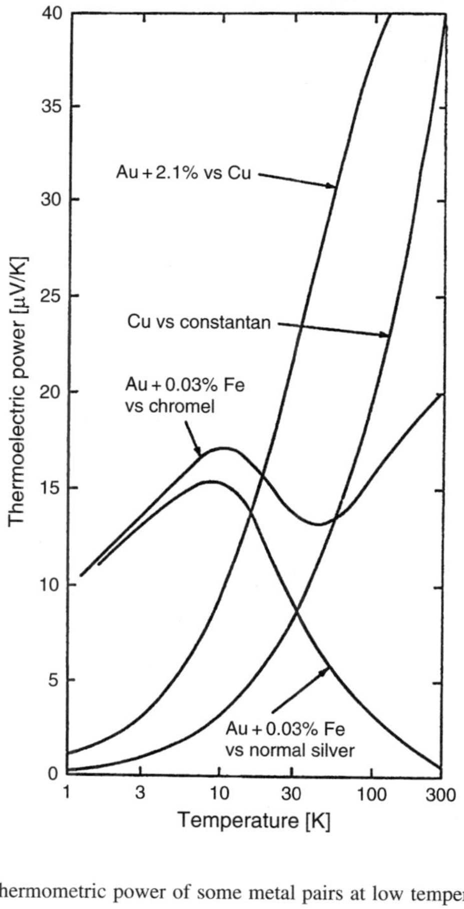

Fig. 9.2. Thermometric power of some metal pairs at low temperature [19].

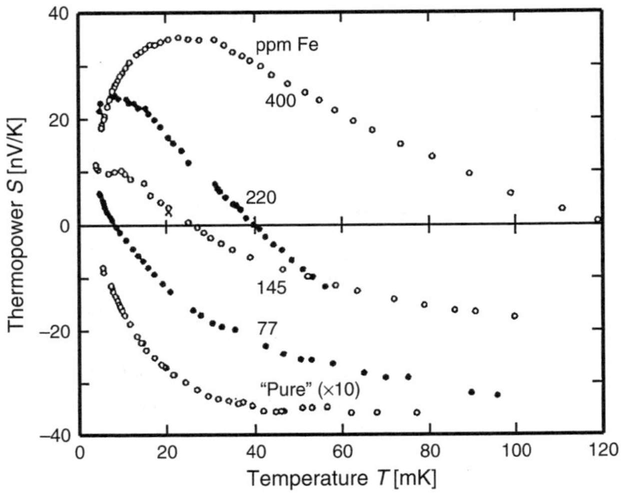

Fig. 9.3. Thermoelectric power of Pd doped with Fe (in ppm) as a function of temperature. The value for the pure sample has been multiplied by 10 for sake of clarity [64].

The main drawback in this type of thermometry is the presence of spurious thermoelectric powers due to chemical inhomogeneity, stress in conductors, contact effects in switches if present, etc.

The measurements are carried out with an a.c. bridge or with a SQUID at the lower temperatures. The resolution is of the order of 1 μK at 1K.

9.6 Resistance thermometry

The resistance thermometry is based on the temperature dependence of the electric resistance of metals, semiconductors and other resistive materials. This is the most diffused type of low-temperature thermometry: sensors are usually commercial low-cost components. At very low temperatures, however, several drawbacks take place such as the low thermal conductivity in the bulk of the resistance and at the contact surface, the heating due to RF pick up and overheating (see Section 9.6.3)

Laws describing the resistance–temperature dependence are not reliable: hence resistance thermometers are secondary devices.

9.6.1 Metal thermistors

The electric resistance of metals decreases as temperature is lowered down to about 20 K. The most used metal as thermometric material is platinum. Platinum is chemically resistant and can be produced with high purity (minimizing the temperature-independent

Table 9.2
Temperature-resistance values for a Pt-100 thermometer

Temperature [K]	Resistance [Ω]
10	0.09
15	0.19
20	0.44
25	0.94
30	1.73
40	4.18
50	7.54
60	11.45
77	18.65
100	28.63
125	39.33
150	49.85
175	60.23
200	70.50
225	80.66
250	90.72
273.2	100.00
300	110.63

contribution to resistivity = high RRR); it is ductile, hence can be drawn in thin wires, and finally its resistivity temperature coefficient is quite high.

The commercial Pt series resistance are well known. The characteristic of Pt-100 resistor are reported in Table 9.2 and Fig. 9.4.

It is important that the metal wire is strain free so it does not change its properties on repeated thermal cycles.

As we said, the sensitivity of a metallic thermometer drastically falls below about 10 K; as it was the case of the thermoelectric power; it is possible to increase the sensitivity by introducing some magnetic impurities. The most commonly used magnetic alloy is the commercial Rh-0.5% Fe (see Fig. 9.5).

9.6.2 Semiconductors, carbon and metal oxide thermistors

We can notice in Fig. 9.5 that Pt and Rh-Fe thermometers have a positive temperature coefficient, whereas other reported materials have a negative coefficient. They usually are either semiconducting or amorphous materials.

The formula which approximately describes the $R(T)$ dependence is the Mott's law [65]:

$$R = R_0 \cdot \exp\left(\frac{T_0}{T}\right)^n \tag{9.14}$$

where n ranges between 0.25 and 0.6 and T_0 between ~1 and ~100 K (see Section 15.2.1).

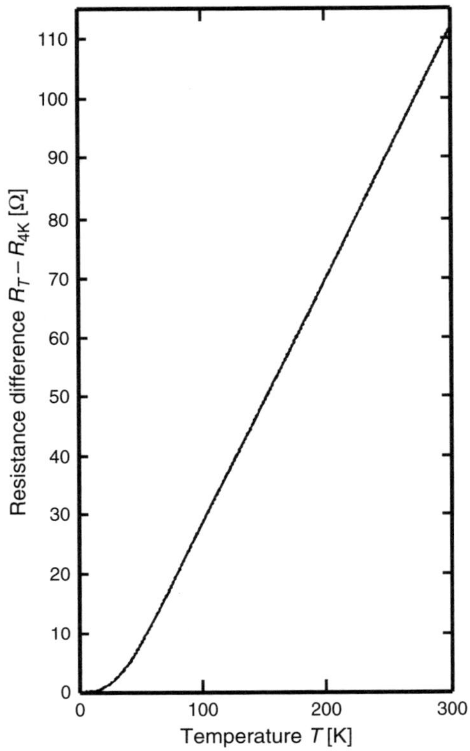

Fig. 9.4. Difference of resistance of Pt at the temperature T and at 4.2 K versus temperature. The curve is linear down to 50 K (data from [19]).

9.6.2.1 Doped germanium resistors

Germanium used for thermometric purposed has doping values of 10^{15}–10^{19} atoms/cm^3. Classic doping methods do not lead to the best results for Ge. In order to obtain a high-doping homogeneity, the so-called NTD (neutron transmutation doping) is used: pure Ge is irradiated by a flux of thermal neutrons produced by a nuclear reactor. Neutrons are captured by some lattice atoms which are transmutated into atoms of the third group. This process (see Section 15.2.1.1) produces a p-type doped Ge.

A precious quality of Ge thermometers is their stability (better than 0.1% after repeated cooling cycles). Problems at very low temperatures are:

1. the electron–phonon decoupling (see Section 15.2.1.3);
2. an excess of noise (see ref. [66,67]);
3. high thermal resistance towards support; this means that measurements must be done with a very low power. This drawback is common to all other resistance thermometers.
4. a high magnetoresistance which depends on the orientation respect to the magnetic field and increases with the sensitivity of the thermometer (that is with T_0) [68,69],

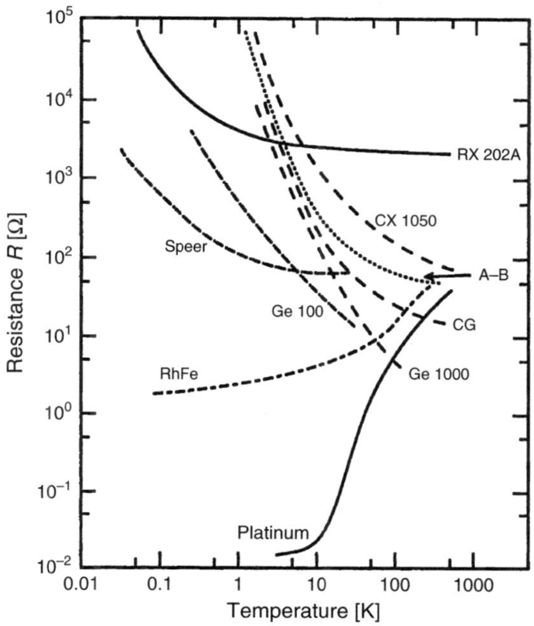

Fig. 9.5. The electrical resistance $R(T)$ for some typical thermometers. A–B denotes Allen–Bradley carbon resistor. Speer is a carbon resistor. CG is carbon-in-glass. CX 1050 is a Cernox and RX 202A is a ruthenium oxide from LakeShore. Ge 100 and Ge 1000 are Cryocal germanium thermometers [45].

see Fig. 9.6 which reports the transversal magnetoresistance of a Ge thermometer. Measurements of magnetoresistance down to 0.1 K are reported in ref. [70,71]. According to theory [72], the exponent of the Mott's law changes, and T_0 becomes proportional to the magnetic field.

We shall discuss the dynamic behaviour of Ge thermistor in Chapter 15.

9.6.2.2 Carbon resistors

Ge resistors are specifically produced for low-temperature thermometry; carbon and RuO_2 resistors are commercial products for electronics. Pure carbon is not a semiconductor. The negative slope $R(T)$ is due to the production process which consists in pressing and sinterization of carbon particles with glue. The resulting resistance is probably determined by the contact resistance between the particles. The cost of the carbon resistor thermometer is very low. Manufacturers such as Speer, Allen–Bradley and Matsushita have produced in the past carbon resistors for many years. Most of firms have now ceased manufacture, although their products may still be found in the storerooms of research laboratories.

Another useful device is the carbon-glass thermometer (CGT) [73] usable between $\sim$1 K and room temperature where it loses sensitivity. A nice property of CGT is its insensitivity to large magnetic fields [74,3, p. 269]. CGT are commercially available,

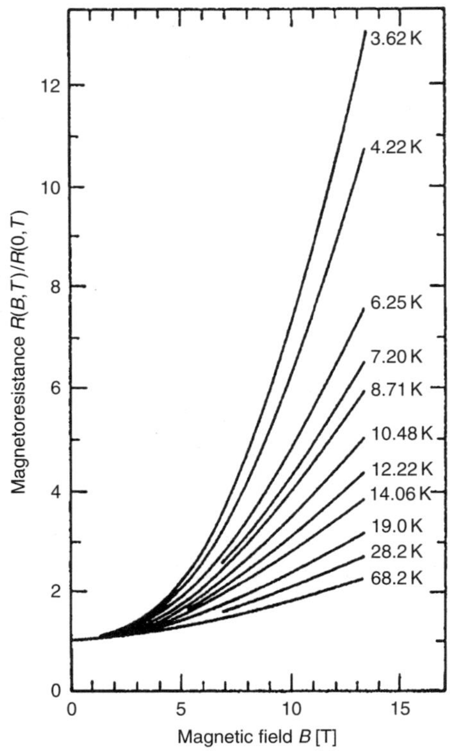

Fig. 9.6. Dependence on the magnetic field (perpendicular to current) of the electrical resistance of a Ge thermometer ($R_{4K} = 856\,\Omega$) at the indicated temperatures [68].

for example, from LakeShore and Cryotronics. As shown in Fig. 9.5, the $R(T)$ curve of CGT falls more steeply than carbon resistors, but eq. (9.14) can be used to represent their behaviour.

It must be noted (also for RuO_2 resistors) that resistors best suited for thermometry are those with the highest temperature coefficient, i.e. the worst for electronics (usually an high room temperature coefficient means also an high low-temperature coefficient). Of course, electronic firms tend to produce more stable resistors; hence the future will see the onset of an antiquary market for thermometers.

Detailed description of carbon resistor is found in some low-temperature text books [2,19,20]. Nowadays, they are usually replaced by thick-film resistors (RuO_2), which are also much easier to mount.

9.6.2.3 Thick-film RuO_2 resistors

RuO_2, zirconium oxinitride and metal oxide bead resistors are called oxide resistors.

RuO_2 resistors are commercial small-power (volume of few mm^3) electronic components. They belong to the group of smd (surface mount device) thick-film resistors and are

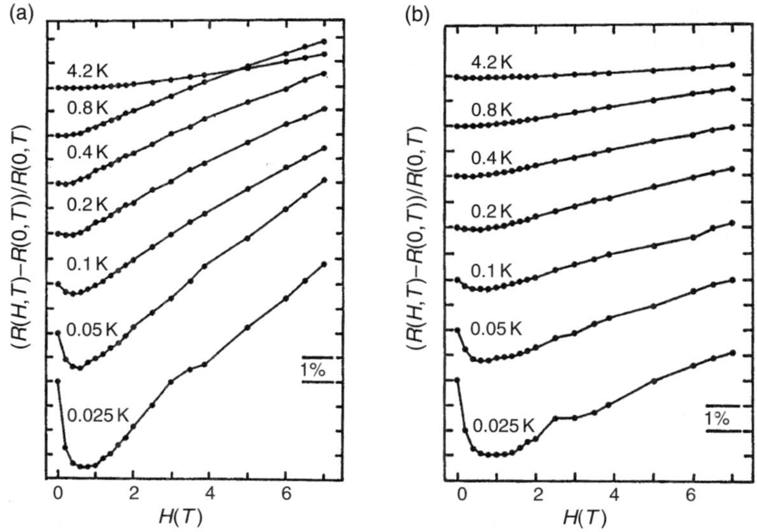

Fig. 9.7. (a) Magnetic field induced relative resistance change $(R(H, T)–R(0, T))/R(0, T)$ versus magnetic field for a 1 kΩ Dale RCW 575 resistor for seven different temperatures. The curves for each temperature are offset for clarity. (b) Magnetic field induced relative resistance change $(R(H, T)–R(0, T))/R(0, T)$ versus magnetic field for 2.7 kΩ Philips RC-01 resistor for seven different temperatures. The curves for each temperature are offset for clarity [78].

sold in wheels containing 2000–10 000 pieces for the automatized mounting of electronic circuits. The resistive material is a 'ceramic' consisting in a mixture of a component containing Ru and a Pb silicate, deposited onto an alumina support. The ratio between the two components defines the resistivity of the material. Below 1 K, the resistance of these thermometers follows approximately the Mott's law, with exponents ranging between 0.5 and 0.25 [75].

Small dimensions, easy mounting, good stability over thermal cycling and low cost are the qualities of these resistors. Magnetoresistivity is usually positive [76–78] for $T > 1$ K. For $T < 1$ K, both positive and negative values have been found (see Fig. 9.7) [78].

Due also to their (amorphous) composition, the heat capacity of a ruthenium oxide resistor is much higher than that of a Ge thermistor of equal mass [61]. This negative property prevents the use of RuO_2 resistors as detector sensors (see Chapter 15).

Note that the electronic industry produces smd components called 'thin-film resistors' of the same shape and size of RuO_2 thermistors. They are usually made of a thin layer of Ni-Cr. Their resistance is extremely stable with temperature and hence they find application as inexpensive (punctual) heaters (see e.g. Chapter 11).

9.6.2.4 Zirconium oxinitride

Zirconium oxinitride resistors (best known are Cernox) are thin-film resistors. Typical commercial sensors have resistance values of about 10^5 Ω at 1 K falling to ~100 Ω at

100 K. Their useful range is from about 0.3 to 300 K. The stability is a few mK at 4 K. For more information about Cernox stability see ref. [79].

The magnetoresistance is less than 0.3% above 10 K; at 4.2 K a field of 13 T has a 2% effect (see ref. [80,81]. About 5000 cernox thermometers have been used at CERN for the control of Large Hadron Collider (LHC) [82].

Another commercial thermistor, Thermox (by LakeShore), is useful between 77 K (~80 kΩ) and room temperature (~0.3 Ω). It consists of a glass-encapsulated metal oxide bead. The short-term stability is about 25 mK in the useful range.

9.6.2.5 Junction diodes

Junction diodes are not strictly resistance thermometers but two terminal devices which are used in a similar way. The forward voltage drop is measured across a p–n junction carrying a constant d.c. of a few μA. Such voltage increases as the temperature falls. Junction diodes are Si, GaAs or AlGaAs commercial electronic components. They can be used above 1 K, but the voltage is almost linear above ~20 K. Typical sensitivities for Si diodes are 30 mV/K at 4 K and 2 mV/K at 77 K. Overheating (see Section 9.6.3) at low temperature is strong in this type of device. GaAs devices are less sensitive to both temperature and magnetic fields [[3], p.269]. The use of junction diodes for low-temperature thermometry has been reviewed in ref. [83,84].

9.6.3 Traps in resistance thermometry

Low-temperature resistance measurements do not differ in principle from room temperature ones: they can be carried out both in d.c. (carachteristographers) or in a.c. (bridges).

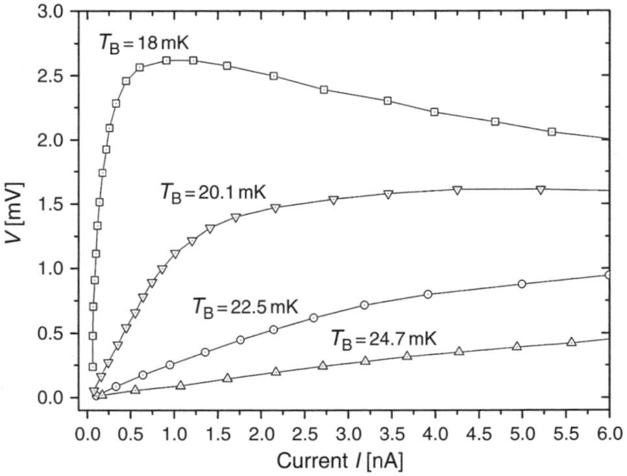

Fig. 9.8. *I–V* curve for a Ge-NTD#31 (see Section 15.2.1.1) resistor at four different temperatures T_B of the mixing chamber. Note the strong nonlinearity at low temperature.

Carachteristographers are d.c. instruments made up by a current (voltage) power supply and a voltage (current) meter. They give a two quadrant V–I relation for the resistive device. The information supplied by such instruments is more exhaustive than that supplied by a.c. bridges, but the measurement is time consuming, and an inversion of polarity is necessary to detect the possible presence of partially rectifying contacts. In Fig. 9.8, an example of a V–I relation in the first quadrant for a Ge-NTD#12 resistor (see Section 15.2.1.1) is shown. Note that the power used in the measurements can be as low as 10^{-14} W.

Bridges give a mean resistance value around $I = 0$. Strictly, they should be used only for linear components. In both cases (d.c. or a.c.), the resistance measurements are made in four-wire configuration, since the resistor to be measured is at low temperature whereas the measuring instrument is at room temperature: the electric connection is usually made by low thermal conductivity wires which are also poor electrical conductors (remind the Wiedeman–Franz law).

The power which must be supplied to the thermistor to measure its resistance depends on the noise level and on the detection system. The latter is the main responsible for the total noise (a few nV/$\sqrt{}$Hz), since the resistors are at low temperature and, hence, their thermal noise can be usually neglected (see eq. (9.17)).

The frequency response of the detection system is of low-pass type for characteristographers and band-pass for bridges (see Section 10.4). In both types of measurements the narrowing of the bandwidth corresponds to a longer time of measurement. Depending on the chosen detection system, several problems (true traps) may be encountered in resistance thermometry.

The main ones are:

1. Thermistor overheating

The small power needed to measure the resistance of the thermometer brings the thermometer at a temperature over that of the support surface. Such 'overheating' is due to the contact resistance at the interface (see Chapter 4). A typical value of the contact resistance is:

$$R_C \cdot T^3 = 8 \times 10^4 \ \left[\mathrm{K^2 m^2/W}\right] \tag{9.15}$$

For example, if we want a thermometer overheating $\Delta T/T$ less than 1%, with a contact surface $A = 10^{-4}\,\mathrm{m}^2$, the power $P(T)$ to be supplied to the thermistor must be:

$$P(T) \leq T^4 \cdot 10^{-4}\,\mathrm{W} \tag{9.16}$$

i.e. at 20 mK $P_{20\,\mathrm{mK}} \leq 2 \times 10^{-11}$ W. With a typical value of resistance of 80 kΩ, the corresponding current is 15 nA.

Equation (9.15) is sometimes optimistic. Thermometer overheating as a function of temperature and contact resistance can be experimentally found, varying the power supplied to the thermometer while keeping constant the support temperature. This is shown in Fig. 9.9 for an RuO_2 thermometer. Note that a measuring power of 5×10^{-12} W at 20 mK produces an overheating of 12 mK!

The contact resistance may change when the thermometer is moved from a position to another. Hence the accuracy of resistance temperature measurements below about 25 mK

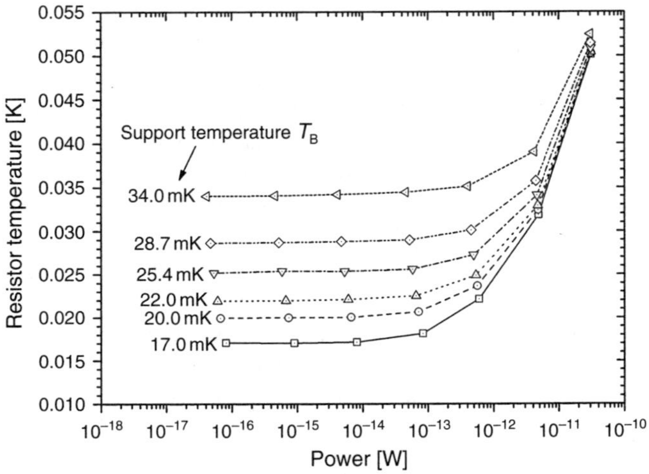

Fig. 9.9. Overheating of an RuO_2 thermometer as a function of temperature for six different support temperatures.

can be illusory. Moreover, in a.c. measurements of high-value resistances (more than $100\,k\Omega$), another phenomenon may be present: since the measuring signal is d.c. coupled to the resistor, a current offset may be present in the output stage of the bridge. This is a random slow varying d.c. signal which supplies a small unpredictable additional power on the thermometer. An 1% current offset produces a 2% power offset. Also for this reason, the use of a secondary resistance thermometer driven with a.c. signal is not advisable below about 25 mK.

Last, we wish to remind that thermal cycling may spoil the thermometer calibration. The frequent check of the calibration by means of reference fixed points (see Section 8.5) is advisable.

2. Filtering

The four lines which electrically connect a thermometer at low temperature with the bridge at room temperature are usually filtered in order to reduce the interference of external noise sources. A low-pass filter may attenuate the signal detected by the bridge, thus giving an erroneous reading of the resistance. As example, let us simply schematize the filter with a capacitance $C = 10^{-8}$ F in parallel with the thermometer resistance R. If the bridge supplies a sinusoidal signal at 16 Hz (e.g. LR700 bridge.) and $R = 10^5\,\Omega$, the error due to the RC filter is $\Delta R/R = 5 \times 10^{-3}$. If $R = 5 \times 10^5\,\Omega$, $\Delta R/R = 12\%$. The situation is much more critical in the case of a square wave signal (e.g. Picowatt) or at higher frequencies. From this simplified example, we see that resistors of more than about $100\,k\Omega$ are to be usually avoided when working with a.c. bridges.

A suggestion: calibrate at room temperature your a.c. bridge connected to your filtered lines using resistors of increasing values to detect the real cut-off frequency of your apparatus.

9.7 Noise thermometry

The noise thermometer is based on the temperature dependence of the mean square noise voltage V^2 developed in a thermistor (Nyquist theorem, 1928):

$$V^2 = 4 \cdot k_B T \cdot R \cdot B \tag{9.17}$$

where B is the amplifier bandwidth.

The main problem encountered in low-temperature noise thermometry is the very low level of signals. For example, with $R = 1\,\mathrm{k\Omega}$ and $B = 1\,\mathrm{kHz}$:

- at 4 K, $V \approx 10^{-8}$ volt and $P = 10^{-19}$ W;
- at 10 mK, $V \approx 10^{-9}$ volt and $P = 10^{-22}$ W.

The latter values are not measurable by means of semiconductor amplifiers and a SQUID system is necessary.

Theoretically, the noise thermometer is a primary thermometer and as such has been used [85,86]. In practice, besides the low level of the signals to be detected, there are other problems [87,88], such as:

(a) noise generated by the resistive elements of the amplification devices;
(b) definition of the bandwidth B of the electronics;
(c) knowledge of the gain $A(f)$ of the electronics.

An additional drawback of this type of thermometry is the very long time needed for the measurement (more than 1 h for an 0.1% accuracy).

In the past, except for the low-temperature range, the uncertainties of noise thermometry were not comparable to those of the gas thermometry due to the non-ideal performance of detection electronics. Up to now, the most successful technique is the switched input digital correlator proposed by Brixy et al. in 1992 [89]. In this method, the noise voltage is fed via two separate pairs of leads to two identical amplifiers whose output signals are multiplied together, squared and time averaged (see Fig. 9.10).

This eliminates the amplifier and transmission line noise superimposed on the thermal noise, since the respective noise voltages are uncorrelated.

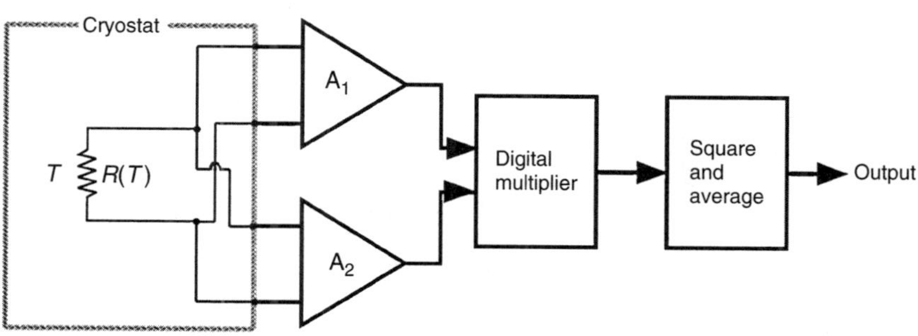

Fig. 9.10. Block diagram of the switched input digital correlator.

The possibility of using the Brixy technique for low-temperature thermometry is under study (see e.g. ref. [90]).

9.8 Dielectric constant thermometry

Capacitance measurements are quite simple. A typical drawback is the need of coaxial cables that introduce a thermal load which is not negligible in low-power refrigerators. On the other hand, capacitance bridges null the cable capacitance. Multiplexing is more difficult than for resistance thermometers. In principle, capacitors have low loss due to Joule heating. This is not always true: losses can be important, especially at very low temperatures. Dielectric constant thermometers have a high sensitivity: capacitance differences of the order of 10^{-19} F can be measured.

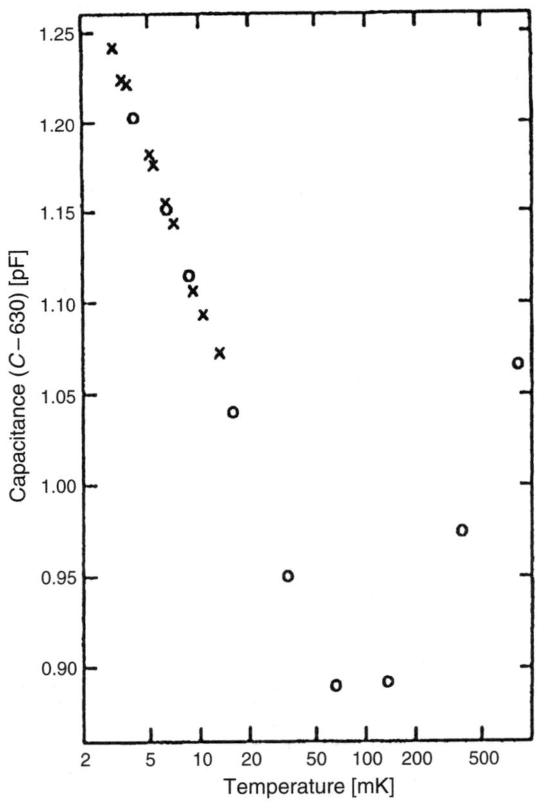

Fig. 9.11. Capacitance (minus 630 pF) of a capacitive glass thermometer as a function of temperature. The data points were obtained at excitation frequencies of 1.0 and 4.7 kHz at zero magnetic field (x) and at $B = 9$ T (o) [92].

Several materials show an appreciable temperature dependence of the dielectric constant down to a few millikelvin. In Fig. 9.11, such dependence is shown for a glass (SiO_2 with 1200 ppm of OH^- [91,92]). Note the presence of a minimum which usually depends on the measuring frequency.

An interesting aspect of dielectric constant thermometry is the small influence of a magnetic field. On the other hand, measurements depend on both the measuring frequency and voltage (see Figs 9.12 and 9.13). Figure 9.13 shows an example of the dependence on frequency of both the dielectric constant and loss for Upilex S [93].

Figure 9.14 shows the low-temperature dependence of the dielectric constant for an epoxy (Emerson & Cuming, Stycast 1266) [94].

Many commercial capacitors can be used as thermometers since they show a strong dependence of their capacitance on temperature. Figure 9.15 shows the low-temperature $C(T)$ curve for a commercial smd capacitor (ceramic capacitor KEMET-Y5V).

Above the minimum, the mean slope of $C(T)$ is about 0.6 pF/mK up to room temperature. Such commercial capacitor would be a thermometer with a very large operating range, a property not very common in other thermometers.

Unfortunately, most capacitance thermometers are not stable and must be recalibrated at every cool down. They may also present problems of 'heat release' [95]: moreover, their thermalization times at low temperatures may be long since the materials used present low thermal conductivity and high specific heat.

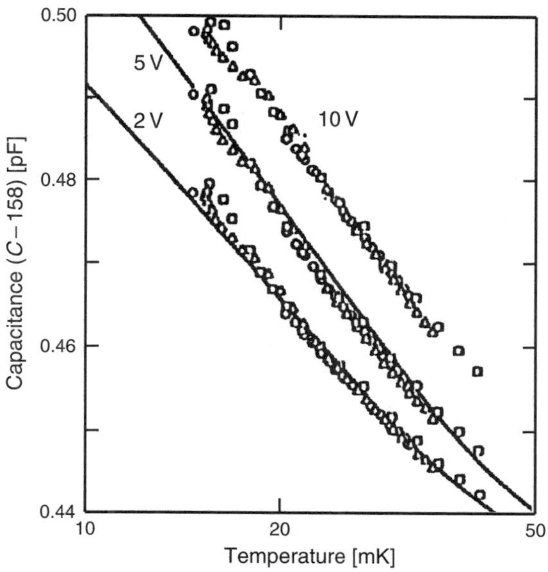

Fig. 9.12. Capacitance of a capacitive glass thermometer as a function of temperature at the same frequency (4.7 kHz) for three measuring voltages. The measurements were carried out in magnetic field of: 0.0 T (o), 0.25 T (Δ) and 6.0 T (□) [91].

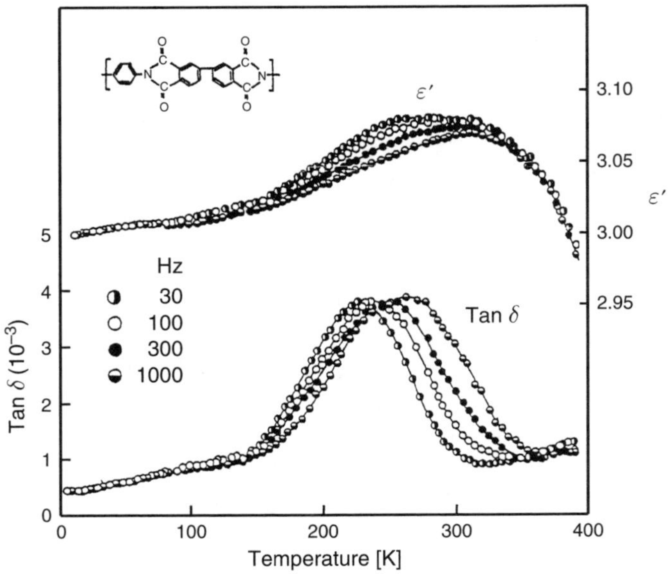

Fig. 9.13. Dielectric properties of Upilex-S as a function of temperature at four measurement frequencies [93].

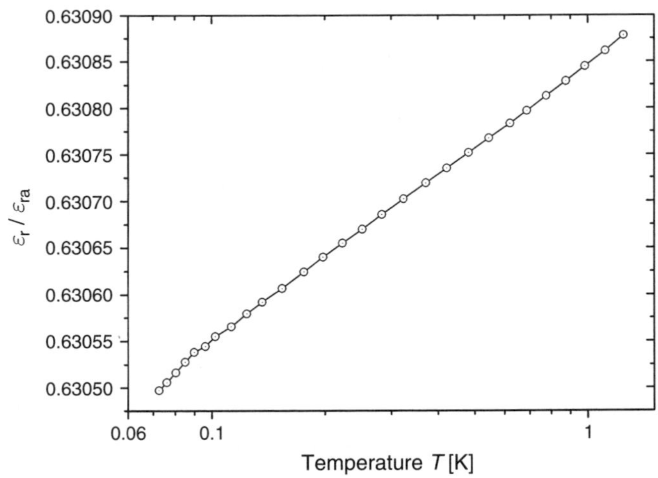

Fig. 9.14. Temperature dependence of the dielectric constant ε_r of Stycast 1266 below 1 K. ε_{ra} is the dielectric constant in vacuum at room temperature [94].

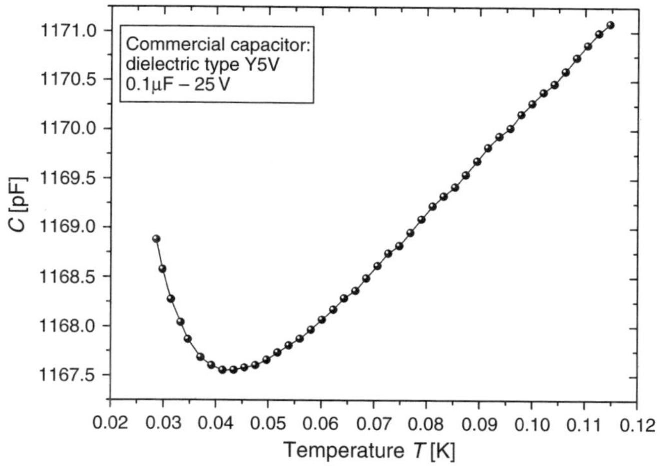

Fig. 9.15. Capacitance of an smd capacitor (ceramic capacitor KEMET-Y5V) as a function of temperature.

9.9 Paramagnetic salt thermometry

The starting relation is the Curie–Weiss law:

$$\chi = \chi_0 + \frac{\lambda}{T - \Delta} \tag{9.18}$$

where χ is magnetic susceptibility of the paramagnetic material and λ is the Curie constant. The Weiss constant Δ depends on the shape of the thermometric sample, on the crystal symmetry and on the interaction between the magnetic moments. Since the parameters of the Curie–Weiss relation (9.18) cannot be defined theoretically, a calibration of the thermometer is necessary. Hence, this is a secondary thermometer.

The paramagnetic salts useful for this type of thermometry are the same as that used in the ADR (adiabatic demagnetization refrigerator, see Section 7.3 and Fig. 7.7). The most commonly used salts are:

- MAS: $(Mn^{2+}\ SO_4 \cdot (NH_4)_2\ SO_4 \cdot 8H_2O)$, $T_c \sim 0.17\,K$;
- FAA: $(Fe^{3+}\ (SO_4)_3 \cdot (NH_4)_2SO_4 \cdot 24H_2O)$, $T_c \sim 0.03\,K$;
- CPA: $(Cr_2^{3+}(SO_4)_3 \cdot K_2SO_4 24H_2O)$, $T_c \sim 0.01\,K$;
- CMN: $(2Ce^{3+}(NO_3)_3 \cdot 3Mg(NO_3)_2 \cdot 24\ H_2O)$, $T_c \sim 0.002\,K$.

where T_c is the minimum temperature usable for each salt.

Note that CMN is the best suited salt for very low-temperature thermometry. The value of T_c can be further decreased to 0.2 mK, partly substituting ions Ce^{3+} with ions La^{3+}.

All these salts contain crystallization water which assure a large distance (about 1 nm on CMN) among the magnetic ions giving a low T_c [1].

We wish to remind that these salts cannot be kept under vacuum at room temperature since they lose the crystallization water.

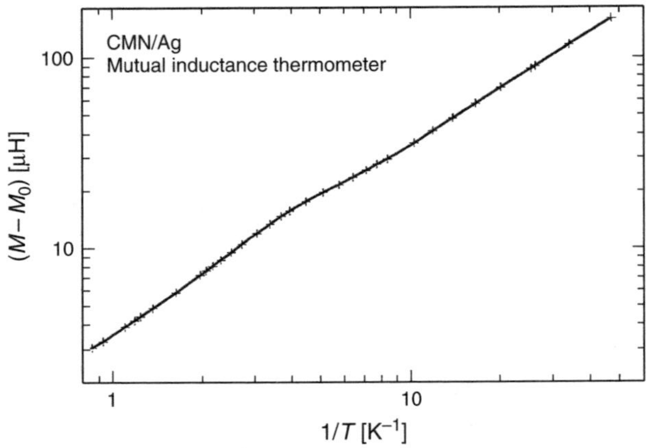

Fig. 9.16. Mutual inductance M (minus electrical connection contribution M_0) versus $1/T$ for a CMN thermometer.

For this type of thermometry a mutual inductance meter of high sensitivity (10^{-9} H) is needed.

Deviations from the $1/T$ dependence are generally observed; these are due to magnetic parasitic contributions (see example of Fig. 9.16 where M_0 is the mutual inductance of the electrical connections).

9.10 Nuclear orientation thermometry

In this type of thermometry, long mean-life radioactive nuclei are used. The latter perform a $\beta^{\pm}$ decay into an excited state which emits γ to return to the fundamental state as depicted in Fig. 9.17.

If radioactive nuclei (^{60}Co and ^{54}Mn are commonly used) are polarized by a strong magnetic field, the direction of emitted γ rays is anisotropic: the emission probability is a function of the angle with the field. Such angular distribution depends on the $(2I+1)$ levels of the nuclear spin in the local magnetic field and is a function of temperature T. In fact, the relative population of such levels is determined by the Boltzmann factor $\exp(-\beta_m \Delta_{\text{hf}})$, with $\beta = 1/k_B T$, m the I component along the local field and Δ_{hf} the hyperfine splitting between two adjacent levels. For $T >> \Delta_{\text{hf}}/k_B$ all levels are equally populated; hence, the γ emission is isotropic; for $T << \Delta_{\text{hf}}/k_B$ only the fundamental level is occupied, hence the anisotropy is practically not dependent on temperature.

There is, however, a temperature range in which the γ emission is a function of temperature. In this range, the anisotropy of γ emission can be used as thermometric property. The advantage of using γ emitters is that the detection can be done from outside the cryostat (no wiring necessary inside the cryostat).

Figure 9.18 shows, for ^{60}Co, the $W(\theta, T)$ function, i.e. the ratio of γ ray intensity for an angle θ with the field direction and the 'isotropic' intensity. On a smaller scale, also the sensitivity $dW/d\theta$, which has a maximum for $\theta = 90°$, is shown: the useful temperature range is about 2–60 mK.

Since an appreciable polarization of nuclear magnetic moments requires high fields ($10 \div 30$ T), ^{60}Co or ^{54}Mn atoms are embedded in ferromagnetic crystals (usually Fe, Ni or ^{59}Co) to get the same local field for all nuclei. However, since also the magnetic

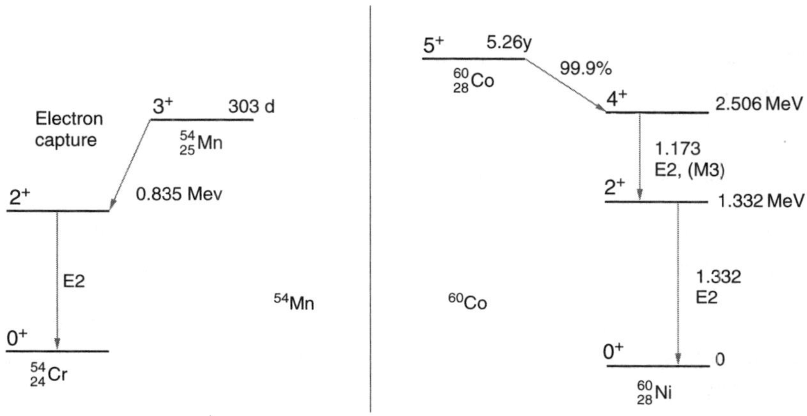

Fig. 9.17. Energy level diagrams for ^{54}Mn and ^{60}Co (slightly simplified). These are the two decay favourite schemes used for nuclear orientation thermometry.

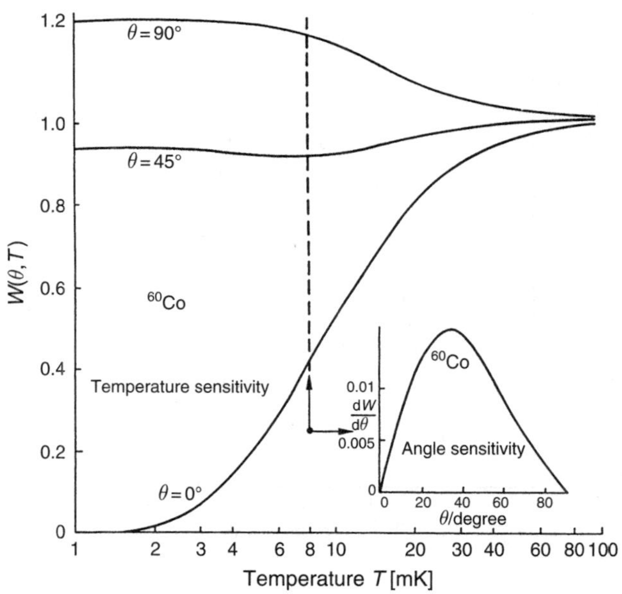

Fig. 9.18. $W(\theta, T)$ function for ^{60}Co (see text) [96].

domains must be aligned, an external field of $0.1 \div 1\,T$ is applied; in calculations also, this external field must be taken into account. The radioactive atoms are usually introduced into the crystal by a surface deposition followed by high-temperature diffusion; ^{60}Co may be otherwise produced by neutron irradiation of ^{59}Co.

A significant source of error in nuclear orientation thermometers is due to the possible absorption of the radioactive emission in the cryostat, with a temperature increase; as a consequence, low-intensity source must be used, with long counting periods to get a good statistics.

Typical power values released by the radioactive substances used in nuclear orientation thermometers are:

- 1 μCurie ^{60}Co: 0.57 nW due to β emission;
- 1 μCurie ^{54}Mn: 0.03 nW due to 5 keV peak of ^{54}Cr after electron capture [97–105].

A drawback of nuclear orientation thermometry is the narrow range where the sensitivity $T(\mathrm{d}W/\mathrm{d}T)$ is adequate for measurements; it happens around $T \approx \Delta_{hf}/k_B$ (see Figs 9.18 and 9.19). $\Delta_{hf}/k_B = 9.14\,\mathrm{mK}$ for ^{54}Mn in Fe and $\Delta_{hf}/k_B = 7.965\,\mathrm{mK}$ for ^{60}Co in Fe. Moreover, since the statistic error on the measurement is $\Delta T/T = (2/n)^{1/2}$ (for n counts), long counting times, typically 10^3 s, but potentially up to 10^5 s for 1% precision at the lowest temperatures, are needed. The temperature should be stable over these timescales.

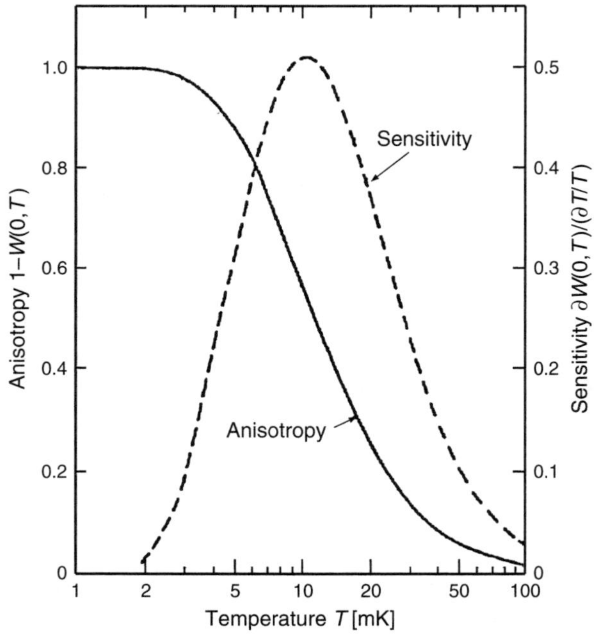

Fig. 9.19. Temperature dependence of the anisotropy of the γ radiation of ^{54}Mn in Ni host along $\theta = 0$. The quantity $\partial W/\partial T$ is a measure of the temperature sensitivity of the thermometer which, in this case, has its maximum value at about 10 mK [1].

An advantage of this technique, instead, is the good thermal contact with the (metallic) radioactive material, and, more important, the good thermal coupling between lattice and nuclear spins. Last, the γ ray energy is usually quite high, and no special window in the cryostat is needed. Nuclear orientation thermometer is a primary thermometer.

Comparisons among various thermometers and the nuclear orientation thermometer can be found in ref. [100–105] Further information may be obtained from ref. [97,106,107].

9.11 Magnetic thermometry with nuclear paramagnets

The low-temperature limit of magnetic thermometry with paramagnetic salts (see Section 9.9) is given by the ordering temperature of the electronic magnetic moments. Such ordering temperature is around 1 mK (example the CMN thermometer).

For a lower temperature thermometry, one has to move from electronic paramagnets to nuclear paramagnets. Magnetic thermometry with nuclear paramagnets extends the range of thermometry down to a few microkelvin.

Since the magnetic moments are smaller, now we have a smaller susceptibility and therefore much smaller signal, requiring more sensitive detection systems. These are resonance or SQUID (see Section 14.5) techniques. Thermal response time are shorter, since pure metals can be used with good thermal conductivity and fast spin-lattice relaxation. The parameter to be measured is the nuclear susceptibility:

$$\chi_n = \frac{\lambda_n}{T_n} \tag{9.19}$$

with:

$$\lambda_n = \frac{N_0 I \cdot (I+1)\,\mu_0 \mu_n^2 g_n^2}{3k_\mathrm{B}} \tag{9.20}$$

In many metals, this law (Curie law) should be valid for at least few μK. For example, for Cu and Pt, it seems to be true with a 1% confidence for fields below 1 mT [2].

Nevertheless deviations from eq. (9.19) have been observed for the intermetallic compound $AuIn_2$ [108,109] and for Tl [110,111]. Requirements for the validity of eq. (9.19) are the absence of changing internal fields due to nuclear magnetic or electronic magnetic ordering in the relevant temperature range, the absence of nuclear electronic quadrupole interactions and no superconductive transition.

Nuclear magnetic thermometry is fundamental for the microkelvin range, but at higher temperatures the signal can be rather weak and may be overcome by contributions from electronic magnetic impurities. For the methods of detection of nuclear magnetization, see e.g. ref. [2].

9.12 Coulomb blockade thermometry

Coulomb blockade thermometry (CBT) is based on the electric conductance characteristics of tunnel junctions. This type of thermometer has been developed at Jyväskylä University. The basic results are reported in ref. [112].

The conductance of a tunnel junction array is determined by three energy contributions: the thermal energy $k_B T$, the electric potential energy eV at bias voltage V and the charging energy $\varepsilon_c = e^2/2C_{eff}$, where C_{eff} is the effective capacitance of the array and $\varepsilon_c = [(N-1)/N]e^2/C$ with N the number of junctions in series.

If $k_B T >> \varepsilon_c$, the dynamic conductance of a junction array is given by [111]:

$$\frac{G}{G_T} = 1 - \left(\frac{\varepsilon_c}{k_B T}\right) \cdot g\left(\frac{eV}{Nk_B T}\right) \tag{9.21}$$

where G_T is the asymptotic conductance at high bias voltage. The function g, which is nearly Gaussian shaped around zero, is given by:

$$g(x) = \frac{\left[x \cdot \sinh(x) - 4\sinh^2(x/2)\right]}{8\sinh^4(x/2)} \tag{9.22}$$

The full width at half minimum of the conductance dip described by eq. (9.21) (see Fig. 9.20) is proportional to the temperature T.

The useful temperature range of CBT is defined by the number and size of the tunnel junctions. These are realized by vacuum evaporation of ~100 nm Al layers (which are

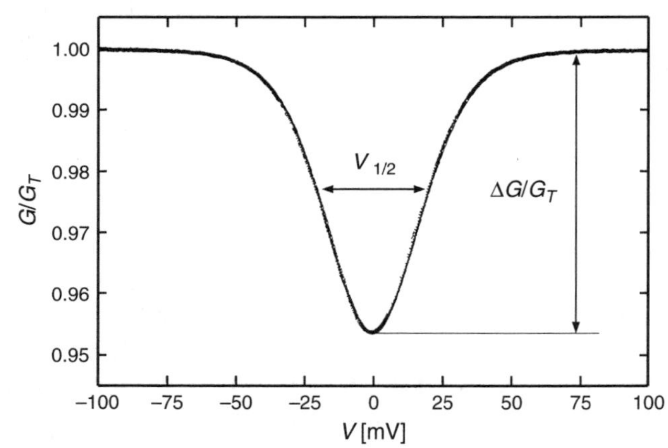

Fig. 9.20. Normalized conductance G/G_T of a coulomb blockade thermometry sensor versus bias voltage V.

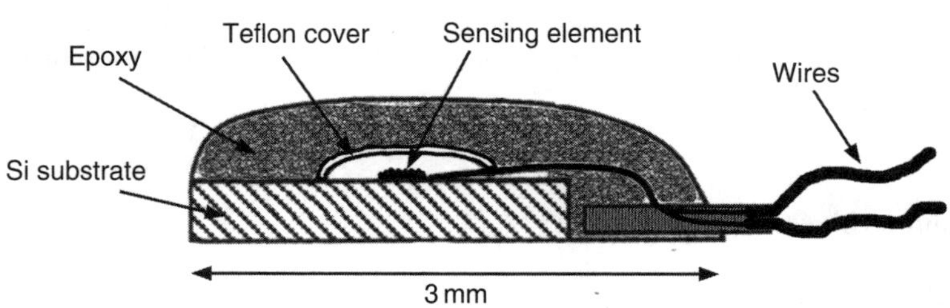

Fig. 9.21. Scheme of a sensor for Coulomb blockade thermometry.

successively oxidized) and ~200 nm of Cu. The substrate is Si oxide. Junctions have a surface of ~1.5 μm^2. A scheme of the sensor is shown in Fig. 9.21.

The useful range of CBT is 0.02 ÷ 10 K [113]. For temperatures below ~1.18 K, a week magnetic field (0.1 ÷ 0.2 T) must be applied to the sensor to suppress Al superconductivity. For temperatures below ~20 mK, electron–phonon decoupling is present. CBT is useful when working with high magnetic fields. The sensor in fact shows a negligible magnetoresistance: tests in the 0.05 ÷ 4.2 K temperature range and with field up to 31 T have been performed [114].

The CBT is in principle a primary thermometer, but its absolute accuracy is at present too low for metrological applications [115].

References

[1] O.V. Lounasmaa: *Experimental Principles and Methods Below 1 K*, Academic, London (1974)
[2] F. Pobell: *Matter and Methods at Low Temperatures*, Springer, Berlin (1995)
[3] T. Quinn: *Temperature*, Academic Press, London (1990)
[4] R.E. Bentley: *Theory and Practice of Thermoelectric Thermometry*, Springer Verlag, Singapore (1998)
[5] L.G. Rubin: Cryogenics **10**, 14 (1970)
[6] L.G. Rubin, B.L. Brandt, H.H. Sample: Cryogenics **22**, 491 (1982)
[7] L.G. Rubin: Cryogenics **37**, 341 (1997)
[8] G. Schuster, D. Hechtfisher, B. Fellmutt: Rep. Prog. Phys. **57**, 187 (1994)
[9] R.C. Kemp, W.R.G. Kemp, L. Besley: Metrologia **23**, 61 (1987)
[10] G.T. McConville: *Proc. LT-17*, p. 401, Elsevier Science (1984)
[11] F.C. Matacottaer et al.: Jpn. J. Appl. Phys. **26**, 1679 (1987)
[12] F.C. Matacotta et al.: Metrologia **24**, 61 (1987)
[13] P.P.M. Steur, M. Duriex, G.T. McConville: Metrologia **24**, 69 (1987)
[14] R.A. Aziz, M.J. Slaman: Metrologia **27**, 211 (1990)
[15] F. Pavese, G. Molinar: *Modern Gas-Based Temperature and Pressure Measurements*, Plenum Press, New York and London (1992)
[16] K.D. Hill: Metrologia 39, 41 (2004)
[17] D.N. Astrov et al.: Metrologia, **26**, 151 (1989)
[18] P.P.M. Steur, M. Durieux: Metrologia **23**, 1 (1986)
[19] G.K. White: *Experimental Techniques in Low-Temperature Physics*, Clarendon, Oxford (1979)
[20] A.C. Rose-Innes: *Low-Temperature Laboratory Techniques*, English University Press, London (1973)
[21] T.J. Quinn: Physica Scripta 41,730 (1990)
[22] P.P.M. Steur: Jpn. J. Appl. Phys. **26**, 1685 (1987)
[23] P.P.M. Steur and F. Pavese: Cryogenics **29**, 135 (1989)
[24] Y. Iye: Cryogenics **28**, 164 (1988)
[25] K. Nara, R.L. Rusby, D.I. Head: Cryogenics **30**, 952 (1990)
[26] H. Sakurai: *Temperature: Its Measurement and Control in Science and Industry*, vol. 6, p. 127, ed. by J.F. Schooley, American Institute of Physics, New York (1992)
[27] K. Grohmann, H. Kock: Proc. LT 17, eds. U. Eckern, A. Schmid, W. Weber, and H. Wuhl, Elsevier Press, Amsterdam (1984)
[28] K.S. Izrailov, R.E. Taimonov: Meas. Tech. **28**, 857 (1985)
[29] D. Gugan, G.W. Michel: Metrologia **16**, 149 (1980)
[30] D. Guganl: Metrologia **28**, 405 (1991)
[31] K. Grohmann, H. Luther: *Temperature: Its Measurement and Control in Science and Industry*, vol. 6, p. 21, ed. by J.F. Schooley, American Institute of Physics, New York (1992)
[32] M.R. Moldover, J.P.M. Trussler: Metrlogia **25**, 165 (1988)
[33] M.R. Moldover et al.: J. Res. Natl. Bur. Stand. **104**, 11 (1999)

[34] B.W. Petley: *Phys. Scr.* **T40**, 5 (1992)
[35] W. Cencek, K. Szalewicz, B. Jeziorski: Phys. Rev. Lett. **86**, 5675 (2001)
[36] G. Lach, B. Jeziorski, K. Szalewicz: Phys. Rev. Lett. **92**, 233001 (2004)
[37] D.Gugan, G.W. Michel: Metrologia **16**, 149 (1980)
[38] H. Luther, K. Grohmann, B. Fellmuth: Metrologia **33**, 341 (1996)
[39] B. Fellmuth et al.: *Dielectric-Constant Gas Thermometry and Determination of the Boltzmann Constant* TEMP-MEKO 2004, Cavtat–Dubrovnik Croatia, ISBN number 953-6313-73-1
[40] E.F. May et al.: Rev. Sci. Instrum. **75**, 3307 (2004)
[41] R.T. Jacobsen, S.G. Penoncello, E.W. Lemmon: *Thermodynamic Properties of Cryogenic fluids*, Plenum Press, New York (1996)
[42] H. Preston-Thomas: Metrologia **27**, 3 (1990)
[43] B.W. Magnum: J. Res. NIST **95**, 69 (1990)
[44] BIPM *Techniques for Approximating the International Temperature Scale of 1990*, Bureau International des Poids et Mesures, Sevres (1990)
[45] G.K. White, P.J. Meeson: *Experimental Techniques in Low-Temperature Physics*, 4th ed., Clarendon Press, Oxford (2002)
[46] V. Steinberg, G. Ahlers: J. Low Temp. Phys. **53**, 255 (1983)
[47] D.S. Greywall, P.A. Bush: Rev. Sci. Instrum. **51**, 509 (1980)
[48] W.P. Halperin et al.: J. Low Temp. Phys. **31**, 617 (1978)
[49] E.R. Grilly: J. Low Temp. Phys. **11**, 243 (1973)
[50] M. Barucci et al.: Cryogenics **40**, 437 (2000)
[51] D.S. Greywall: Phys. Rev. B **31**, 2675 (1985)
[52] R.A. Scribner, E.D. Adams: in *Temperature, Its Measurement and Control in Science and Industry*, vol. 4, p. 37, ed. by H.H. Plumb, Instrument Society of America, Pittsburgh, PA (1972)
[53] J.R. Thompson, H. Meyer: Cryogenics **7**, 296 (1967)
[54] G.C. Straty, E.D. Adams: Rev. Sci. Instrum. **40**, 1393 (1969)
[55] W.P. Halperin et al.: Phys. Rev. Lett. **34**, 718 (1975)
[56] R.J. Soulen, Jr., W.E. Fogle, J.H. Colwell: J. Low Temp. Phys. **94**, 385 (1994)
[57] W.E. Fogle, R.J. Soulen Jr., J.H. Colwell: in *Temperature: Its Measurement and Control in Science and Industry*, vol. 6, p. 91, ed. by J.F. Schooley, American Institute of Physics, New York (1992)
[58] R.J. Soulen, Jr: Physica B **109&110**, 2020 (1982)
[59] G. Schuster, A. Hoffmann, D. Hechtfischer: Czech. J. Phys. **46**, 481 (1996)
[60] D.S. Greywall, P.A. Busch: J. Low Temp. Phys. **46**, 451 (1982)
[61] G. Schuster et al.: Physica B **165&166**, 31 (1990)
[62] G. Frossati: J. Physique **41** (C7), 95 (1980)
[63] C.C. Kranenburg et al.: Jpn. J. Appl. Phys. **26**, 215 (1987)
[64] D.J. Bradley et al.: J. Low Temp. Phys. **45**, 357 (1981)
[65] N.F. Mott: Phil. Mag. **19**, 835 (1969)
[66] B.I. Shklovskii: Phys Rev B **67**(4), 045201 (2003)
[67] D. McCammon et al.: NIM A **436**, 205 (1999)
[68] L.J. Neuringer et al.: Rev. Sci. Instrum. **42**, 9 (1971)
[69] H.H. Sample, L.G. Rubin: Cryogenics **17**, 597 (1977)
[70] P. De Moor et al.: J.Appl.Phys. **79**, 3811 (1996)
[71] E. Pasca et al.: *Bolometers in Magnetic field: Use of NTD Ge Sensors*, NIM A 2007 in print, available online: http://dx.doi.org/10.1016/j.nima.2007.02.092
[72] B.I. Shklovskii, A.L. Efros: *Electronic Properties of Doped Semiconductors*, Springer, Berlin (1984)
[73] W.N. Lawless: Rev. Sci. Instrum. **43**, 1743 (1972)
[74] L.G. Rubin, B.L. Brandt: Adv. Cryogenic. Eng. **31**, 1221 (1986)
[75] A. Briggs: Cryogenics **31**, 932 (1991)
[76] W.A. Bosh et al.: Cryogenics **26**, 3 (1986)
[77] M.L. Sigueira, R.J. Viana, R.E. Rapp: Cryogenics **31**, 796 (1991)
[78] K. Uhlig: Cryogenics **35**, 525 (1995)
[79] D. Giraudi et al.: *Proc. of TEMP-MEKO 1996*, ed. by P.Marcarino, p.155, Levrotto and Bella, Turin (1997)

[80] G. Heine, W. Lang: Cryogenics **38**, 377 (1998)
[81] H.D. Ramsbotton, S. Ali, D.P. Hampshire: Cryogenics **36**, 61 (1996)
[82] T. Junquera et al.: *Neutron Irradiation Tests of Calibrated Cryogenic Sensors on Low Temperatures*, CERN-LHC Project Report n°153 (1997)
[83] J.K. Krause, P.R. Swinehart: Adv. Cryog. Eng. **31**, 1247 (1986)
[84] M.G. Rao, R.G. Scurlock, Y.Y. Wu: Cryogenics **23**, 635 (1983)
[85] R.P. Giffard, R.A. Webb, J.C. Wheatley: J. Low Temp. **6**, 533 (1972)
[86] R.P. Giffard, R.A. Webb, J.C. Wheatley: J. Low Temp. **13**, 383 (1973)
[87] D.R. White et al.:Metrologia **33**, 325 (1996)
[88] L. Crovin, A. Actis: Metrologia **14**, 69 (1978)
[89] H. Brixy et al.: in *Temperature: Its Measurement and Control in Science and Industry*, vol. 6, p. 993, Toronto, ed. by J.F. Schooley, American Institute of Physics, New York (1992)
[90] H. Sakurai: in *Proc. of SICE Annual Conference in Fuhi*, p. 3029, Fukui University, Japan (2003)
[91] P.J. Reijntjis et al.: Rev. Sci. Instrum **57**, 1413 (1986)
[92] S.A.J. Wiegers et al.: Rev. Sci. Instrum. **58**, 2274 (1987)
[93] O. Yano, H. Yamaoka: Prog. Polym. Sci. **20**, 585 (1995)
[94] G. Ventura et al.: Cryogenics **39**, 963 (1999)
[95] K. Gloos et al.: J. Low Temp. Phys. **73**, 101 (1988)
[96] D.S. Betts: *Refigeration and Thermometry Below One Kelvin*, Sussex University Press, Brighton (1976)
[97] R.P. Hudson et al.: J. Low Temp. Phys. **20**, 1 (1975)
[98] D.S. Parker, L.R. Corruccini: Cryogenics **15**, 499 (1975)
[99] R.C. Richardson: Physica B **90**, 47 (1977)
[100] S.R. de Groot, H.A. Tolhoek, W.J. Huiskamp: *Alpha-, Beta-, and Gamma-Ray Spectroscopy*, vol. 2, p. 1199, ed. by K. Siegbahn, North-Holland, Amsterdam (1985)
[101] H.J. Rose, D.M. Brink: Rev. Mod. Phys. **39**, 306 (1967)
[102] H. Marshak: J. Low Temp. Phys. **6**, 769 (1972)
[103] P.M. Berglund et al.: J. Low Temp. Phys. **6**, 357 (1972)
[104] J.R. Sites, H.A. Smith, W.A. Steyert: J. Low Temp. Phys. **4**, 605 (1971)
[105] N.J. Stone, H. Postma eds.: *Low Temperature Nuclear Orientation*, North-Holland, Amsterdam (1986)
[106] R.J. Soulen Jr., H. Marshak: Cryogenics **20**, 408 (1980)
[107] G. Shuster, D. Hectfisher, B. Fellmuth: Rep. Prog. Phys. **57**, 187 (1994)
[108] T. Herrmannsdörfer et al.: Phys. Rev. Lett. **74**, 1665 (1995)
[109] T. Herrmannsdörfer, F. Pobell: J. Low Temp. Phys. **100**, 253 (1995)
[110] U. Angerer, G. Eska: Cryogenics **24**, 515 (1984)
[111] G. Eska, E. Schubert: Jpn. J. Appl. Phys., Suppl. **26(3)**, 435 (1987) (*Proc. 18th Int'l Conf. Low Temp. Phys.*)
[112] J.P. Pekola et al.: Phys. Rev. Lett. **73**, 2903 (1994)
[113] J.P. Pekola et al.: J. Low Temp. Phys. **128**, 263 (2002)
[114] J.P. Pekola et al.: J. Appl. Phys. **83**, 5582 (1998)
[115] A.J. Manninen et al.: Physica B **284–288**, 2010 (2000)

10 Instrumentation for Cryogenics

Contents

10.1 Magnets

Small-size fixed magnets with fields up to ~1 T are nowadays commercially available; they usually retain their characteristics at low temperature. Variable field magnets are used to operate superconducting heat switches or for NMR thermometers.

Magnets producing fields more than 50 T are used to apply quantizing fields for experiments on electrons in metals and in semiconductors.

We will not discuss resistive magnets, capable of producing 2–3 T, but only superconducting magnets.

10.1.1 Superconducting magnets

The discovery of superconductivity is due to Heike Kammerlingh Onnes (1911) (see Section 2.1). Zero electrical resistance means that there is no ohmic heating: the wires can carry a higher current density than Cu wire and may be packed closely together, since there is no heat to remove. A power supply is used only to initially energize the magnet. Superconductors turn back resistive metals above a critical magnetic field, even at $T = 0$. Thus, the field produced by a solenoid may destroy superconductivity. The field producing such a phenomenon is known as critical field and depends on a critical current. Superconducting magnets are operated close to the temperature of liquid helium. They are usually immersed in a He bath, but also cooled by a pulse tube low-temperature stage. In the latter case, the practical construction of the magnet is more sophisticated.

Small superconducting magnets are often used to operate superconducting heat switches or for NMR thermometry. They may be installed in a helium bath or directly, e.g. onto the mixing chamber of a dilution unit. Larger magnets, for adiabatic demagnetization refrigeration, are operated in the helium bath. Magnets producing fields up to 1 T are often self-built in the laboratory, especially when specific field geometries or other design factors are required. Further information may be found in ref. [1–6].

10.1.2 Magnet wires

The superconducting wires used in commercial magnets are usually made of NbTi alloys and Nb_3Sn. Superconducting wire is formed from filaments of superconducting alloy in a matrix, usually of Cu. The presence of the matrix is very important: it acts to thermally stabilize the superconductor. When carrying large currents, any unwanted flux variation may result in a local heat production in the wire. A local transition to the normal state is possible. If the normal state propagates along the wire, 'quenching' may occur. By surrounding the superconductor with a Cu matrix, the heat from the initial flux change is removed. Also, the Cu matrix shorts the resistance created in the superconductor, acting as a parallel channel for current flow.

Single or multifilament wire with a matrix of Cu or CuNi is commonly available in several diameters. Single-filament wire is usually reserved for low current applications. Multifilament wire has a higher critical current, since 'supercurrents' flow only in the surfaces of superconductors, and in multifilamentary wire there is more superconducting surface.

For more information about superconducting wires, their metallurgical production and magnet construction see ref. [6, p.187–8].

10.1.3 Magnet specifications

When a superconducting magnet is needed, the best is to buy it. The main specifications are:

(a) *The maximum field.*

This governs the choice of material and the size of the magnet. To get fields above 10 T will require Nb_3Sn wire.

(b) *The homogeneity.*

The field is specified to be uniform over some volume. A typical example is to ask for 10^{-3} variation over a 1 cm diameter spherical volume in the centre of the magnet. For a typical solenoid of length about 5–10 times the bore diameter, this would be the natural homogeneity. Homogeneity may be improved by adding coil sections at either end of the magnet to flatten the field profile. The limit of available homogeneity is about 10^{-9} over 1 cm as requested for NMR magnets. In order to take full advantage of the field homogeneity, the sample must be placed at the centre of the magnet. Differential thermal contractions in the cryostat may change this positioning.

(c) *Access to the field.*

In solenoid magnets, the clear bore diameter (the inner diameter of the innermost magnet former) is the usual information needed. The diameter may range from a few millimetres to a few metres. A diameter around 30–50 mm is typical for the use with a dilution refrigerator.

It may be necessary to reduce the stray field at some distance from the magnet; typically this might mean reducing the field to below 50 mT over some volume where the metallic mixing chamber of a dilution refrigerator might be. This is accomplished by compensation coils. Whether this is possible depends on geometric factors mainly on the distance from the magnet centre.

10.1.4 Persistent mode

Figure 10.1 shows a scheme of a superconducting magnet with the persistent current switch and switch heater. To energize the magnet, the heater is powered and the switch branch becomes resistive. Hence the current delivered by the power supply flows in the superconducting winding of the magnet. When the desired current (magnetic field) is reached, the heater is switched off and the current flows through the persistent current switch which has returned to the superconducting state. The switch is simply a link of superconductor wire embedded in epoxy to provide a thermal resistance to the helium bath. Superconducting links with normal state resistances of 100 ohm are typical. Such a link on a 100 H magnet produces a low-pass filter with a time constant $\tau = L/R = 1$ s. This is a convenient filter against the power supply ripple and noise.

Such switches have a drawback: when the magnet current is sweeping, a back electromotive force (e.m.f.) is present equal to $L \cdot \mathrm{d}I/\mathrm{d}t$. This voltage is also present across the switch and so a current flows through the switch.

Thus, when the magnet is sweeping, the field lags the field calculated from the known current. The lag is sweep rate dependent, but may be rather precisely corrected for.

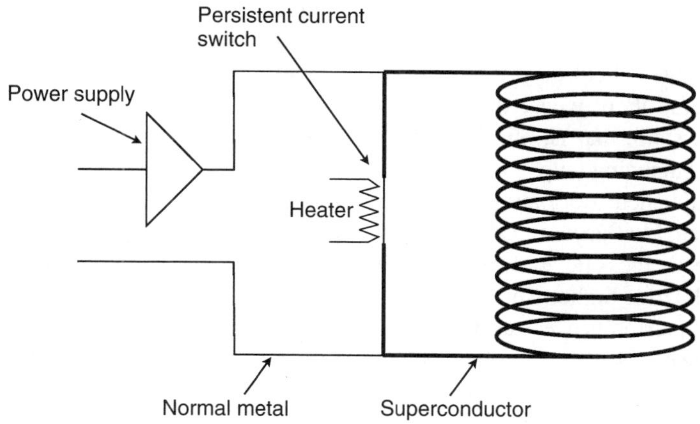

Fig. 10.1. A schematic of a superconducting magnet.

To cancel a persistent mode introduced at high field, the persistent current must be matched by the current in the leads and the persistent current switch heated. The magnet may quench if the current is not accurately matched. Should it be necessary to 'depersist' a magnet with an unknown current, back-to-back diodes (of right size) can be put across the leads; they are a safety when heating the link.

10.1.5 Power supplies for magnets

Power supplies providing large currents at low voltages are required for the operation of superconducting magnets. They are designed to provide precisely known currents with well-defined sweep rates. In a power supply, it is also important to know the maximum available voltage. This voltage is necessary to overcome the back e.m.f. generated across the magnet: low voltages lead to long energization times.

Modern power supplies solve almost all the problems, but unluckily they usually produce high-frequency noise, a dramatic drawback in a low-temperature laboratory.

10.2 Radio frequency shielding and filtering

The shielding of experiments from external interference is of great importance in cryogenics. There are two main reasons for this need:

1. Many of the experiments being performed necessitate the measurement of very small signals;
2. The quest for lower temperatures makes the problem of RF heating more acute as heat capacities decrease.

10.2.1 Electric and magnetic fields

At low frequencies, interference produced by electric and magnetic fields can be treated separately. Electric fields introduce noise via the mutual capacitance between noise sources and signal lines. A circuit with a large load impedance is most susceptible to electric interference. The standard method for suppressing this coupling is to surround the signal lines with a shield held at ground. This is accomplished using a coaxial cable, but one must be careful about where the shield is held to ground. It may be preferable to make a balanced system, that is, one for which both signal lines have the same impedance to ground. The balancing helps in eliminating common mode interference. Balancing is most easily accomplished by floating the source and carrying the signal by twisted pairs with an external shield at ground. Large currents flowing through the shield should be avoided since its effectiveness depends on being at a constant potential. Twisted pairs are used up to about 1 MHz (at higher frequencies, triaxial cables are to be preferred).

Magnetic interference is due to time-varying magnetic fluxes on circuit loops. Since the interference appears as a voltage source, it is more effective on low impedance loops. Low-frequency magnetic isolation is much more difficult to achieve than electrostatic shielding. The simplest precaution against this is the use of twisted pair to null the total flux through the loop (see Fig. 10.2).

Fig. 10.2. Twisted pair suppresses magnetically coupled interference because it tends to null out the effects of an applied field.

An ordinary metallic shield must be very thick to have an effectiveness against inductively coupled low-frequency noise. High permeability materials such as μ-metal are useful for magnetic shielding, but one must be sure that the shielding material has been annealed and degaussed before use. If the shielding is magnetized, the d.c. field inside the shielded area will cause interference when leads vibrate. In one gauss field, 1 cm of signal line vibrating at 1 kHz with an amplitude of 0.1 mm induces about 1 microvolt (to be compared with typical input voltage noise of preamplifiers: few nanovolt/$\sqrt{}$Hz).

The proper configuration for a magnetic field set-up is a relatively low-susceptibility, high-saturation shield as close as possible to the magnetic field and a high susceptibility, lower saturation shield some distance away from the first.

10.2.2 Superconducting shields

An advantage that low-temperature experiments have is that superconducting shields may be used (T_C in Table 8.11). These provide nearly ideal shielding and are generally employed for the most sensitive experiments. Niobium capillaries with twisted pairs and filled with epoxy are sometimes used below 1 K. Niobium tubing are used to shield inductors and SQUIDS. Lead foil is used for more irregularly shaped objects; the seams of these packages must be carefully soldered. The possibility of trapping magnetic fields in a superconducting shield suggests to surround the entire dewar with a μ-metal shield. This reduces the field which is present when the superconducting shield temperature becomes lower than T_C. This does not help, however, when large magnetic fields are applied in the experimental region inside the cryostat. The most sensitive devices should be provided with their own μ-metal shields outside and the superconducting shield inside [7].

The temperature dependence of the μ-metal permeability is not well known, although it decreases at low temperatures. Another material, cryo-perm, seems to retain large permeability even at liquid helium temperatures.

10.2.3 Electromagnetic interference (EMI) filtering

Another aspect of the EMI problem for low-temperature experiments is the injection of RF (at megahertz frequencies and up) energy via the wiring into the cryostat. A common RF-induced problem is a heating caused by local telephone broadcasters. RF transmitters often ride in over resistance bridge wirings whose signal are at very low frequencies.

Since gaps in shielding of the order of $\lambda/20$ are significant, making an enclosure RF tight means making it water tight also. The use of conductive gasket material is required when a totally RF-tight system must be built.

Since it is difficult to achieve effective RF shielding, the principle of shielding as few as different areas as possible should be followed. A standard procedure is to place the cryostat inside a shielded room. Such rooms commonly provide 120 dB of attenuation from 50 kHz to 1 GHz. If there is no shielded room, one can shield known RF sources. When filtering does not produce unwanted change in the signal (see Section 9.6.3), lines should be filtered as heavily as possible. Filters in pi-type (π) configuration are usually used (see Fig. 10.3) mounted in a shield box separated into two compartments as shown in Fig. 10.4.

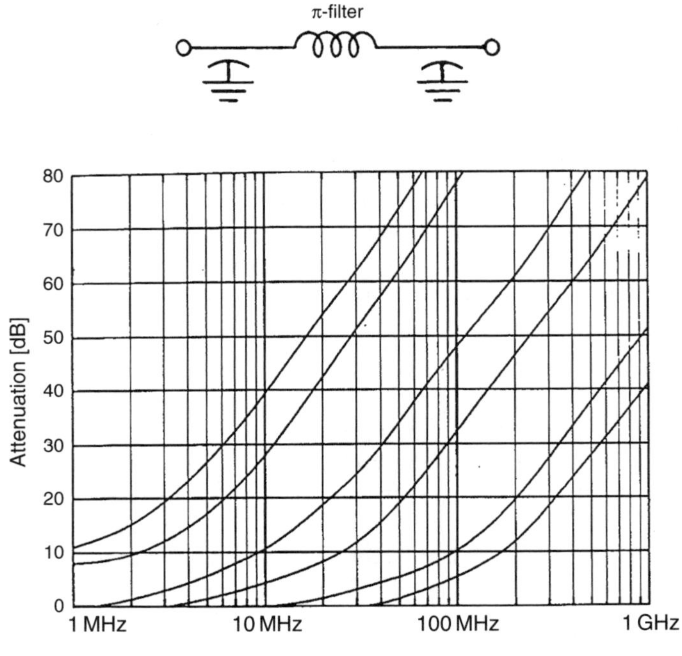

Fig. 10.3. Example of attenuation-frequency plot of electromagnetic interference filters as a function of resistance to ground.

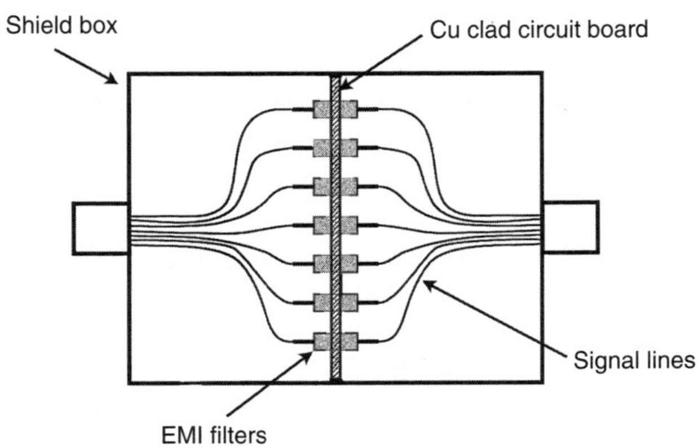

Fig. 10.4. Mounting of electromagnetic interference (EMI) filters inside a shielding box.

10.3 Bridges

The signal supplied by cryogenic sensors is very low (in the order of μV). The white noise voltage is proportional to the frequency bandwidth. To reduce the bandwidth, low-pass filters (carachteristographers with ~0.01 Hz cut-off) or band-pass filters (lock-in) are used. The former method is more precise, but a longer time for the measurements is required.

The measure of the resistance of thermometers, of the capacitance of level detectors, of the magnetic susceptivity for paramagnetic salt thermometers, etc., is nowadays made with commercial ‘bridges’. Bridges for cryogenics allow accurate measurement of component values with a very low power dissipation. In a cryogenic experiment, several bridges may be necessary at the same time. Figure 10.5 shows the block diagram of the LR-400 (from Linear Research), an excellent commercially available bridge. Its measurement frequency is 15.9 Hz and its feedback loop has an effective d.c. gain of 10^5. Both phases are automatically nulled, and the input stage of the preamplifier changes to noise match to the source impedance. Balance occurs when:

$$I \cdot R_x = V \qquad (10.1)$$

where V is voltage across the transformer. The error due to lead resistance is $<10^{-4}$. Resistance changes of one part in 10^5 at 15 mK can be measured with this bridge.

The feedback scheme does not need to involve analogue circuitry.

In several capacitance bridges and a few resistance bridges, a computer is used to control the feedback, as shown in Fig. 10.6. An Anderson bridge is used in this diagram. The ‘set range’ line controls several relays which select the value of R. The computer also controls the ratio transformer.

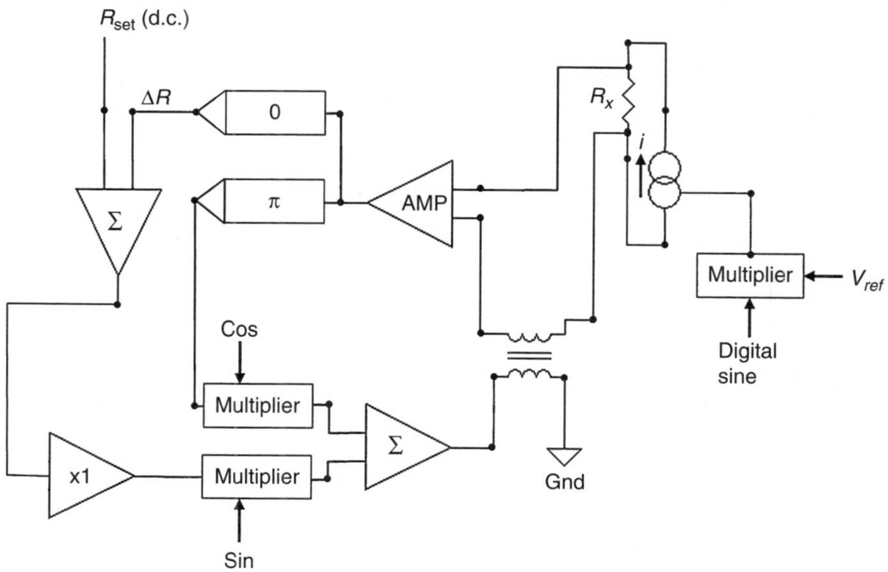

Fig. 10.5. Block diagram of LR-400.

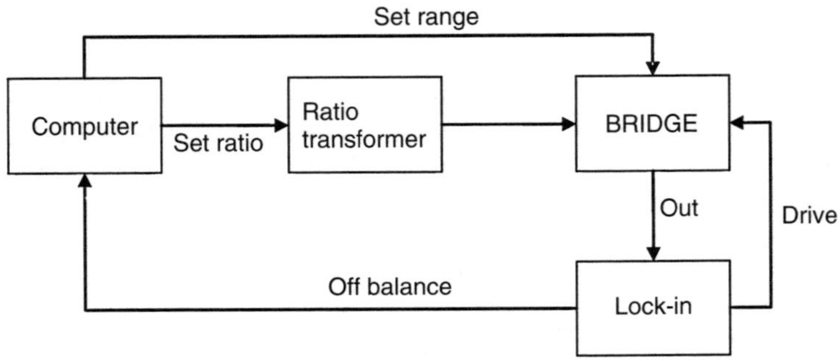

Fig. 10.6. Typical block diagram for the computer control of a bridge.

10.4 The synchronous demodulator (lock-in)

As we have seen in Section 9.5.3, in the case of resistance thermometry, the signal produced by a low-temperature thermometer is very low (microvolt range). Low-pass filters are not sufficient to narrow the detection bandwidth in order to get a suitable signal to noise ratio (S/N). Bandpass filters are needed. The most commonly used method is the synchronous demodulation, usually simply called 'lock-in' technique, as shown in the block diagram of Fig. 10.7.

A lock-in is an electronic apparatus (or computing program) which allows to extract the signal from a dominating noise, taking advantage of the knowledge of the frequency of the signal and of its phase difference with a reference signal at the same frequency.

For example, the light emitted by a faint source may be 'chopped' by means of a chopper producing a square wave-shaped light signal on an optical detector (see Fig. 10.8). Other 'spurious' optical signals which are not modulated at the frequency of the chopper, as we

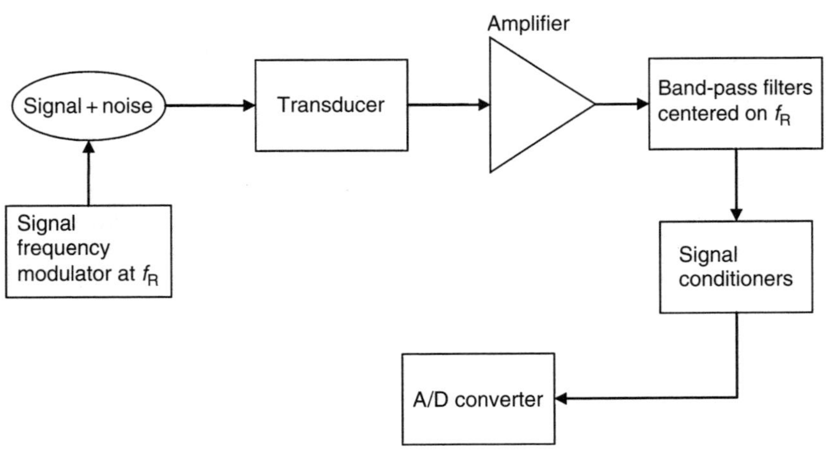

Fig. 10.7. Lock-in block diagram.

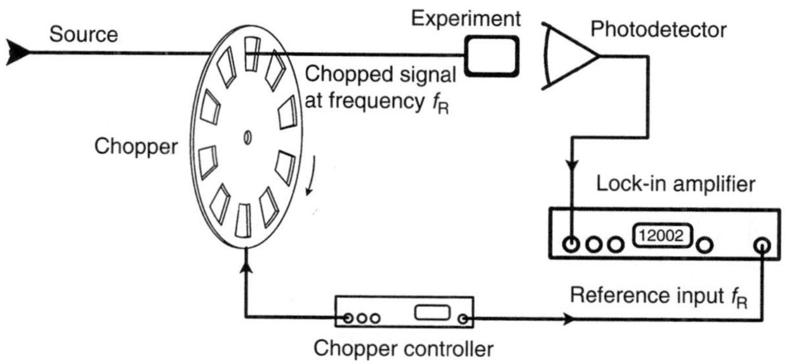

Fig. 10.8. Chopping of a faint source to reject light signals other than that of the source.

shall see, are not detected. In this case, the reference signal for the lock-in is produced by an optical transmitter/receiver pair (LED/photodiode) across the chopper wheel. In a resistance bridge, the reference signal is created by the instrument itself.

The lock-in has two (electrical) inputs: the first consists of the signal $s(t)$ to be measured with noise $n(t)$; the second is the square wave reference signal $r(t)$.

The operation carried out by a lock-in is the product:

$$v_p(t) = r(t) \cdot [s(t) + n(t)] \tag{10.2}$$

followed by a mean with a fixed time constant τ. The output is a constant (really very slow varying) voltage signal:

$$v_{\text{out}}(t) = \langle v_p(t) \rangle_\tau = \langle r(t) \cdot s(t) \rangle_\tau + \langle r(t) \cdot n(t) \rangle_\tau \tag{10.3}$$

The term $r(t) \cdot s(t)$, because of the correlation between $r(t)$ and $s(t)$, produces a constant voltage (positive if $r(t)$ and $s(t)$ are in phase, negative in the case of opposite phases, but in any case proportional to $s(t)$). Instead, the term $r(t) \cdot n(t)$ fluctuates around zero since $r(t)$ and $n(t)$ are not correlated. Obviously, τ must be smaller than the typical time of variation of $s(t)$ and much larger than $2\pi/\omega_s$, with $\omega_s/2\pi$ frequency of the reference signal. In such condition, the response to a signal of varying amplitude will be an output signal with some noise. If the amplitude varies slowly, it will be possible to use a low-pass filter with very low cut-off, to integrate the signal $v_p(t)$, thus eliminating most of noise.

Let us first examine the lock-in input in the simplest case, i.e. when both the reference and the signal are sinusoidal, attributing for the moment different frequency to the signal and to the reference:

$$s(t) = \sqrt{2} \cdot V_s \cdot \cos(\omega_s t + \phi_s) \quad \text{and} \quad r(t) = \sqrt{2} \cdot V_R \cdot \cos(\omega_R t + \phi_R) \tag{10.4}$$

hence:

$$v_p(t) = V_s \cdot V_R \cdot \cos[(\omega_s + \omega_R) \cdot t + \phi_s + \phi_R] + V_s \cdot V_R \cdot \cos[(\omega_s - \omega_R) \cdot t + \phi_s - \phi_R] \tag{10.5}$$

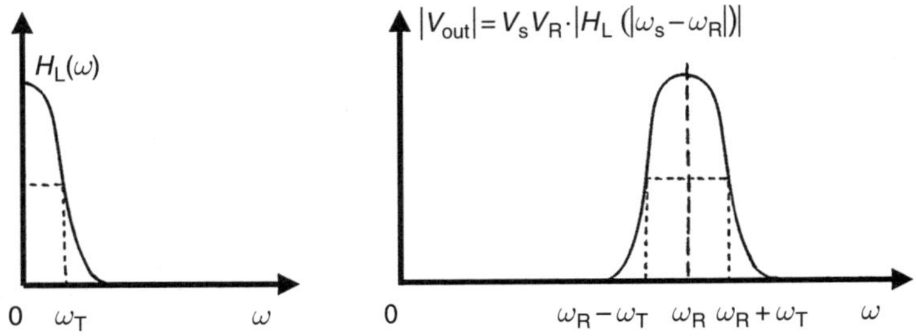

Fig. 10.9. Transfer function of a low-pass filter (on the left) and band-pass transfer function (on the right), obtained through the multiplication process.

We see that the multiplier output signal is made up of a term at the sum frequency and another at the difference frequency. The signal of eq. (10.5) passes through a low-pass filter of response $H_L(j\omega)$ and high-frequency cut-off well below the reference frequency, in order to eliminate the sum frequency term. The output will be:

$$|V_{out}| = V_s V_R \cdot |H_L(|\omega_s - \omega_R|)| \tag{10.6}$$

Since the filter is low pass, it will attenuate all the frequencies far from the reference frequency, that is, the multiplier together with the low-pass filter acts as a band-pass filter centred at the reference frequency ω_R. The bandwidth of such band-pass filter is the double of the bandwidth of the low-pass filter (see Fig. 10.9). This result will be used in the evaluation of the residual noise in the output signal.

In reality, $\omega_R = \omega_s$ and $\phi_s - \phi_R = \phi =$ constant. Hence from eq. (10.6):

$$V_{out} = V_R \cdot |H_L(0)| \cdot \cos(\phi) \cdot V_s = \alpha \cdot V_s \tag{10.7}$$

The lock-in output is thus proportional to the amplitude of the signal V_s through a constant α which is only weakly dependent on ϕ (for small ϕ).

The noise in the output signal is that present in the bandwidth $(1/2\tau)$ of the band-pass filter. If the power noise spectrum at the input is white (flat) and given by $w_n(f)$, the lock-in output noise will be simply:

$$\langle \Delta V_{out}^2 \rangle = \alpha^2 \int w_n \mathrm{d}f = \alpha^2 \frac{2w_n}{\tau} \tag{10.8}$$

If the filter time constant is τ, a datum independent of the preceding one will be got after a time $\approx\tau$. The *S/N* ratio for a time of measurement τ is:

$$\frac{S}{N} = \frac{V_{out}}{\sqrt{\langle \Delta V_{out}^2 \rangle}} = V_s \cdot \sqrt{\frac{\tau}{2w_n}} \tag{10.9}$$

The total duration of a measurement is usually longer than τ. As a consequence, a mean over many independent data is done.

Equation (10.7) is important because it gives the integration time needed to get a requested *S/N* ratio, once the noise and signal levels are known.

There are three possibilities to carry out the multiplication process:

1. by analogic electronics which unfortunately introduces distortions in the signal;
2. by conversion of the signal and reference to numerical form and the use of a digital signal processing;
3. by using of a phase-sensitive detector (PSD).

In the last method, which is the most commonly used, the multiplier is replaced by an electronic switch controlled by the reference signal. The switch changes the amplification of the signal $s(t)$ from $+1$ to -1 (for example $+1$ when the reference is positive and -1 when it is negative), as shown in Fig. 10.10(a). When there is no phase delay between the

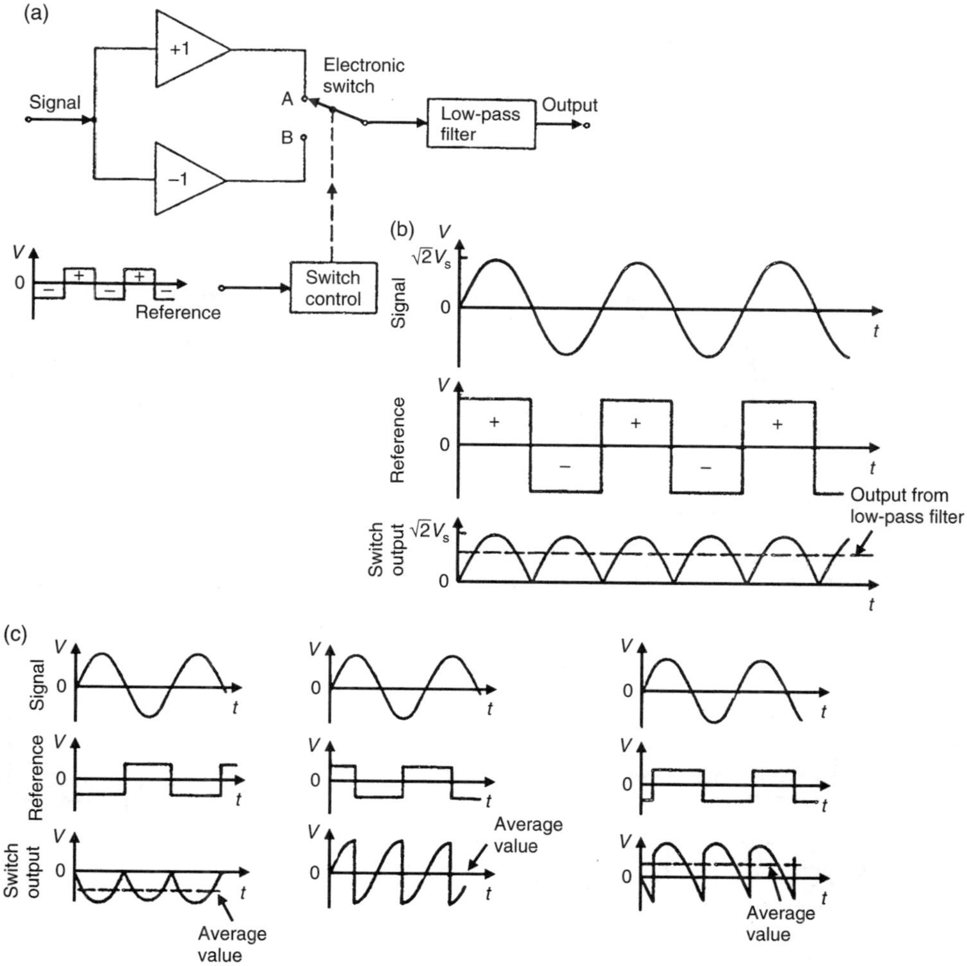

Fig. 10.10. (a) Scheme of a PDS, (b) output waveforms (in phase signals), (c) output waveforms (out-of-phase signals).

signal and the reference, the output is a positive average signal as shown in Fig. 10.10(b). However, there is usually a phase delay between the signal and the reference (due to filters). In this case, the resulting waveforms are shown in Fig. 10.10(c).

In Fig. 10.11, the transmission window of a PSD is shown.

In Fig. 10.12, the block diagram of a real lock-in is shown.

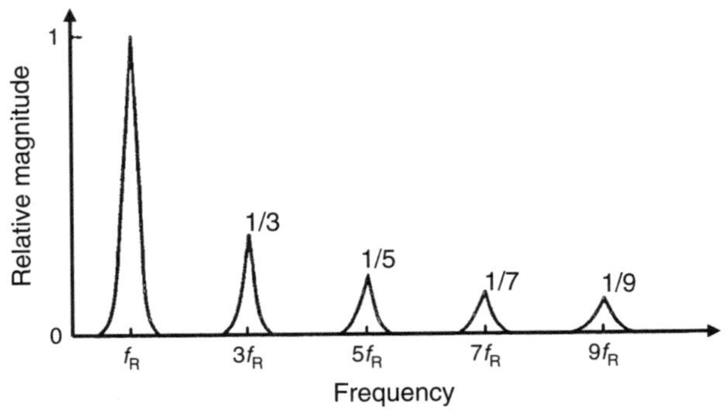

Fig. 10.11. Transmission window of a phase-sensitive detector.

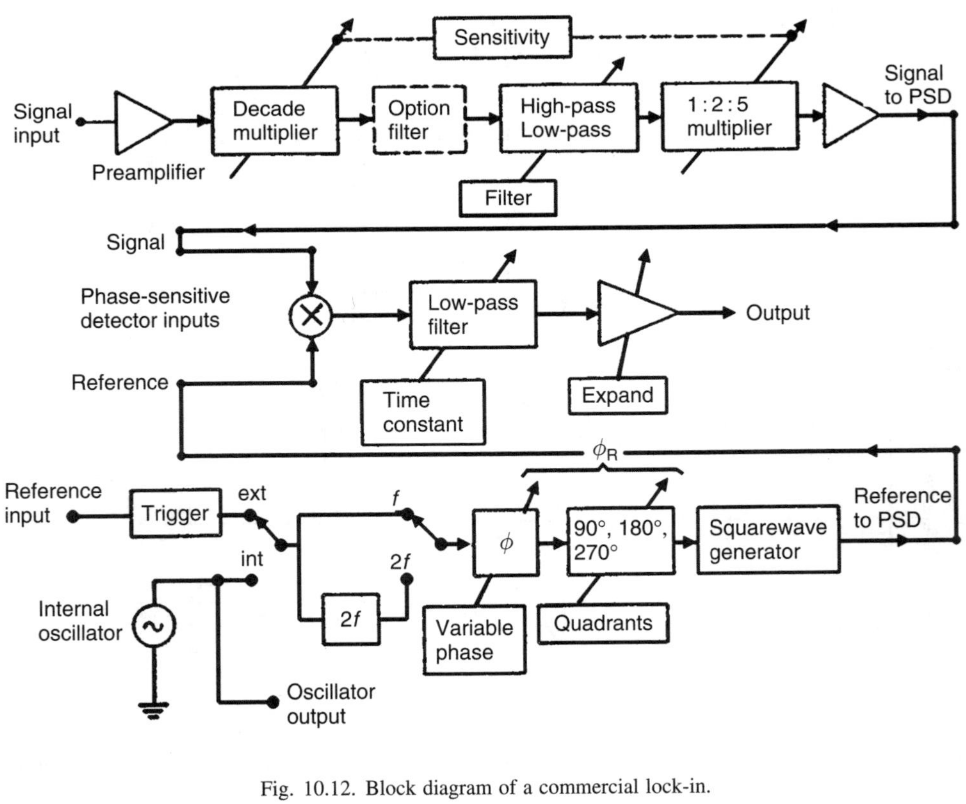

Fig. 10.12. Block diagram of a commercial lock-in.

In a lock-in used in resistance bridges for cryogenics, the reference signal may be either sinusoidal (as in LR-700) or a square wave (as AVS-47 Picowatt). The square wave reference (easier to produce at low frequencies) increases the output signal level by a factor 1.23, thus compensating the noise, increased by $(1.23)^{1/2}$. A sinusoidal reference allows to use lower cut-off filters on the electric wiring of the cryostat.

A lock-in present in a bridge for cryogenics has usually a reference at frequencies between 10 and 1000 Hz. Electrical power delivered to the resistive sensors are between $10^{-16} \div 10^{-14}$ W. The maximum measuring rates are usually less than 10 readings for second. The latter fact poses severe limits in dynamic measurements as, e.g., the measurement of heat capacity (see Section 12.6.4).

Resistance bridges sometimes contain a multiplexer to carry out measurement of several thermometers with the same bridge. However, the multiplexing procedure presents several drawbacks (e.g., the temperature readings are not taken at the same time). It is safe, when possible, to use one bridge for each thermometer. The reference frequencies of the bridges must differ for at least a few hertz and of course must not be at the frequency of the power line or motors of the turbomolecular pumps.

10.5 Temperature control

To carry out measurements at a fixed temperature, the refrigerator temperature must be kept constant for a suitably long time. The problem of the temperature control depends not only on the refrigerator itself, but on the thermal characteristics of the experiment. Let us now consider an oversimplified case in which heat capacities are neglected: the mixing chamber temperature of a dilute refrigerator (DR) is to be controlled by a resistive heater H_R and a d.c. power supply.

The cooling power of the refrigerator is:

$$P_{DR} \approx A \cdot T^2; \quad A = 1.5 \times 10^{-3} \text{ [W K/2]} \tag{10.10}$$

We wish to fix the temperature of the mixing chamber at 0.1 K. Let us consider four cases:

1. $H_{Rc} = R_{cost.} = 1\,k\Omega$ constant resistor (i.e. Ni-Cr resistor);
2. $H_{R1} = R_1(T) = R_0 \cdot \exp\left(T_0/T\right)^{1/4}$ with $R_0 = 500\,\Omega$, $T_0 = 3$ K, constant current biased;
3. $H_{R2} = R_2(T) = R_1(T)$ but constant voltage biased;
4. $H_{R3} = R_3(T) = R_0 \cdot \exp\left(T_0/T\right)^{1/2}$ with $R_0 = 3.54\,\Omega$, $T_0 = 5$ K, constant voltage biased.

The cooling power of the refrigerator at 0.1 K is: $P_{DR}\,(0.1\,\text{K}) = 1.5 \times 10^{-5}$ W, then:

$$\begin{aligned} &V(H_{Rc}) = 0.122\,\text{V};\\ &I(H_{R1}) = 5.37 \times 10^{-5}\,\text{A};\\ &V(H_{R2}) = 0.280\,\text{V};\\ &V(H_{R3}) = 0.250\,\text{V}; \end{aligned} \tag{10.11}$$

The operating temperature point (0.1 K) is stable if $\mathrm{d}P_{DR}/\mathrm{d}T > \mathrm{d}P_R/\mathrm{d}T$.

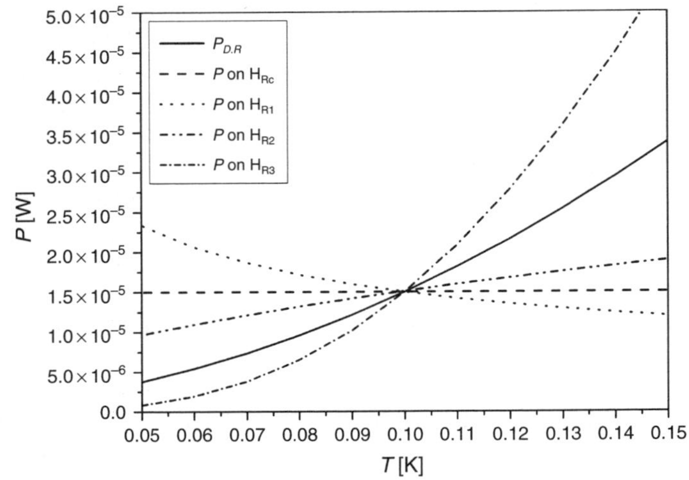

Fig. 10.13. Stability diagram in the operating temperature point (0.1 K) for the four cases.

From Fig. 10.13, we see the latter condition is fulfilled in the first three cases, but not in the fourth case. The most stable situation is obtained with R_1. The choice $R = R_{cost.}$ is however usually adopted when the power supplied to the resistor must be measured. The control of temperature in the real (dynamic) case is much more complex. The problem is similar to that encountered in electronic or mechanical systems. The advantage in the cryogenic case is the absence of 'thermal inductors'. Nevertheless, the heat capacities and heat resistances often show a steep dependence on temperature (i.e. $1/T^3$ of Kapitza resistance) which makes the temperature control quite difficult. Moreover, some parameters vary from run to run: for example, the cooling power of a dilution refrigerator depends on the residual pressure in the vacuum enclosure, on the quantity and ratio of $^3He/^4He$ mixture, etc.

For all these reasons, commercial instruments, based on the classical P + PD + PI control (see for example ref. [8]), are of little utility except for $T > 1$ K. Their heater control methods sometimes employ pulse width modulation techniques of a square wave. Thus, they introduce vast amounts of noise to sensitive regions of the experiment.

For a further discussion about the temperature control see ref. [5, pp. 118–133].

10.6 Low-noise cold amplifiers

The information about a cryogenic experiment usually consists of electrical signals coming from the cold region via electrical wires. The length of such wires is typically 1–2 m. The choice of the type of wire depends on several factors (see Section 4.1), but the main goal is to minimize the noise level. Long wires introduce noise signals due to mechanical vibrations, RF pick-up by high impedance loops and excess capacitance. Preamplifiers working at low temperature, placed close to the cold experiment, reduce the length of the high impedance line. GaAs devices work down to 4 K (see Section 9.6.2.5), but they are noisy. Silicon J-FET are low-noise devices that are often used in cooled preamplifiers. Around 100 K, the transconductance of Si J-FET increases by about a factor of two. Also the thermal noise shows a decrease, but the $1/f$ contribution increases. The $1/f$ contribution can be reduced using an a.c. bias (of a resistive sensor, see Section 15.4) as schematically shown in the example of Fig. 10.14.

Figure 10.15 shows an example of a low-noise amplifier. The cold module works at about 100 K. If the cold module is mounted on a heat sink at temperature below 100 K

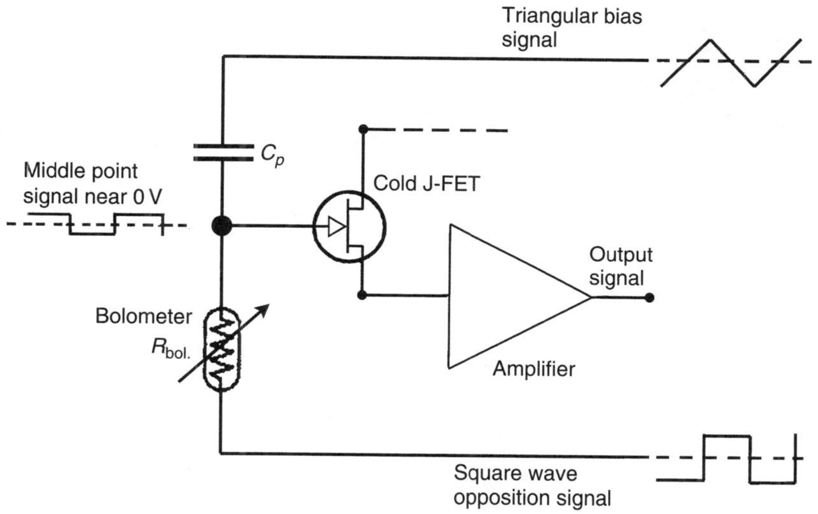

Fig. 10.14. Bolometer a.c biased to reduce the $1/f$ noise of the preamplifier.

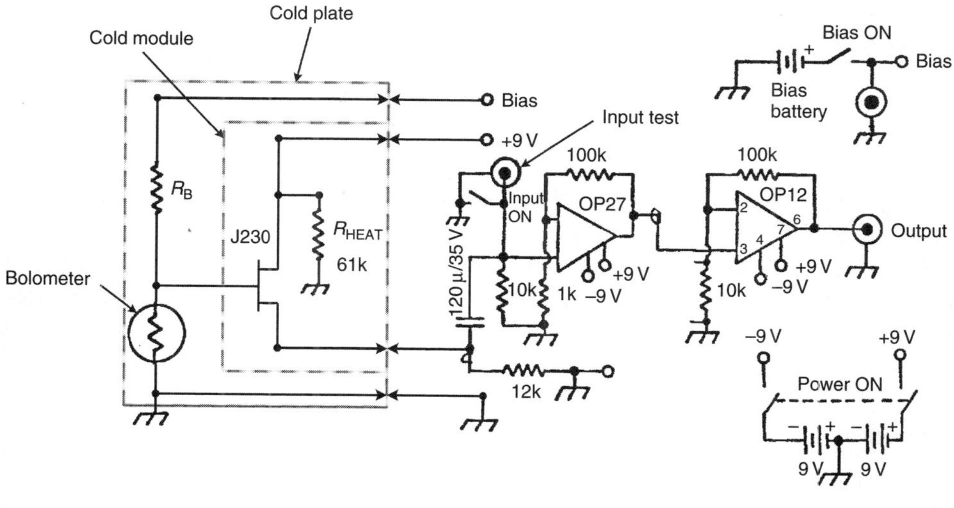

Fig. 10.15. Example of preamplifier with cold input stage. R_B = bolometer load resistor, R_{HEAT} = cold module heater.

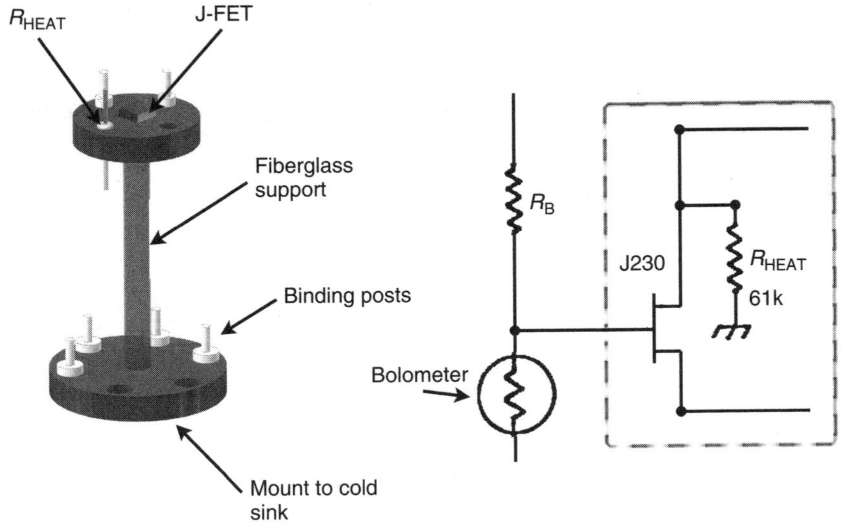

Fig. 10.16. Example of the realization of the thermal impedance for the cold stage of Fig. 10.15 with sink at 4.2 K.

Table 10.1
Overall amplifier characteristic

First stage	Remote at cold surface
Voltage gain	1000
Input impedance	$>10^{13}\,\Omega$
Output impedance	$<500\,\Omega$
Frequency response	0.75 Hz to 30 kHz
Maximum output	± 7 V
Power requirements	± 9 V (batteries included) ± 3 mA
Shorted input noise	$<5\,\mathrm{nV/Hz^{1/2}}$
Bias voltage	Switched, to suit application

(typical 4.2 K), a thermal impedance (see Section 11.2) tailored for the dissipation of the J-FET is needed.

An example of the realization of such impedance by a thin-walled fibreglass tube for a sink at 4.2 K is shown in Fig. 10.16. The overall amplifier characteristics are reported in Table 10.1.

References

[1] F.M. Asner: *High field superconducting magnets*, Clarendon Press, Oxford (1999)
[2] M.N. Wilson ed.: *Superconducting magnets*, Clarendon Press, Oxford (1983)
[3] D.B. Montgomery: *Solenoid magnet design*, Wiley, New York (1969)

[4] H.J. Schneider-Muntau ed.: *High magnetic fields*, World Scientific Pub. Co, Singapore (1977)
[5] G.K. White, P.J. Meeson: *Experimental techniques in low-temperature Physics* p. 182, Clarendon Press, Oxford (2002)
[6] R.C. Richardson, E.N. Smith eds.: *Experimental techniques in condensed matter physics at low temperatures*, p. 22, Addison-Wesley Pub. Co. (1988)
[7] F. Pobell: *Matter and methods at low temperatures*, 2nd ed., Springer Verlag, Berlin (1995)
[8] U. Tietze, Ch. Schenk: *Advanced electronic circuits*, p. 465, Springer Verlag, Berlin (1978)

PART V

11 Measurement of the Properties of Solids at Low Temperature

Contents

11.1 Introduction

In this and in the two following chapters, we shall describe in detail some measurements of the properties of solids at low temperatures. As we shall see, each measurement is a real experiment. This is due not only to the complex operations needed to reach and maintain the desired low temperature, but also to the number of parameters involved in a measurement. Such parameters are never perfectly reproducible in two nominally identical experiments: a typical example is that of the thermal contacts (see Section 4.3).

In the following sections, we will describe in detail some measurements of thermal conductivity of:

- an alloy and an almost pure metal in the 4–77 K range;
- a pure metal at very low temperatures;
- a polymer between 80 mK and room temperature.

11.2 Measurement of the thermal conductivity

For an isotropic material, the instantaneous power flow $P_x(x, y, z, t)$ through a surface $A(x)$, perpendicular to the flow direction, is given by:

$$P_x(x, y, z, t) = -k(T) \cdot A(x) \cdot \frac{\partial T(x, y, z, t)}{\partial x} \tag{11.1}$$

where $k(T)$ is the thermal conductivity. Two equations equivalent to eq. (11.1) can be written for the directions y, z. If the material is anisotropic, $k(T)$ depends on the direction. As we saw in Section 3.10, the thermal conductivity k is a function both of the type of material and of the temperature range.

The most common technique for the measurement of $k(T)$ at low temperatures is the 'steady longitudinal flow method': a steady thermal power flow along a sample of section A (i.e. a cylinder or a bar) is produced by a temperature gradient ΔT. If the power flows only in the x direction, eq. (11.1) becomes:

$$P = -k(T) \cdot A(x) \cdot \frac{\mathrm{d}T}{\mathrm{d}x} \tag{11.2}$$

Hence the power flowing through a sample of length $L = (x_2 - x_1)$, is:

$$P = \frac{-\int_{T_1}^{T_2} k(T) \cdot \mathrm{d}T}{\int_{x_1}^{x_2} \frac{\mathrm{d}x}{A(x)}} \tag{11.3}$$

where T_1 and T_2 are the temperature at x_1 and x_2 respectively.

If the section $A(x)$ of the sample is constant between x_1 and x_2, eq. (11.3) can be written as:

$$P = -\frac{A}{L} \cdot \int_{T_1}^{T_2} k(T) \cdot \mathrm{d}T = -g \times \int_{T_1}^{T_2} k(T) \cdot \mathrm{d}T \tag{11.4}$$

where $g = A/L$ is called geometrical factor of the sample.

If $\Delta T = T_2 - T_1$ is small, the mean thermal conductivity k_{m} in the T_2–T_1 interval is given by:

$$k_{\mathrm{m}} = \frac{P}{g \cdot \Delta T} \tag{11.5}$$

The product $g \cdot k(T) = G(T)$ is called thermal conductance of the sample.

The choice of g in each measurement, as we shall see, depends on several factors, in particular on the range of temperature. In any case, for the measurement of the thermal conductivity with the longitudinal flux method, one end of the sample is put in contact with a 'thermal bath' (which must not appreciably change its temperature while absorbing heat), whereas the other end is heated by an electrical resistor (but sometimes by other means, for example light). In the hypothesis of longitudinal flux, the measured quantities are the power P and the temperatures T_1 and T_2, in two points of the sample at a know distance L. For the determination of $k(T)$ eq. (11.4) or eq. (11.5) are used. The former method leads to the 'integrated thermal conductivity', the latter to the 'mean thermal conductivity'. In the integrated thermal conductivity method, the temperature T_1 at the colder end of the sample is held constant while T_2 is changed by means of the applied power P. A $P(T)$ curve is thus obtained, and the thermal conductivity $k(T)$ can be got by derivation of eq. (11.4). In the mean conductivity method, the application of eq. (11.5) is immediate.

The main experimental difficulties in both methods are due to the presence of shunt thermal conductances (i.e. electrical wires) and series thermal conductances (thermal contacts). The latter problem can be overcome by the 'potentiometric' method (see Fig. 11.2).

Among the shunt spurious contribution, the power exchanged through the residual gas in the vacuum chamber does not represent usually a problem. The shunt conductances are always to be compared with the thermal conductance of the sample. At temperatures above ~77 K, the thermal contact drawback vanishes (see Section 4.4), but the problem of the radiative exchange ($\propto T^4$) becomes very important.

The simplest experimental arrangement (two-probe method) uses two thermometers: one on the thermal bath at T_s, the second on the warm end of the sample together with the heater (see Fig. 11.1). Such configuration can be used when one is sure that contact resistances are negligible compared with the sample thermal resistance. This is seldom the case at very low temperature. A sample bath (and sample support) temperature drop:

$$\Delta T = R_C \cdot P \tag{11.6}$$

is to be expected (R_C is the thermal contact resistance and P the heat power delivered by the heater). Usually R_C, and hence ΔT, varies as T^3 (see Section 4.4).

At very low temperatures and for high conductivity materials, R_C may become of the order of the thermal resistance of the sample. To overcome this problem, for samples of

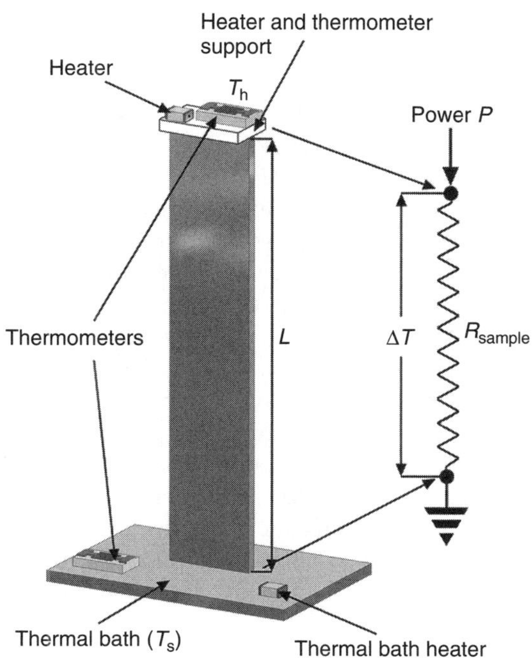

Fig. 11.1. Scheme of the two-probe method for the measurement of the thermal conductivity.

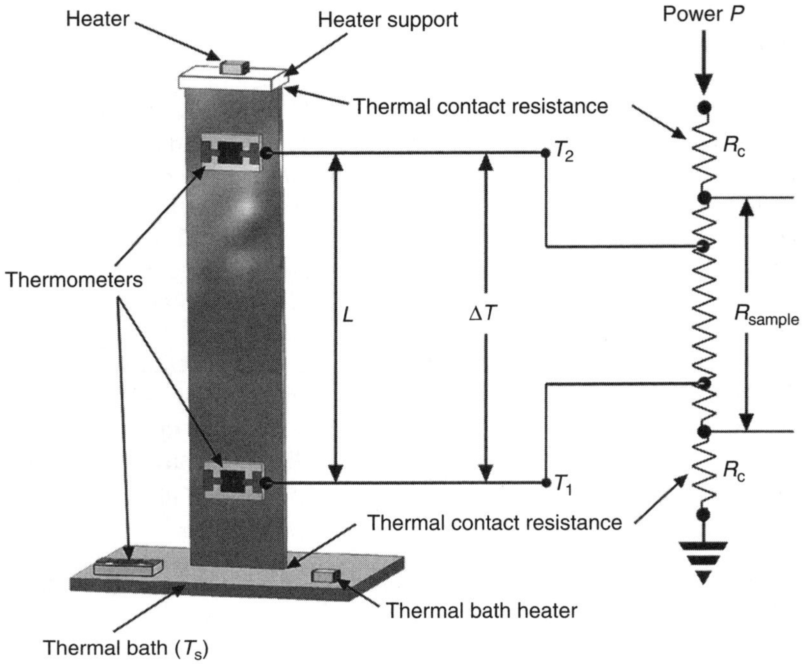

Fig. 11.2. Scheme of the four-probe method for the measurement of the thermal conductivity. R_C is the contact thermal resistance.

large conductance and at low temperatures, the 'four-probe method' is used. The analogy with the potentiometric method used for the measurements of small electrical resistances is evident (see Fig. 11.2). In the four-probe method, two thermometers measure T_1 and T_2 in two points, along the sample, at a distance L. The heater is at one end of the sample, the other end being in contact with the thermal bath at T_s. In principle, the measure of T_s is not indispensable. With such method, the power shunted by the wires of the thermometers T_1 and T_2 can be made negligible, and the measurement does not depend on R_C. However, to keep T_1 constant instead of T_s becomes more difficult.

It is impossible to give a general recipe to carry out the measure of the thermal conductivity of a material. The following general considerations must be kept in mind:

Temperature range: At very low temperatures (below 1 K), contact thermal resistance becomes important (see Section 4.4). In general, the power which must be used in the measurement of the thermal conductivity is low; spurious power contributions such as radiative input and mechanical vibration must be sometimes taken into account. At higher temperatures, the radiative power impinging onto the sample becomes the most worrying phenomenon; an accurate thermal shielding is necessary and the geometrical factor of the sample (see Section 11.5.3) must be usually larger than that used in the very low-temperature range.

Conductivity of the sample material: For low-conductivity materials, the contact resistance is sometimes negligible in comparison with the sample thermal resistance, whereas in the case of high-conductivity materials 'a four-probe measurement' is usually necessary. Remember that electrical wiring to thermistors and heaters form a thermal shunt to the sample. The contribution of such shunt must be carefully evaluated, in particular for low-conductivity samples. For $T << 9\,\text{K}$, the use of bare superconducting wires is a frequently adopted solution (see Section 4.2). A technique which requires no electrical wire in thermal contact with the sample has also been proposed [1]: the temperature gradient in the sample is generated via optical heating and detected by two magnetization thermometers with inductively coupled dc-SQUID readouts (see Section 14.5). The parasitic heat leak and power dissipation of the set-up can be arranged to be below $10^{-15}\,\text{W}$. Therefore, this contact-free method is well-suited for measuring samples with very small thermal conductivity at very low temperatures.

Mechanical properties of the sample: In the simplest case, the sample stands alone (it is fixed by a screw as in measure of Section 11.5.2), or a support may be necessary as in the examples of Section 11.3.2 and Section 11.4.2. In the latter case, the support represents again a thermal conductance in parallel with the sample. Such shunt conductance must be taken into account for low-conductance samples. Contractions (or dilatations, see Chapter 13), which take place when the sample is cooled, not only change the geometrical factor, but also influence the mechanical joints and hence the thermal contact resistance. Various mechanical solutions, some of them very 'exotic', have been used in the measure of the thermal conductivity [2–4].

11.3 Measurement of the thermal conductivity of A6061-T6 and A1050 between 4.2 and 77 K

11.3.1 Introduction

In the last decades, aluminium and aluminium alloys have received an increasing interest in cryogenic applications [5–7]. Pure aluminium is used in the building of superconducting heat switches (see Section 4.3) because of its high switching ratio [8,9] and convenient critical field. Satisfactory procedures for welding have also been found [10,11].

Aluminium alloys are very seldom used below their transition temperatures ($\sim$1.2 K) [12], but find application at temperatures above 4 K [13] where the thermal conductivity is lower than that of the pure metal [14], and the mechanical characteristics are much better. In particular, because of their lightness, good mechanical properties and high thermal conductivity [15–17], aluminium alloys are often used in the realization of structures and shields in space applications [18–20].

We describe here the measurement of the thermal conductivity between 4.2 and 70 K of A6061-T6, a tempered Al/Mg/Si alloy, with good corrosion resistance and mechanical properties, used for heavy-duty structural applications, and A1050, an almost pure Al/Fe/Si alloy with good workability.

11.3.2 Experiment and results

Both samples were in the form of a strip. The geometrical factor $g = A/L$ was $(5.7 \pm 0.2) \times 10^{-6}$ [m] for A6061-T6 and $(1.5 \pm 0.05) \times 10^{-6}$ [m] for A1050. An integrated

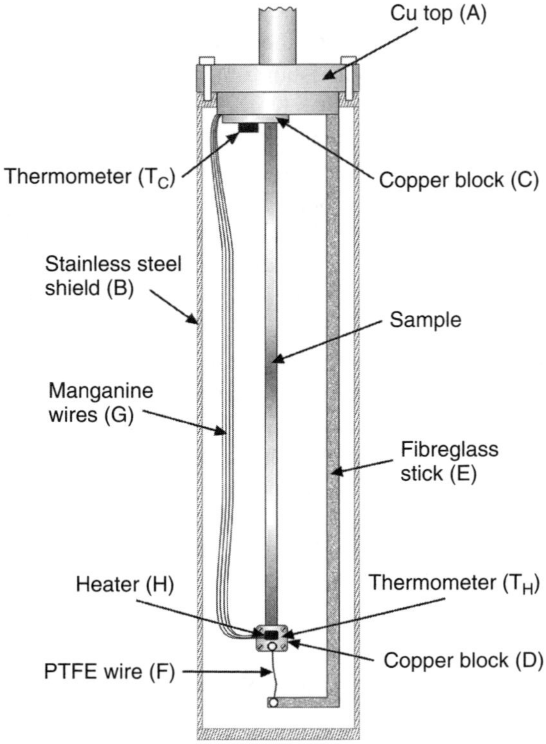

Fig. 11.3. Scheme of the experiment set-up.

thermal conductivity technique was used. The experiment (see Fig. 11.3) was housed in a vacuum chamber made by a Cu top (A) supporting the experiment and a stainless steel shield (B).

The tightness of the two parts was assured by an indium gasket, and the vacuum chamber was immersed in liquid helium. A small volume (about $1\,cm^3$) of helium gas was inserted into the vacuum chamber for the thermalization at 4.2 K of all the parts of the experiment. The exchange gas was successively pumped out, and the pressure during the measurements was kept below 10^{-5} torr to reduce any contribution from convection. Each end of the sample strip was tightened between two copper blocks (C, D of Fig. 11.3) by means of brass screws; one copper block was anchored to the Cu top (A) and the other (the 'hot end') held the heater (H) and the thermometer (T_H).

Therefore, the strip was sticking out from the Cu top; to avoid oscillations of the sample, the second copper block (D) was tied to a fibreglass stick (E) attached to the copper top by means of a thin PTFE wire (F) of negligible thermal conductance. The electrical connections of the thermometer (T_H) and the heater (H) was made by eight manganin wires (G), 100 μm in diameter. Another thermometer (T_C) was used to monitor the temperature of the cold end of the sample. An LR700 AC resistance bridge was used to read the resistance of the thermometers.

The cold end of the sample was kept at a constant temperature $T_C \sim T_s = 4.2\,K$. The hot end was warmed at different temperatures T_h (ranging from 5 to 77 K) by means of a known power P_h supplied to the Ni-Cr heater (H). For $T_h > T_C$, a power P produced a temperature gradient along the sample. The power flowing through the sample can be written as:

$$P = P_h - P_{res.} \tag{11.7}$$

where P_h is the supplied power and $P_{res.}$ can be written as:

$$P_{res.} = P_{mang.} + P_{PTFE} + P_r \tag{11.8}$$

where $P_{mang.}$ is the power through the eight manganin wires, P_{PTFE} is the power through the thin PTFE wire and P_r is the power radiated by the sample. In this experiment $P_{res.} < 0.05\,P_h$ over the whole temperature range of the measurements. In evaluating P_r, an emissivity factor $\varepsilon_r = 0.2$ was used. Thus, neglecting $P_{res.}$, the supplied power P_h is equal to the heat power P flowing through the sample, and $P(T_h)$ can be written as:

$$P(T_h) = g \cdot \int_{T_C}^{T_h} k(T)\,dT \tag{11.9}$$

where g is the geometrical factor of the sample. In Fig. 11.4, a $P(T_h)$ graph for the A6061-T6 sample is reported. From experimental data $P(T_h)$, the integral of thermal conductivity is obtained and by derivation of eq. (11.9), we get $k(T)$ which can be well represented (for metal alloys) by:

$$k(T) = \frac{1}{\frac{a}{T} + b \cdot T^2} = \frac{T}{a + b \cdot T^3} \tag{11.10}$$

where a/T is the contribution to the thermal resistivity due to the electron-impurity scattering and $b \cdot T^2$ comes from the electron–phonon scattering process (see Chapter 3). The numerical values of a and b obtained for the two materials are reported in Table 11.1.

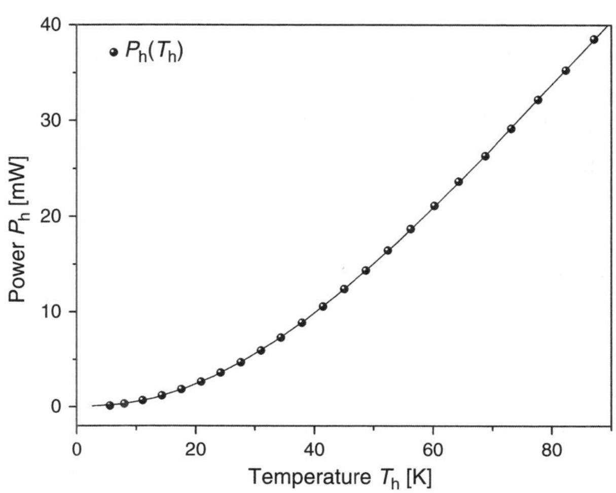

Fig. 11.4. $P(T_h)$ curve for the A6061-T6 sample.

Table 11.1
Numerical values of the fit parameters *a* and *b* from eq. (11.10)

Material	a [K^2 m/W]	b [m/KW]
A6061-T6	0.445 ± 0.005	$(4.9 \pm 0.5) \times 10^{-7}$
A1050	$(5.46 \pm 0.05) \times 10^{-2}$	$(5.3 \pm 0.2) \times 10^{-7}$

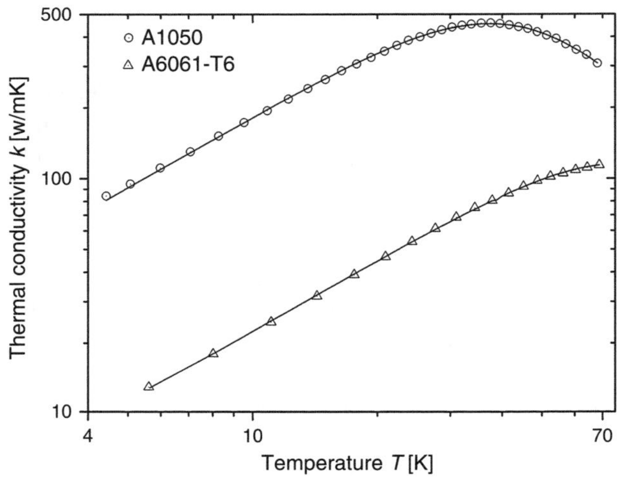

Fig. 11.5. Thermal conductivity of A6061-T6 and A1050 as a function of temperature.

The main contribution to the error is due to the uncertainty on the geometrical factor g and on the calibration of the thermometer T_H.

Figure 11.5 shows a plot of the thermal conductivity of the two alloys. Note that the presence in A6061-T6 of several impurities (about 1% of Mg and 0.5% of Si) is responsible for a conductivity k of A6061-T6 about one-tenth of that of A1050.

11.4 Thermal conductivity of copper at very low temperatures

11.4.1 Introduction

Only few measurements of the thermal conductivity of copper at very low temperatures have been published. Suomi et al. [21] reported about measurements carried out on Cu wires down to 20 mK; more recently Gloos et al. [22] measured the thermal conductivity of rod and foil samples down to even lower temperatures.

As a matter of fact, since the low-temperature heat conduction of a pure metal mostly depends on electron scattering by lattice defects and impurities, one should speak of conductivity of a particular sample. Here we describe the measurement of the thermal conductivity down to 30 mK of three different types of commercial copper, before and after an annealing thermal treatment [23].

The purpose of this experiment was to select the best thermal link to cool down sensitive masses of the order of 100–1000 kg for the detection of rare decays [24] and gravitational waves [25].

11.4.2 Experiment

We have carried out the measurement of the thermal conductivity of six copper samples, whose characteristics are shown in Table 11.2.

The samples were shaped as strips 1 m long and 2 mm wide. The longitudinal steady heat flow method was used to measure the integrated thermal conductivity in the 30–150 mK range.

A known power P_h was supplied to one end of the sample to establish a difference of temperature $\Delta T = T_h - T_C$ between the ends of the sample (see Fig. 11.2).

By derivation of the integrated power:

$$P(T_h) = g \cdot \int_{T_C}^{T_h} k(T)\,\mathrm{d}T \tag{11.11}$$

where $g = A/L$, the geometrical factor of the samples, and the thermal conductivity $k(T)$ can be obtained. Due to the large thermal conductivity k of the copper, long samples were chosen to increase the accuracy of measurements. Since the working space in the dilution refrigerator was quite small, each sample was wound around a cylindrical support made of a 100 μm thick Kapton foil (see Fig. 11.6). A special care was taken to avoid kinks in the sample which could cause lattice defects. Annealing was carried out before winding the sample on the holder, and the cold work is not known. The sample was kept in place by means of some spots of GE-Varnish. The top end of the sample was in thermal contact with the mixing chamber of a dilution refrigerator.

Table 11.2
Copper sample characteristics

Name	Type	Condition
CV 540 (Copper 1)	Cu OFHC, thickness 75 μm ± 10%, purity 99.95%.	Raw
CV 500 (Copper 2)	Cu OFHC, thickness 50 μm ± 15%, purity 99.95%, phosphor deoxidized	Raw
CV 491 (Copper 3)	Cu OFHC, thickness 50 μm ± 10%, purity 99.99%.	Raw
CV 540 (Copper 1A)	Cu OFHC, thickness 75 μm ± 10%, purity 99.95%.	Annealed in vacuum for 12 h at 450°C
CV 500 (Copper 2A)	Cu OFHC, thickness 50 μm ± 15%, purity 99.95%, phosphor deoxidized	Annealed in vacuum for 12 h at 450°C
CV 491 (Copper 3A)	Cu OFHC, thickness 50 μm ± 10%, purity 99.99%.	Annealed in vacuum for 12 h at 450°C

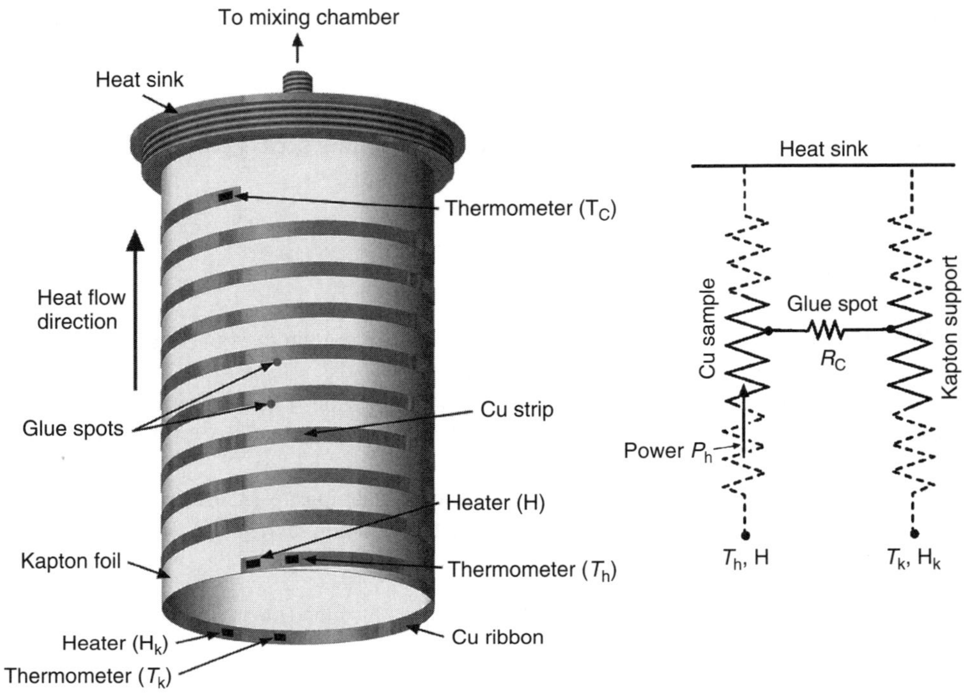

Fig. 11.6. Experimental set-up.

An NiCr heater (H) was glued onto the bottom end of the sample, and two RuO_2 calibrated thermometers (for the measurement of T_h and T_C) were glued near the bottom end of the sample (see Fig. 11.6).

The thermal conductance of each glue spot below 150 mK was very low because of the two contact resistances R_C (Kapton-glue and glue-copper), and the power P_h delivered to the copper sample did not flow through the Kapton foil. To be sure of that, however, 1 mm large, 56 μm thick copper ribbon was internally glued around the upper end of the Kapton support. A heater H_k and a thermometer T_k (Fig. 11.6) were fixed on the ribbon and a power P_k was delivered to the Kapton support in such a way that $T_k = T_h$.

Since the thermal conductivity of Kapton is linear in T below 150 mK as happens for copper [26], a negligible temperature difference was supposed to be present across the glue spots. In practice, we found an extremely high value of R_C, since the temperature T_h was not influenced at all by changes of T_k up to about ten times T_h.

The low-temperature working space was surrounded by a copper thermal shield at the temperature of the mixing chamber, hence, the radiative power was always negligible. For electrical connections, RF filters both at room temperature and at 4 K were used: the total spurious power on the sample was estimated to be below 10^{-11} W.

11.4.3 Results

At very low temperatures, the thermal conductivity, for non-superconducting metals, can be written as:

$$k(T) = k_{\mathrm{el.}} + k_{\mathrm{ph.}} = a \cdot T + b \cdot T^2 \tag{11.12}$$

where $k_{\mathrm{el.}}$ and $k_{\mathrm{ph.}}$ represent the electronic and phonon contributions respectively. Since at low temperatures, electrons are mainly scattered by lattice defects and impurities, and this scattering mechanism is temperature independent, $k_{\mathrm{el.}}$ depends linearly on T as the specific heat, within the usual kinetic formulation. Instead $k_{\mathrm{ph.}}$ is due to the electron–phonon scattering and hence quadratic.

Low-temperature ($T < 1\,\mathrm{K}$) heat conduction of a pure metal, like copper of our experiment (Cu Debye temperature $\theta_{\mathrm{D}} \approx 340\,\mathrm{K}$), is mostly electronic [27] and the phonon contribution should be negligible. With the latter hypothesis, in the 30–150 mK temperature range:

$$k_{\mathrm{Cu}}(T) = K \cdot T \tag{11.13}$$

and:

$$P_{\mathrm{h}} = g \cdot \int_{T_{\mathrm{C}}}^{T_{\mathrm{h}}} K \cdot T \mathrm{d}T = \frac{g}{2} K \cdot T_{\mathrm{h}}^2 - P_{\mathrm{C}} \tag{11.14}$$

where P_{C} is a constant which depends on T_{C}.

From a linear fit on a plot of T^2 as a function of P_{h}, K is immediately obtained from the slope of the curve. Such plots for two of the six samples are shown in Fig. 11.7. In Fig. 11.8, the plots of the measured thermal conductivity for all samples are reported.

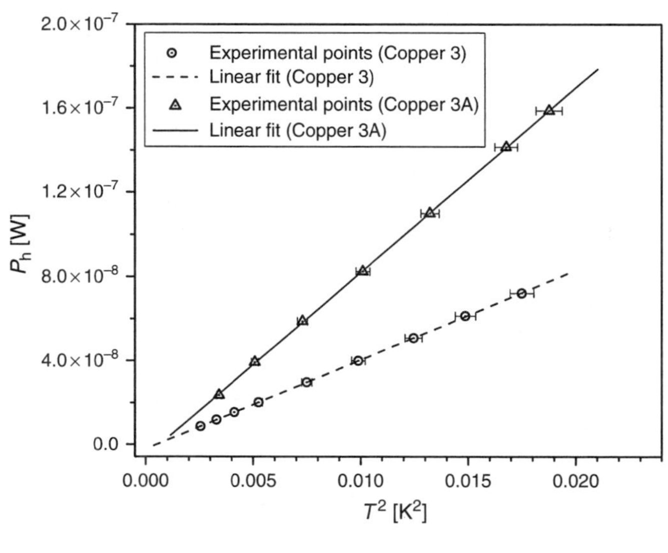

Fig. 11.7. Experimental data: P_{h} as a function of T^2 for the samples Copper 3 and Copper 3A.

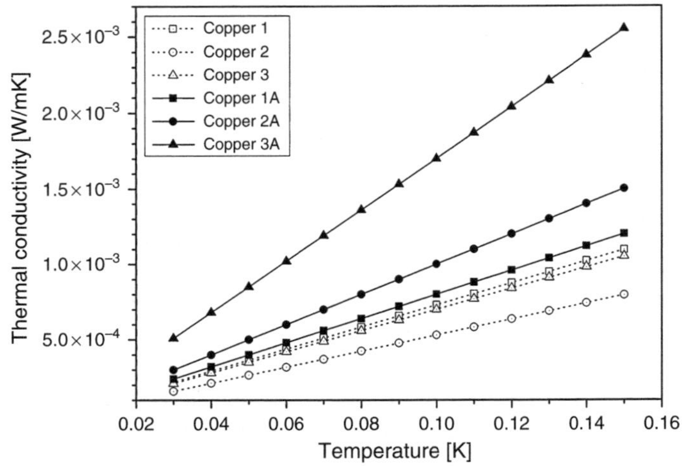

Fig. 11.8. Thermal conductivity $k(T) = K \cdot T$ for the six samples.

Table 11.3
Thermal conductivity parameter K of the six Cu samples, before (first three items) and after (last three items) annealing at 450°C

Sample name	Parameter K [W/mK^2]
CV 540 (Copper 1)	$(7.3 \pm 0.9) \times 10^{-3}$
CV 500 (Copper 2)	$(5.3 \pm 1.4) \times 10^{-3}$
CV 491 (Copper 3)	$(7.0 \pm 1) \times 10^{-3}$
CV 540 (Copper 1A)	$(8.0 \pm 2) \times 10^{-3}$
CV 500 (Copper 2A)	$(1.0 \pm 0.3) \times 10^{-2}$
CV 491 (Copper 3A)	$(1.7 \pm 0.5) \times 10^{-2}$

In evaluating the geometrical factor g of the strips, a contraction factor of 0.33% from room temperature has been introduced [28].

The relative errors on P_h, of the order of 0.1%, have been disregarded in the computation of K. We have checked for the presence of non-linear terms in the curves of T^2 vs P_h, but the contribution of higher-order terms was found to be within the errors due to thermometer calibration.

The thermal conductivity of the six samples is summarized in Table 11.3.

The uncertainty on $k(T)$ is mainly due to the error in the evaluation of the geometrical factor g.

We can see from Fig. 11.8 that the annealing procedure has produced, as expected, an increase of the conductivity because of the reduction of lattice defects. The increase in conductivity after annealing is large in the case of Copper 2 and Copper 3 (a factor of 1.9 and 2.4 respectively) but is negligible in case of Copper 1. This behaviour can be attributed to the large impurity content and to the larger thickness of the sample that implies larger bending stress in the assembly operation.

Using the standard scaling for copper properties, we find values of RRR (see Section 3.10.2) from around 40 (Copper 2) to 120 (Copper 3A) which are typical of commercial Cu.

A comparison can be done with the few published data [21,22,29,30] of different purity copper and different investigated temperature ranges. Suomi et al. [21] report about measurements in 20–200 mK temperature range and obtain a thermal conductivity close to that reported in these measurements. The value of K of Copper 3A is rather good compared with ref. [29,30] (measurements of commercial copper in a temperature range of, respectively, 400–900 and 200–700 mK) where K varies between 1.1 and 1.6 W/cm K^2. Gloos et al. [22] find a much higher value of K. This is due to the higher purity of the samples (5N) as well as to the higher annealing temperature (950°C).

11.5 Measurement of the thermal conductivity of Torlon

11.5.1 Introduction

Torlon (polyamide-imide) is a thermoplastic polymer which exhibits excellent mechanical properties and good chemical resistance. These characteristics enable Torlon to compete with metals and other engineering plastic resins in a growing number of automotive, aerospace and electronics applications. Torlon is capable of performing throughout a wide temperature range. In particular, it retains its high strength at low temperatures, so that it is often used in cryogenic applications.

As we shall see, it is impossible to carry out measurements of thermal conductivity in a very extended range of temperature on the same sample (with the same g factor) and with the same cryogenic apparatus. Hence the measurements were performed in two overlapping ranges, as reported in the following sections.

11.5.2 Thermal conductivity of Torlon 4203 in the 0.08–5 K temperature range

The sample was a cylinder with a geometrical factor $g = 0.106$ cm (see Fig. 11.9). The lower end of the sample was fixed onto a copper holder (heat sink) in good thermal contact with the mixing chamber of a dilution refrigerator. A RuO_2 thermometer (R_s) monitored the temperature T_s of the holder. A small block of copper carrying another RuO_2 thermometer (R_u) and also a NiCr heater (H_u) was screwed onto the upper end of the specimen. At the ends of the sample, there are two thermal guard rings made with a copper ribbon glued to the sample. Such rings protrude along the sample for a length equal to that of the screws, creating an isothermal zone around the screws and defining an effective length L of the sample. Four bare NbTi wires, 25 μm in diameter, were used to make the electrical connections of R_u and H_u. Their thermal resistance was about four orders of magnitude greater than that of the sample [31].

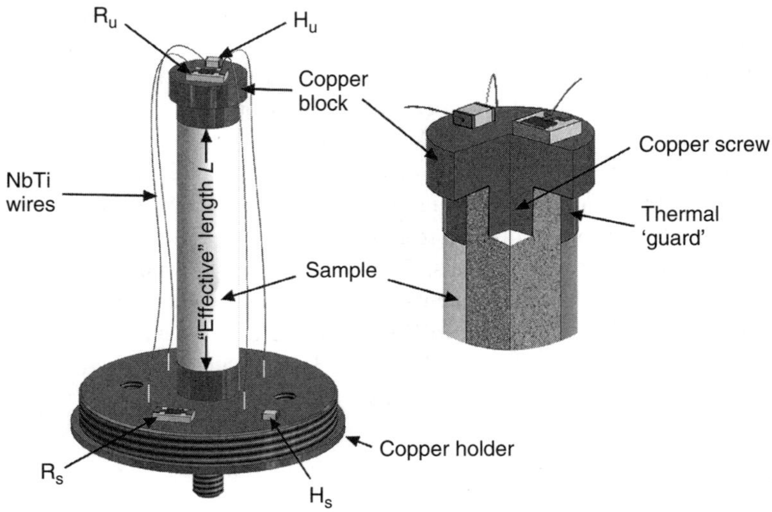

Fig. 11.9. Experimental set-up for the measurement of the thermal conductivity of Torlon.

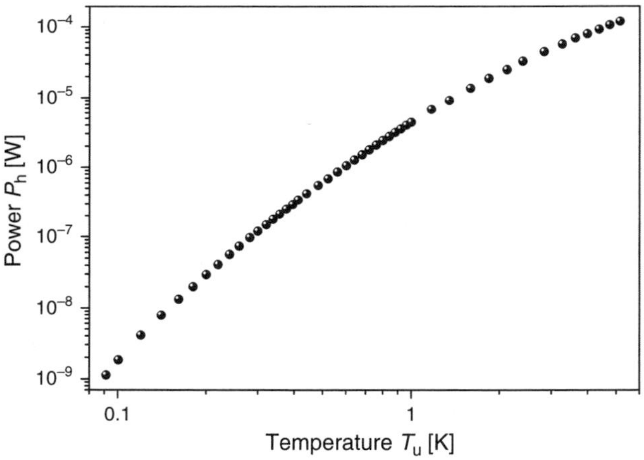

Fig. 11.10. Heating power supplied to the sample.

Thermal conductivity was measured by a steady-state technique: the measurements below 1 K (above 1 K) were carried out with the mixing chamber maintained at a constant temperature $T_s \sim 70\,\text{mK}$ ($T_s \sim 300\,\text{mK}$) by controlling the power dissipated in a heater (H_s) glued to the copper holder.

Electrical power P_h was supplied to H_u and, once the thermal equilibrium was reached, the temperature T_u was read by R_u. The heating power P_h was then changed, in order to obtain a set of data, shown in Fig. 11.10.

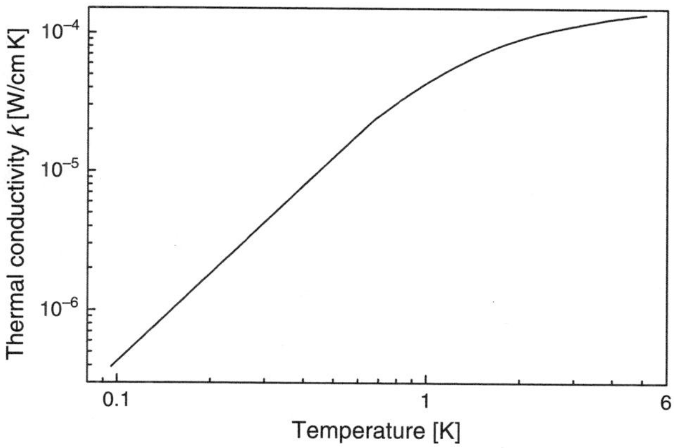

Fig. 11.11. Thermal conductivity of Torlon 4203.

The conductivity, obtained as the derivative of P_h/g fit curve, is shown in Fig. 11.11. The estimated error of k is less than 3%.

Below 0.8 K, thermal conductivity can be represented by a 'typical' power law $k(T) = \alpha \cdot T^n$ (see Section 3.10.1), and the fit gave:

$$k(T) = (6.13 \pm 0.07) \times 10^{-5} T^{2.18 \pm 0.01} \quad [\text{W/cm K}] \tag{11.15}$$

This behaviour is in good agreement with the tunnelling model [32,33], which for polymers and $T < 1$ K predicts a quadratic temperature dependence of thermal conductivity, due to the resonant scattering of phonons by two-level systems. We wish to point out that Torlon has a very low thermal conductivity: for instance, at 0.1 K, k is only a factor of four greater than that of graphite AGOT [34] (which is one of the best thermal insulators known up until now). Torlon is also easily machined and is much stronger than most insulators. This property, together its low linear thermal expansion coefficient (see Section 13.3), makes Torlon particularly suitable as construction material for low-temperature apparatus.

11.5.3 Thermal conductivity of Torlon 4203 between 4.2 and 300 K

For this range of temperature, the experimental apparatus, the shape and geometrical factor g of the sample and the method of measurement (mean conductivity method) are different from those used for the very low-temperature range.

The measurement of the thermal conductivity of Torlon has been carried out on a sample whose shape and dimensions are shown in Fig. 11.12.

The part of the sample along which the gradient of temperature is produced has a cylindrical form of length $L = 5.75 \pm 0.03$ mm, and radius $r = 4.00 \pm 0.01$ mm, giving a geometrical factor $g = A/L = 8.74 \pm 0.09$ mm at room temperature. The thermal contacts at the end of the sample have been realized by means of two gold-plated copper screws (Sc_1, Sc_2) 4 mm in diameter. The threading in the sample had a depth of 5 mm. Since

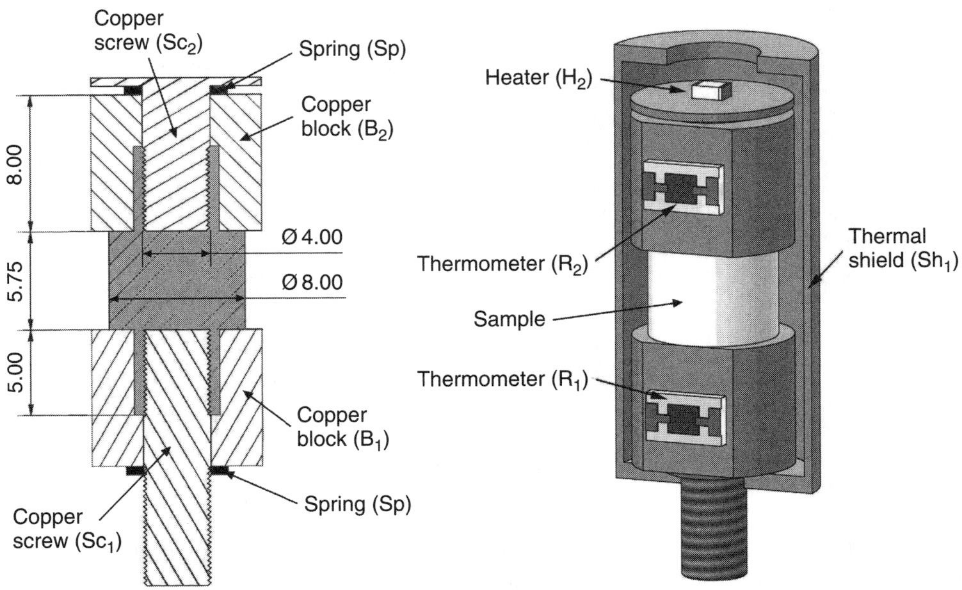

Fig. 11.12. Shape and size of the sample (mm) and view of the sample holders.

the thermal contraction of Torlon is slightly greater than that of copper (see Section 3.9 or ref. [[35], p. 50]), the thermal contact between the screws and the threaded parts of the sample becomes better on cooling. On the other hand, to ensure the thermal contacts on the two flat surfaces of the sample, two gold-plated copper blocks (B_1, B_2), on which thermometers (R_1, R_2) were mounted, were pressed by stainless steel springs (Sp) against the two ends of the sample.

Thermal conductivity was measured by a steady-state technique. One end of the sample was fixed (see Fig. 11.13) onto a gold-plated copper platform (Pf) whose temperature T_1 can be set by means of a heater (H_1). The thermometer (R_1), glued on the copper block (B_1), measured T_1. The copper block (B_2) held a carbon thermometer (R_2), which measured T_2, and a NiCr heater (H_2) was glued on the top of the copper screw (Sc_2) (see Fig. 11.12). Electrical connections were made of Ø 50 μm, ~35 cm long manganin wires.

A cylindrical gold-plated copper thermal shield (Sh_1) enclosed the sample and an outer gold-plated copper thermal shield (Sh_2) enclosed the experiment (see Fig. 11.12 and Fig. 11.13).

A power P_h was supplied to the heater (H_2) in order to create a temperature gradient $\Delta T \sim 2\% \cdot T_1$ along the sample. The thermal conductivity at a temperature T was evaluated from:

$$\begin{cases} k(T) = \dfrac{P_h}{g \cdot \Delta T} \\ T = \dfrac{T_1 + T_2}{2} \text{ and } \Delta T = T_2 - T_1 \end{cases} \tag{11.16}$$

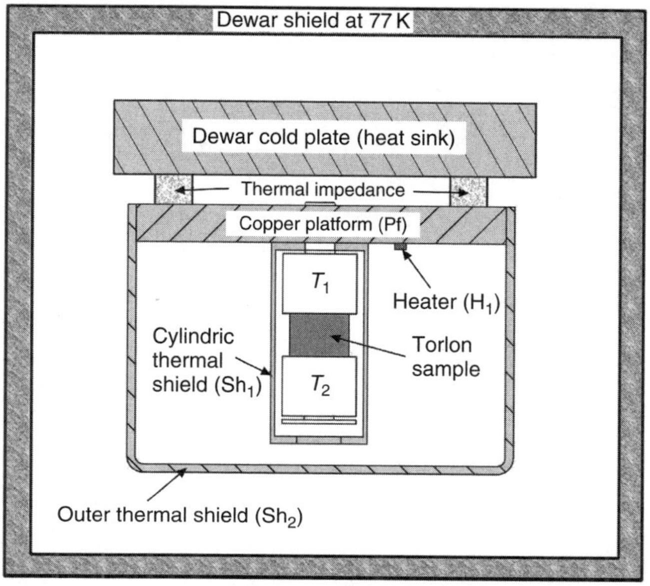

Fig. 11.13. Schematics of the experimental set-up for the measurement of the thermal conductivity of Torlon between 4.2 and 300 K.

Three runs of measurements were carried out in a ^{4}He cryostat (such as that described in Section 5.4):

1. 8–25 K: thermal bath at 4.2 K and dewar shield at 77 K;
2. 30–80 K: same as previous run, with a greater thermal impedance to reduce helium consumption;
3. 80–300 K: heat sink and dewar shield at 77 K, with a greater heat sink platform (Pf) thermal impedance.

The thermal conductivity of Torlon 4203 in the 4.2–300 K range is shown in Fig. 11.14. In Fig. 11.15, $k(T)$ is shown over the range 80 mK–300 K using data of Section 11.5.2. The correction due to the thermal contraction (max $\Delta g/g = 0.4\%$) was neglected [36].

The thermal conductivity data for Torlon, presented in Fig. 11.15, are typical of amorphous polymers. In particular, they show:

1. a T^2 dependence of k below 1 K, in agreement with the 'tunnelling model' [32,33];
2. a plateau between 5 and 10 K, as predicted by the 'soft potential model' [34];
3. a steep rise of conductivity after the plateau;
4. an almost linear behaviour of $k(T)$ between 30 K and room temperature.

A similar behaviour has been found for the thermal conductivity of other polymers, e.g. polymetilmethacrylate (PMMA) [36].

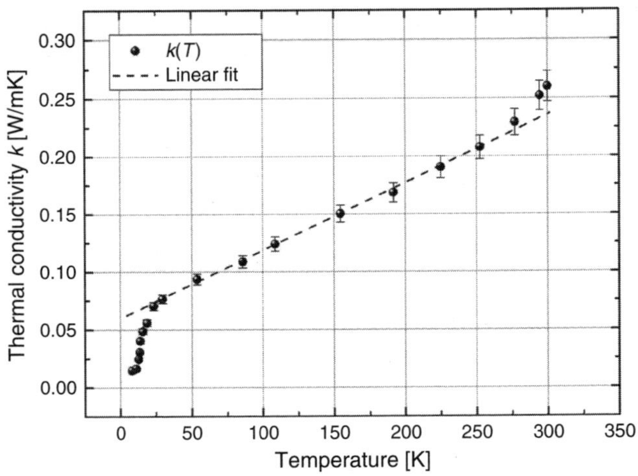

Fig. 11.14. Thermal conductivity of Torlon 4203 in the temperature range 4.2–300 K. The dashed line represents formula (11.17).

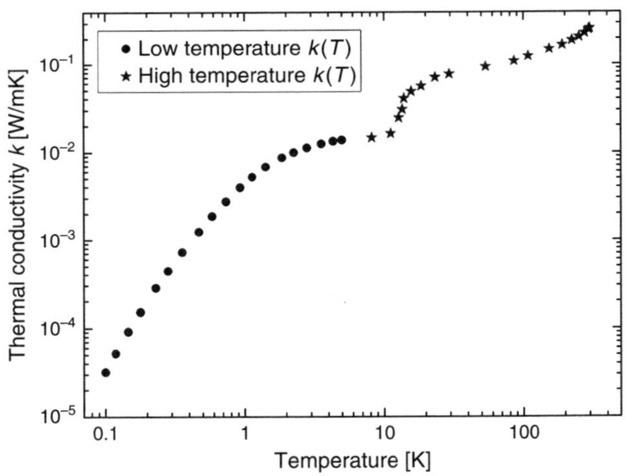

Fig. 11.15. Thermal conductivity of Torlon in the 0.08–300 K range.

As regards these data, in the 30–250 K range, the thermal conductivity of Torlon can be well represented by:

$$\begin{cases} k(T) = b + a \cdot T \\ a = (0.58 \pm 0.03) \times 10^{-3} \ \left[\mathrm{W/mK^2}\right] \\ b = (0.60 \pm 0.03) \times 10^{-1} \ [\mathrm{W/mK}] \end{cases} \tag{11.17}$$

Data at the lowest temperatures of Fig. 11.14 match well with data at $T < 5\,\mathrm{K}$ of Fig. 11.11.

Several assumptions in the preceding section deserve a more detailed explanation.

11.5.3.1 Comparison among the power passing through the sample and the spurious power contributions

Table 11.4 shows a comparison among:

- the electrical power P_h delivered to the warmer end of the sample;
- the power P_M shunted by the eight manganin wires;
- the power P_R radiated by the sample toward the cylindric shield.

In Table 11.4, P_R was calculated considering the thermal exchange for radiation between two cylindrical coaxial surfaces with the following hypotheses:

1. neglecting the contribution from the gold-plated supports (B_1) and (B_2);
2. assuming emissivity $\varepsilon_T = 0.5$ for Torlon and an emissivity $\varepsilon_S = 0.03$ [37] for the cylindrical gold-plated copper shield;
3. considering the sample at its mean temperature $T = (T_2 + T_1)/2$;
4. assuming a ratio between the surfaces $A_T/A_S = 0.5$, with A_T lateral surface of the sample and A_S surface of the cylindrical gold-plated copper shield that the sample 'sees'.

From Table 11.4, at 300 K, the estimated P_R becomes about $\sim$1.1% of P_h.

11.5.3.2 Thermal contacts to the sample

The thermal resistance between the ends of the sample and the copper blocks must be negligible compared with the thermal resistance of the sample. This assumption must be verified especially for short samples at low temperature where the contact resistance is higher. For this reason, a second measurement of the thermal conductivity of Torlon in the 4.2–25 K range was carried out. The second sample had a different length ($L = 24.51$ mm) and the same section A. This additional measurement gave the same value of k within 2%. Moreover, we see from Fig. 11.15 that data of thermal conductivity at 4.2 K well join data at lower temperatures (within 3%) obtained on a sample of much smaller geometrical factor and with a different method (integrated thermal conductivity method) and a different apparatus [38]. Finally, at room temperature, we find $k = 0.26$ W/mK, which is the data sheet value.

Table 11.4
Comparison among the electrical power P_h delivered to the warmer end of the sample, the power P_M shunted by the eight manganin wires and the power P_R radiated by the sample toward the cylindrical shield

Temperature T [K]	P_h [W]	P_M [W]	P_R [W]
10	3.5×10^{-5}	5.7×10^{-9}	1.8×10^{-10}
80	1.5×10^{-3}	3.0×10^{-7}	7.5×10^{-7}
300	1.4×10^{-2}	1.9×10^{-6}	1.5×10^{-4}

11.5.3.3 Error budget

From eq. (11.16) we obtain:

$$\frac{\Delta k}{k} = \frac{\Delta P_h}{P_h} + \frac{\Delta g}{g} + \frac{\Delta(\Delta T)}{\Delta T} \tag{11.18}$$

There are three main contributions to the maximum relative error in $k(T)$:

1. The power supplied to the sample: from Table 11.4 we can estimate that the relative error of P_h is of the order of ~1%;
2. The measurement of g: the error in the measurements of L and A has been estimated as 0.5%;
3. The uncertainty on the temperature gap ΔT due to the sensitivity of the thermometers in the temperature range 4.2–300 K: a conservative value of $\Delta(\Delta T)/\Delta T$ is ~2%.

Taking into account the uncertainty on T, that is about 1%, we conclude that the maximum (at 300 K) relative error in $k(T)$ over the 4.2–300 K is 5%.

References

[1] H.Y. Hao et al.: Rev. Sci. Instrum. **75**, 2718 (2004)
[2] G. Ventura et al.: Cryogenics **38**, 917 (1998)
[3] G. Ventura et al.: Cryogenics **40**, 489 (2000)
[4] L. Risegari et al.: Cryogenics **44**, 875 (2004)
[5] A.L. Woodcraft: Cryogenics **45**, 626 (2005)
[6] A.L. Woodcraft: Cryogenic **45**, 421 (2005)
[7] M. Barucci et al.: *Proceedings of 8th International Conference Advanced Technology and Particle Physics*, p. 541, ed. by M. Barone et al., Como 2003, World Scientific (2003)
[8] R.M. Bueller et al.: Rev. Sci. Instrum. **49**, 515 (1978)
[9] R.M. Bueller et al.: Cryogenics **20**, 395 (1980)
[10] K.W. Wittekers et al.: Cryogenics **29**, 904 (1989)
[11] Yu.M. Bunkov: Cryogenics **29**, 938 (1989)
[12] G. Ventura et al.: Nucl. Phys. B, **56–57**, 415 (1997)
[13] R.W. Kaffky, N.S. Moham, D.H. Damon: Phys Rev. B **11**, 1297 (1974)
[14] C.Y. Ho, R.W. Powell, P.E. Liley: J. Phys. Chem. Ref. Data **1**, 305 (1972)
[15] H.I. McHenry: Adv. Cryog. Eng. **22**, 9 (1977)
[16] R.P. Reed, A.F. Clark: *Materials at Low Temperatures*, American Society of Metals, Metals Park, OH (1983)
[17] H.I. McHenry: Nucl. Eng. Des. **58**, 219 (1980)
[18] B. Collaudin, T. Passvogel: Cryogenics **39**, 157 (1999)
[19] S. Triqueneaux et al.: Cryogenics **46**, 288 (2006)
[20] J.M. Lamarre et al.: in *Proceedings of 9th International Workshop on Low Temperature Detectors*, vol. 605, p. 571, Madison 22–27 July 2001, American Institute of Physics, New York (2002)
[21] M. Suomi, A.C. Anderson, B. Holmstroem: Physica **38**, 67 (1968)
[22] K. Gloos et al.: Cryogenics **30**, 14 (1990)
[23] L. Risegari et al.: Cryogenics **44**, 167 (2004)
[24] R. Ardito et al.: Prog. in Part. and Nuc. Phys. **57**, 203 (2006)
[25] P. Astone: Class. Quant. Grav. **19**, 1227 (2002)
[26] M. Barucci et al.: Cryogenics **40**, 145 (2000)
[27] K. Mendelssohn, H. Rosenberg: *Solid State Physics*, vol. 12, p. 223, ed. by Turnball-Seitz, Academic Press (1961)

[28] R.F. Seligmann, R.E. Sarwinski: Cryogenics **12**, 239 (1972)
[29] G. Davey, K. Mendelssohn: Phys. Lett. **7**, 183 (1963)
[30] A. Dupré, A. Van Itterbeek, L. Michiels: Phys. Lett. **8**, 99 (1964)
[31] J.R. Olson: Cryogenics **33**, 729 (1993)
[32] W.A. Phillips: J. Low. Temp. Phys. **7**, 351 (1972)
[33] W.P. Anderson, B.I. Halperin, C.M. Varma: Philos. Mag. **25**, 1 (1972)
[34] V.G. Karpov, M.I. Klinger, F.N. Ignat'ev: Sov. Phys. JETP **57**, 439 (1983)
[35] F. Pobell: *Matter and Methods at Low Temperatures*, Springer, Berlin (1991)
[36] C.L. Choy: Polymers **18**, 984 (1977)
[37] G.K. White, P.J. Meeson: *Experimental Techniques in Low-Temperature Physics*. 4th ed., Oxford University Press Inc., New York (2002)
[38] G. Ventura et al.: Cryogenics **39**, 481 (1999)

12 Measurements of Heat Capacity

Contents

12.1 Introduction

We saw in Section 3.2 that the knowledge of low-temperature specific heat is extremely important to understand the physical properties of a solid. The measurements of heat capacity are not, conceptually, more difficult than those of thermal conductivity. On the contrary, some problems such as the anisotropy of materials are not present, and the shape of the sample to be measured is usually unimportant. Nevertheless, from a technical point of view, the measurements of heat capacity are more complex, mainly for the two following reasons:

- since thermometers contribute to the addendum heat capacity, it is necessary in most cases to use crystalline thermometers of small heat capacity. In particular, RuO_2 thermometers are usually replaced by doped Si or Ge crystals;
- the characteristic times of the heat capacity measurements are strongly limited by the instrumentation time response (typically 0.1 s).

For a discussion of the measurement of low-temperature specific heat see, for example, ref. [1].

12.2 Measurement methods

The five most common methods used to measure the heat capacities at low temperature are:

- the heat pulse technique [2–9];
- the a.c. calorimetry [10,11];
- the time constant (relaxation) method [12–17];
- the dual slope method [18].
- the method of the thermal bath modulation [19]

12.2.1 Heat pulse technique

In the heat pulse method, the sample of heat capacity C is thermally linked to the thermal bath (at temperature T_B) by a conductance G.

In the simplest case (single time constant), $\tau_s = C/G$ is the sample to thermal bath relaxation time.

If the sample is heated by a pulse of power P_i of duration $\tau_i << \tau_s$:

$$C = \frac{P_i \cdot \tau_i}{\Delta T} \tag{12.1}$$

where ΔT is the corresponding temperature increase. If the system is 'quasi-adiabatic', the measurement of C would seem very simple. Really this is not true.

Let us examine in Fig. 12.1 a schematics of a set up for heat capacity measurement: a support (Sp) of heat capacity C_{Sp} is thermally linked to the thermal bath through a thermal resistance $R_G = 1/G$.

The sample (S), a thermometer (Th) and a heater (H) of heat capacity C_S, C_{Th} and C_H are in thermal contact with the support (Sp) through his contact resistances r_S, r_{Th} and r_H respectively.

In Fig. 12.2, the thermal equivalent circuit is shown: R_{Th} and R_H represent the thermal resistances of the wiring of the thermometer and heater respectively. There are four thermal nodes at the temperatures T_H, T_{Sp}, T_{Th} and T_S. The only experimentally measured quantity is T_{Th}, while we wish to know T_S.

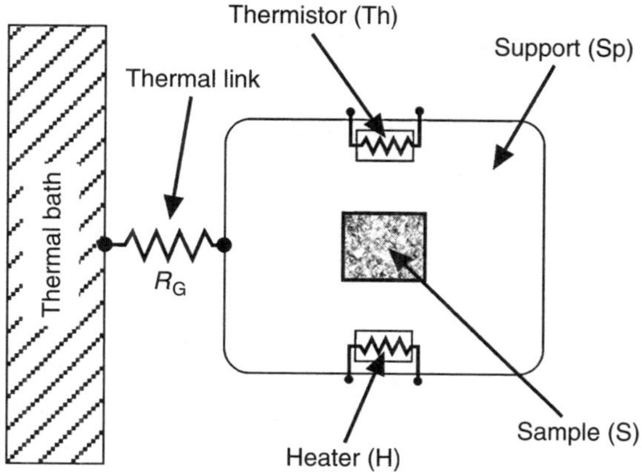

Fig. 12.1. Scheme of the main elements present in a measurement of heat capacity.

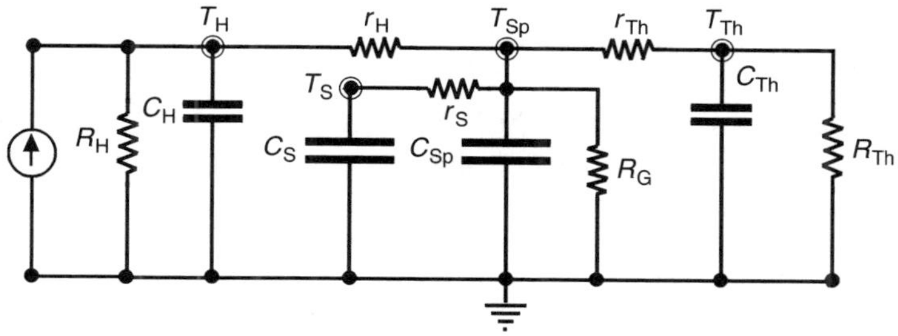

Fig. 12.2. Thermal equivalent circuit for the heat capacity measurement.

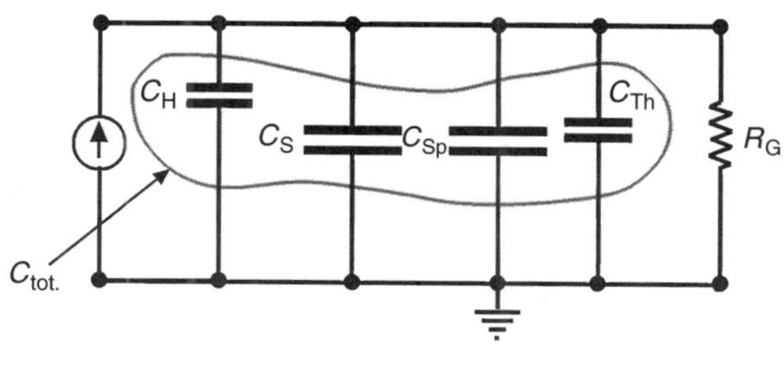

Fig. 12.3. Simplified thermal circuit of the heat capacity measurement.

The relation between $T_{Th}(t)$ and $T_S(t)$ can be formally obtained for small temperature changes if all the circuit parameters, in particular the contact resistances, are known.

Only when r_S and r_{Th} can be neglected $T_S = T_{Th}$. However, the shape of the pulse which charges C_S still depends on the other circuit time constants.

The most idealized configuration is depicted in Fig. 12.3. Even in this simplified situation, it is required that $R_G \cdot C_{tot.} >> \tau_{acq.}$ ($\tau_{acq.}$ = shortest acquisition time of the instrument measuring T_{Th}).

Only in the latter case, the measurement is practically adiabatic. Otherwise, it is necessary to extrapolate data to get the effective ΔT at $t = 0$ (start of the heat pulse) [2,3].

The heat capacity of the addendum ($C_{\text{addendum}} = C_{\text{Th}} + C_{\text{Sp}} + C_{\text{H}}$) can be obtained by a separate measurement without the sample. The heat pulse technique is typically used in the 0.05–4 K temperature range.

12.2.2 AC calorimetry

In AC calorimetry, the experimental set up is still that of Fig. 12.1, but, in this case, P is a sinusoidal power. Amplitude and phase differences between the signal P and the sinusoidal change of the sample temperature are a function of the heat capacity C [10]. This method has been used in the 1–70 K temperature range.

12.2.3 Time constant (relaxation) method

In the time constant (relaxation) method, the waveform of P is a negative step which produces a relaxation of the sample temperature from $T_B + \Delta T$ to T_B. The measure of $P(T)$ may be critical when the power P is comparable with the spurious power or when the thermal conductance G is steeply variable with the temperature (i.e. $G \propto T^3$ in the case of contact conductances).

Another quite delicate step is the determination of the relaxation time constant τ from the relaxation curve $T(t)$. In this method, instead of the electrical heater, other heating sources, such as an optical source, can be used [13].

12.2.4 Dual slope method

In the dual slope method, referring to the Fig. 12.1, the support (Sp), with the thermometer (Th) and the sample (S), is heated by the heater (H) with a constant power P_i during the first phase of an acquisition. In this step, the sample temperature T_S rises from T_1 to T_2. Then, the power P is reduced to a value P_f to bring the sample back to a temperature $T_1' \approx T_1$. During these phases, the temperatures T_{rise} and T_{fall} are both recorded. The two following equations describe the heating and the cooling phase respectively:

$$
\begin{aligned}
C(T) \cdot \dot{T}_{\text{rise}} &= P_i - P_G(T) + P_{\text{sp}}(T) \\
C(T) \cdot \dot{T}_{\text{fall}} &= P_f - P_G(T) + P_{\text{sp}}(T)
\end{aligned}
\tag{12.2}
$$

where $\dot{T}_{\text{rise/fall}}$ are the time derivatives of T in the rising and falling phase, $P_G(T)$ is the power flowing to the bath through the thermal conductance G and $P_{\text{sp}}(T)$ is the spurious power impinging onto the addendum and the sample. If both $P_G(T)$ and $P_{\text{sp}}(T)$ depend only on temperature T and do not vary with time, the equation system (12.2) can be solved, giving:

$$C(T) = \frac{P_i - P_f}{\dot{T}_{rise} - \dot{T}_{fall}} \tag{12.3}$$

Hence, the heat capacity can be obtained with a fit from the derivatives of the temperatures. In the dual slope method, the analysis of the experimental data is quite complex, but there are two advantages: the range of temperature can be wide and the explicit knowledge of $G(T)$ is not necessary.

12.2.5 Thermal bath modulation

In the thermal bath modulation, a thermocouple is used as a weak link to the thermal bath, and the temperature of the bath is modulated sinusoidally in time. This configuration eliminates the need for a separate thermometer and heater on the sample, while retaining the ability to make measurements with minimal addenda.

12.2.6 Measurement constrains

In the methods reported above, the temperature change ΔT used to measure the heat capacity $C(T)$ was supposed to be so small that the time constant $\tau = R \cdot C$ could considered constant in the ΔT interval. Let us consider, for example, the thermal discharge of a system with heat capacity $C(T) \propto T$ and thermal conductance to the bath $G(T) \propto T^3$ (e.g. a metal sample and a contact resistance to the bath at T_B). A $\Delta T/T_B = 10\%$ gives a $\Delta\tau/\tau = 20\%$ over the interval ΔT, that is a 'time constant' definitely not constant.

When possible [20], one should try to compensate the temperature dependence $C(T)$ with the dependence of $G(T)$.

We shall report hereafter three examples of measurement of heat capacity: the first (of a crystal) with a negligible addendum; the second (of a polymer) with a heavy addendum; the third is the measurement of the carrier specific heat of a heavily doped semiconductor.

12.3 Example of 'classical' set up for the measurement of heat capacities

In a 'classical' low-temperature measurement of heat capacity, the sample is placed (usually glued) onto a low heat capacity support which is thermally linked to the thermal bath by a thermal conductance (see Fig. 12.1).

Figure 12.4 shows an example of experimental set up for a 'classical' measurement of heat capacity: the sample is glued onto a thin Si support slab. The thermometer is a doped silicon chip and the heater is made by a (~60 nm thick) gold deposition pattern. Electrical wiring to the connect terminals are of superconductor (NbTi). The thermal conductance to the thermal bath (i.e. mixing chamber of a dilution refrigerator) is made with thin nylon thread. The Si slab, the thermometer and the heater represent the 'addendum' whose heat

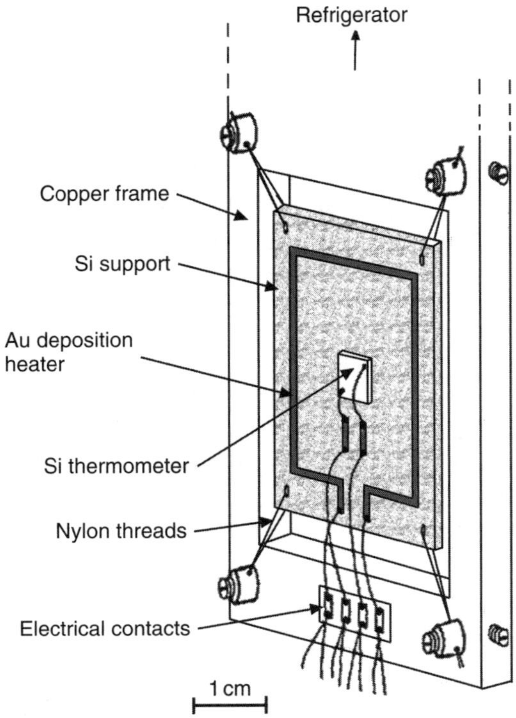

Fig. 12.4. Example of experimental set up for a 'classical' measurement of heat capacity.

capacity $C_A(T)$ must be measured in a separate run. When a sample of heat capacity $C(T)$ is added, a second run of measurements gives $C_A(T) + C(T)$. It is obvious that, if possible, the condition $C_A << C$ should be fulfilled. When measuring heat capacities of crystalline insulating materials at very low temperatures, due to the cubic dependence on temperature of the $C(T)$, the addendum heat capacity may become larger than that of the sample, thus decreasing the accuracy of the measurement. In these cases, one looks for a 'non-classical' set up such as that we describe in the next section [20].

12.4 Heat capacity of a TeO_2 single crystal between 0.06 and 0.28 K

For this measurement, the sample was a TeO_2 optical quality crystal of $2 \times 2 \times 3\,cm^3$, corresponding to a mass of 75.493 g. It was sustained by four pure tin cylinders, which kept the crystal blocked inside a copper frame, as shown in Fig. 12.5. The length of the cylinders was chosen to compensate the thermal contraction of the crystal, down to the lowest reached temperatures.

The frame was in good thermal contact with the mixing chamber of a dilution refrigerator. A RuO_2 thermometer measured the temperature T_B of the frame. A thermal shield at the same temperature T_B of the mixing chamber was used.

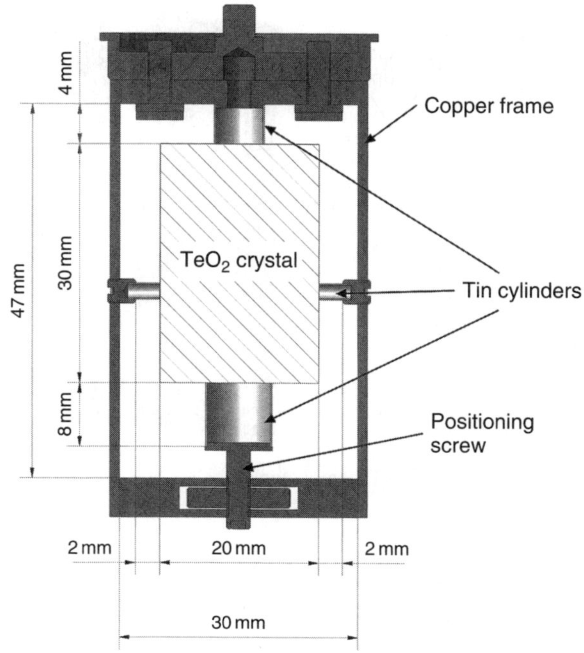

Fig. 12.5. Section view of the mounting of the TeO_2 crystal inside the copper frame. The whole experiment was surrounded by a copper thermal shield, not shown in figure.

A gold film (thickness 0.1 μm, and residual resistance $R \sim 9\,\Omega$) was deposited onto the crystal in order to obtain a heater of negligible heat capacity. It was carefully checked that the film resistance was constant along the temperature range of the specific heat measurements.

A calibrated NTD 31 Ge sensor of $2.95 \times 1.15 \times 0.3\,\text{mm}^3$ (see Section 15.2.2), glued on one face of the crystal with a minimum amount of GE-varnish (to minimize the addendum), was used as a thermometer. The electrical contacts were made by ball-bonding with Au wires. This thermistor was calibrated in the range 0.05–0.3 K. Bare NbTi wires, 25 μm diameter, realized the electrical connections for the thermometer and the heater on the TeO_2 crystal. The wires (two for each resistance) were stretched along the opposite sides of the copper frame (Fig. 12.6). Aluminium capillaries (0.2 mm outer diameter, 0.1 mm inner diameter and length 2 mm) clamped together the NbTi wires and the gold wires (25 μm diameter), connecting both the NTD sensor and the gold deposition.

In Table 12.1, the contributions to the heat capacity C_{sp} of the addendum are shown; specific heat data references are reported in ref. [20]. A factor 1/3 was attributed to the heat capacity contribution of the elements linking the crystal to the frame [15]. Note that the electron heat capacity of the NTD Ge 31 sensor was derived from the electron

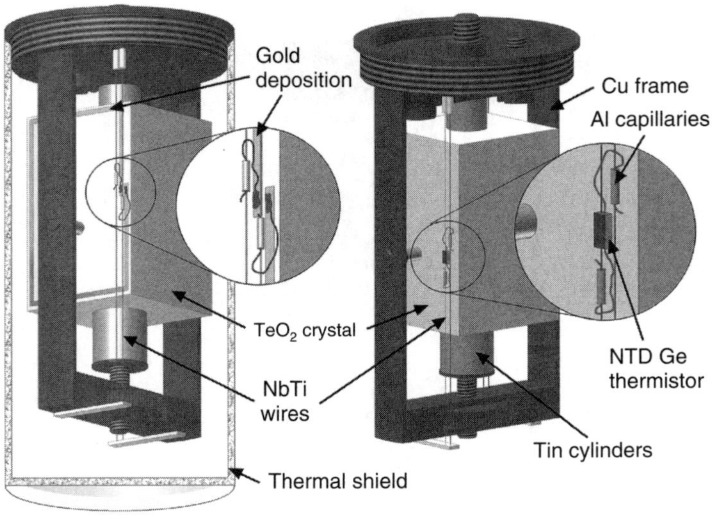

Fig. 12.6. View of the experimental assembly for specific heat measurement of the TeO_2 crystal. The electrical connections of the gold film heater (left) and of the NTD Ge thermistor (right) are also shown.

Table 12.1
Heat capacity contributions C_{sp} of the addendum

Material	Volume [mm^3]	$C = \alpha \cdot T + \beta \cdot T^3$ [μJ/Km^3]		$C_{sp}(T = 0.1\,K)$ [μJ/K]
		α	β	
Sn	57	—	1.478×10^{-2}	2.81×10^{-4} (1/3 contribution)
NTD Ge 31	1.61	6.02×10^{-4}	3.0×10^{-3}	1.02×10^{-4}
Ge-Varnish	0.381	7.031×10^{-3}	2.104×10^{-2}	2.76×10^{-4}
Al	0.263	—	2.83×10^{-1}	7.44×10^{-5}
In	0.215	—	7.77×10^{-2}	1.67×10^{-5}
NbTi	9.44×10^{-2}	—	7.03×10^{-2}	2.21×10^{-6} (1/3 contribution)
Au	1.93×10^{-2}	6.8×10^{-2}	4.5×10^{-2}	1.32×10^{-4}
Pd	0.102×10^{-2}	1.1	1.1×10^{-2}	1.12×10^{-4}

In the last column, the heat capacities at 0.1 K are reported.

heat capacity of the NTD Ge 12 thermistor (close to Ge 31 as doping level [21]), since differences in electron heat capacity for similar sensors are negligible [22]. C_{sp} was between 10^{-12} and 10^{-10} J/K: the main contributions were those due to tin, glue and gold.

12.4.1 Thermal conductance of the sample to the thermal bath

The thermal conductance between the TeO_2 crystal and the thermal bath was measured by a standard integral method (see Section 11.2), supplying a know power P to the gold-deposited heater. When thermal equilibrium was reached, the temperature T_c of the crystal was recorded. By varying the power supplied to the crystal and keeping the thermal bath temperature T_B constant, we obtained a set of data $P(T_c)$, which were differentiated with respect to T in order to obtain the thermal conductance G. The fit of the values obtained is:

$$G(T) = (1.23 \pm 0.3) \cdot T^{(2.54 \pm 0.01)} \, [\mu W/K] \tag{12.4}$$

It can be observed that the bulk thermal conductance of the tin cylinders [23] and of the NbTi wires [24] are respectively about two orders of magnitude higher and four order of magnitudes lower than the measured $G(T)$. Therefore, the main contribution to G is the thermal conductance of the contacts Cu/Sn and Sn/TeO_2. The exponent between 2 and 3 has been already reported for measurements of contact thermal resistances between solids at very low temperatures [25].

12.4.2 Measurement of the heat capacity

The heat capacity of the crystal was measured in the 0.06–0.28 K temperature range using a relaxation method (see Section 12.2.3). A small power supplied by the heater rose the temperature T_c of the crystal above T_B by a few millikelvin. When the thermal equilibrium was reached, the heating power was switched off, and the exponential decay of the crystal temperature was recorded by means of an LR700 bridge (see Section 10.8), at a rate of 5 sample/s.

An example of $T_c(t)$ is reported in Fig. 12.7. A single relaxation time constant τ was always observed.

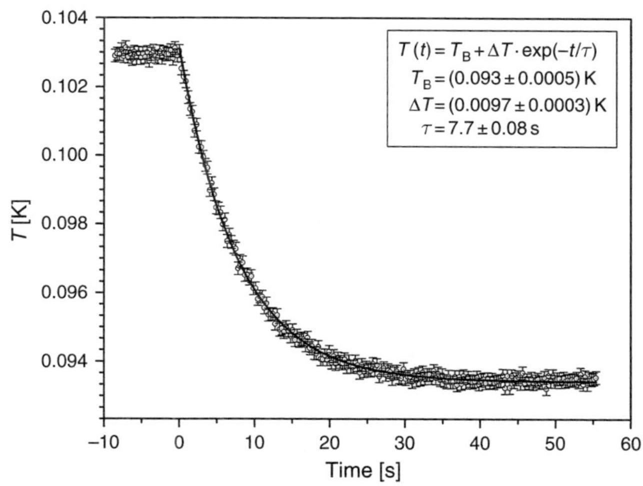

Fig. 12.7. Example of the exponential decay of the crystal temperature after removal of the heating power.

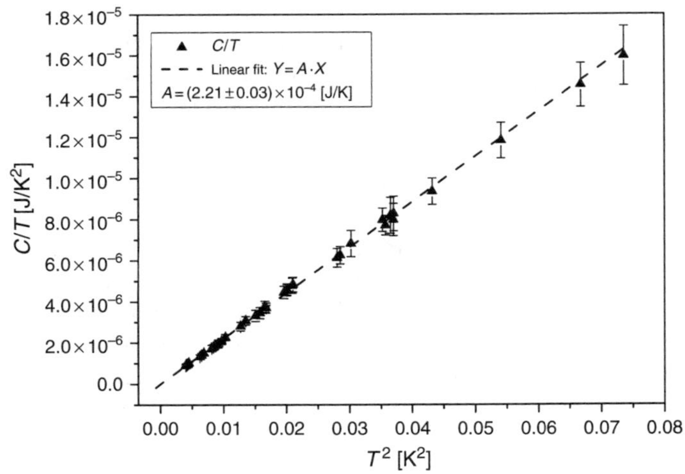

Fig. 12.8. Heat capacity data, showing the linearity of C/T versus T^2.

We repeated the same procedure at different values of T_B, between 0.06 K and 0.28 K, in order to obtain a set of thermal discharges.

The value of the heat capacity C was obtained as $\tau \cdot G$. These values were corrected subtracting the contribution C_{sp} of the addendum. Heat capacity data are shown in Fig. 12.8, in which the ratio C/T is reported as a function of T^2.

Note that the contribution of the addendum to the total heat capacity is between 1.85 and 0.3% over the temperature range.

Since C had a cubic dependence on T and G was mainly due to the contact resistance, τ was almost constant over the measurement temperature range and close to 10 s.

12.4.3 Results

As we can see from Fig. 12.9, the measured specific heat is consistent with White's data [26] above 2 K.

From the slope of the linear fit of C/T versus T^2, it is straightforward to extract the limiting value of the Debye temperature: $\theta_D = (232 \pm 7)$ K.

All values reported in the literature for θ_D of TeO_2 are decreasing with decreasing T, and the lowest value reported is the White's value 265 ± 10 K, extrapolated from heat capacity data at about 1 K. The value $\theta_D = (232 \pm 7)$ K is the lowest ever reported and is in excellent agreement with the value calculated from the extrapolation at $T = 0$ K of Ohmachi and Uchida's [27] and Schweppe's [28] ultrasonic measurements of elastic stiffness constants c_{ij}. In fact, using the Betts, Bhatia and Horton's approximation [29] to evaluate θ_D for a tetragonal crystal, including all tetragonal harmonics up to those of degree six, we obtain: $\theta_D = (232 \pm 8)$ K. The error on the elastic estimation of θ_D is mainly due to the extrapolation of the elastic constant values at low temperatures. Measurements

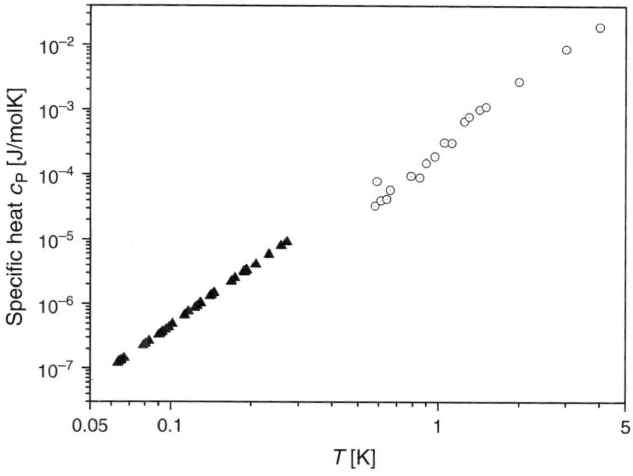

Fig. 12.9. Comparison of White's measurements (open circles [28]) with present data (full triangles).

of θ_D for TiO_2, a crystal with the same structure as TeO_2, reported consistent values for the Debye temperature resulting from calorimetric and elastic measurements [30], excluding the presence of excitations other than acoustic lattice vibrations.

12.5 Measurement of the specific heat of Torlon between 0.15 and 4.2 K

12.5.1 Experimental technique

In this experiment (see also ref. [31]), the relaxation method (see Section 12.2.3), was used for the measurements below 1 K; for higher temperatures, the dual slope technique (see Section 12.2.4) was adopted.

The sample was in the shape of a slab of $10.03 \times 3.38 \times 0.43\,cm^3$ and a mass $m = 19.88\,g$. The Torlon slab was enveloped in a 7 μm Cu foil glued on the whole sample area by means of a thin layer of diluted GE 7031 Varnish in order to get an isothermal sample. The internal thermalization time τ_i of the enveloped sample was evaluated to be [32]:

$$\tau_i = \frac{1}{\pi^2}\frac{c}{k}l^2 \tag{12.5}$$

where c and k are the specific heat for unit volume and thermal conductivity respectively and l is the slab thickness (0.43 cm).

As we shall see (Fig. 12.13), the internal thermalization times are $\tau_i = 0.02\,s$ at $T = 0.1\,K$ and $\tau_i = 0.03\,s$ at $T = 4.2\,K$, much shorter than the external time constants used in the measurements (see Section 12.5.2).

A NiCr heater (H) and a RuO_2 thermometer (T_h) were glued onto the Cu envelope of the sample. The sample was firmly suspended inside a copper frame by four screws, 3 mm

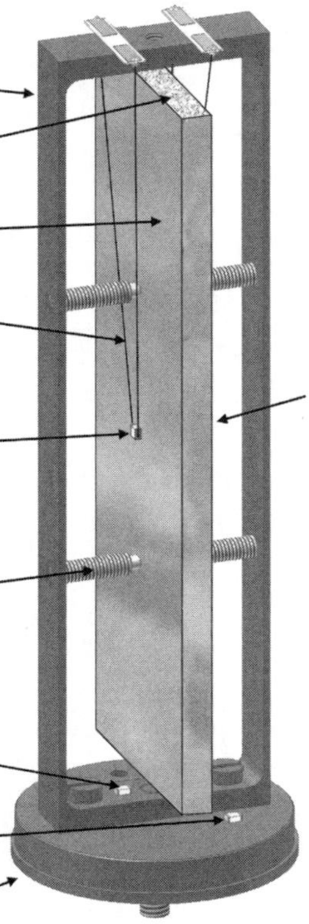

Fig. 12.10. Scheme of the experimental set up.

in diameter (see Fig. 12.10). The screws were made of copper with Torlon tips, fastened at room temperature just to cause a small bending of the frame in order to compensate the slight difference in thermal contraction [33,34] between copper and Torlon.

A Cu wire, 20 μm in diameter, was added for the measurements in the 0.7–4.2 K range in order to get a sample suitable to bath thermal conductance.

The electrical connections to the heater and thermometer were made with NbTi wires. These wires (Ø25 μm) were connected at the ends by tiny crimped Cu tubes (2 mg). At the copper frame upper ends of the NbTi wires a four lead connection was adopted. The frame, surrounded by a copper shield, was in thermal contact with the mixing chamber of a dilution refrigerator.

An AVS 47 AC resistance bridge for the thermometer and a four-wire *I–V* source (Keitley 236) for the heater were used. The contributions to the addendum are reported in Table 12.2.

The addendum contributions are evaluated from data in the literature [35,36]. The heat capacity data for Torlon were corrected only for the Cu contribution.

12.5.2 Results

Two runs of measurement have been carried out.

In the first run, the heat capacity of the sample was measured in the 150–900 mK temperature range, with the relaxation method and the thermal conductance G made by the Torlon tips.

To measure $G(T)$, an electrical power P_H was supplied to the heater (H), keeping the frame temperature constant. $G(T)$ was obtained as the derivative of the $P_H(T)$ curve. The heat capacity of the sample was evaluated simply as $C = \tau \times G$, where τ was obtained by the fit of the thermal discharges around a set of fixed temperatures. A single time constant was always found. An example of thermal discharge at $T \approx 160\,\text{mK}$ is shown in Fig. 12.11.

In the second run, the heat capacity of the sample was measured in the 0.7–4.2 K temperature range, with the dual slope method and the conductance G made by a Cu wire. Data of heat capacity were obtained with a more complex method described in ref. [37].

Table 12.2
Main contributions to the addendum at different temperatures

Temperature [K]	C_{Tor} [J/K]	C_{NiCr} [J/K]	C_{RuO_2} [J/K]	C_{glue} [J/K]	C_{add}/C_{Tor} %	C_{Cu} [J/K]
0.2	1.76×10^{-5}	3.0×10^{-8}	4.5×10^{-8}	3.3×10^{-8}	0.61	1.0×10^{-6}
0.5	9.83×10^{-5}	5.0×10^{-8}	8.0×10^{-8}	1.3×10^{-7}	0.26	2.5×10^{-6}
1.0	5.20×10^{-4}	7.0×10^{-8}	1.0×10^{-7}	5.9×10^{-7}	0.15	5.0×10^{-6}
4.0	2.5×10^{-2}	1.5×10^{-7}	2.2×10^{-7}	2.9×10^{-5}	0.12	2.0×10^{-5}

The heat capacity of the copper envelope and capillaries was subtracted from the heat capacity of the Torlon sample, whereas the other heat capacity contributions were neglected.

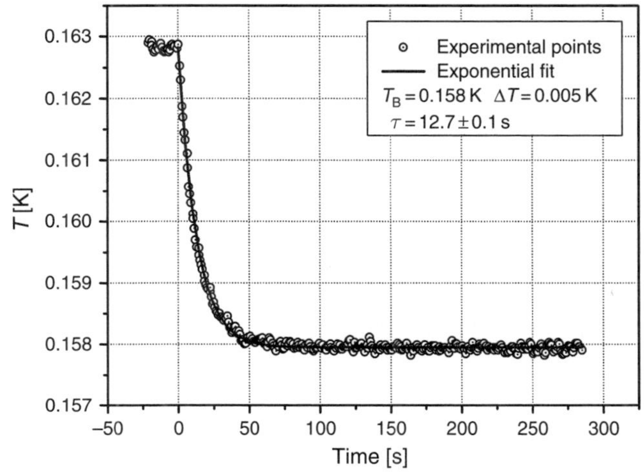

Fig. 12.11. Example of thermal discharge at $T \approx 160\,\text{mK}$.

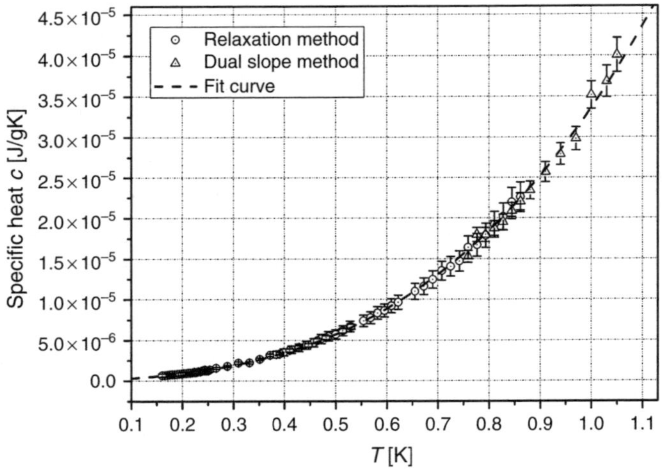

Fig. 12.12. Specific heat of Torlon in the range 0.15–1 K and fitting curve of eq. (12.6) and eq. (12.7).

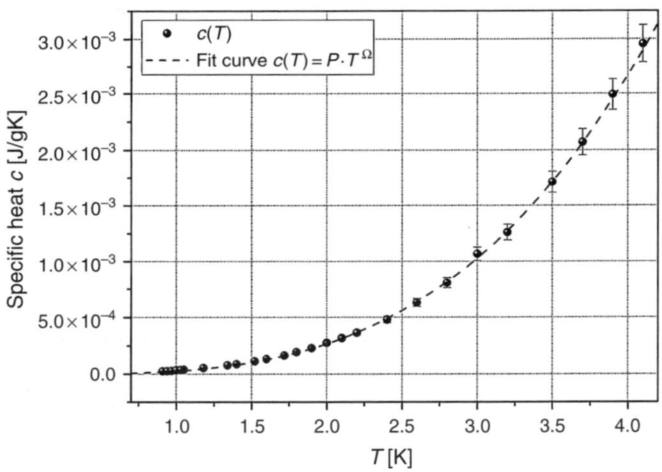

Fig. 12.13. Specific heat of Torlon in the range 0.9–1 K and fitting curve: $c(T) = P \cdot T^{\Omega}$ (see eq. (12.8) and eq. (12.9)).

Figure 12.12 shows the specific heat of Torlon in the low-temperature range (0.15–1 K), while Fig. 12.13 in the range (0.9–4.2 K).

12.5.3 Discussion

The specific heat of Torlon in the 0.15–1 K can be represented by:

$$c(T) = P_1 \cdot T^{1+\delta} + P_2 \cdot T^3 \tag{12.6}$$

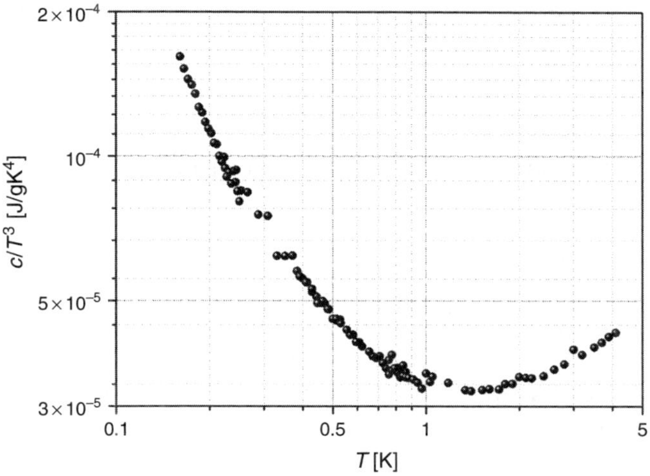

Fig. 12.14. The specific heat c of Torlon divided by T^3 vs T.

with:

$$\begin{cases} P_1 = (5.41 \pm 0.08) \times 10^{-6}\ [\mathrm{J/K^{(2+\delta)}g^1}] \\ P_2 = (2.82 \pm 0.03) \times 10^{-5}\ [\mathrm{J/K^4 g^1}] \\ \delta = 0.28 \pm 0.01 \end{cases} \tag{12.7}$$

Equation (12.6) is in the shape predicted by the tunnelling theory for the amorphous materials [38,39] and δ of eq. (12.7) is within the range of values obtained for other disordered solids [40].

Above 1 K, the specific heat of Torlon cannot be represented by eq. (12.6).

In Fig. 12.14, we report c/T^3 as a function of T in a logarithmic scale in the full temperature range. Note the minimum at $\sim$1.5 K.

In the case of a perfect crystalline solid, for temperature well below the Debye temperature θ_D, a c/T^3 versus T graph would give a constant value: $c/T^3 \propto 1/\theta_D$. However, most crystals show deviations from the Debye's law, in particular c/T^3 versus T presents a maximum. This behavior is present also in amorphous solids where the maximum is more evident and appears at temperatures higher than in crystals [40].

The maximum of c/T^3 is probably due to localized vibrational modes such as excitations in the molecular structure [40]. We can see from Fig. 12.14 that the maximum has not been reached in these measurements which are limited to 4.2 K. For the similarity with polypropylene, the maximum in c/T^3 for Torlon can be expected at about 10 K.

In the case of these measurements, a fit of data in the higher temperature region gives (see Fig. 12.13):

$$c(T) = P_1 \cdot T^{\Omega} \quad (1\,\mathrm{K} \le T \le 4.2\,\mathrm{K}) \tag{12.8}$$

with:

$$\begin{cases} P = (2.68 \pm 0.07) \times 10^{-5} \quad [\mathrm{J/K^{(\Omega+1)}g}] \\ \Omega = 3.32 \pm 0.02 \end{cases} \tag{12.9}$$

A dependence of specific heat $c \propto T^{\Omega}(3 < \Omega < 3.5)$ in the 1–10 K temperature range has been observed in most amorphous solids. For example $\Omega = 3.45$ for some epoxies [40].

12.6 Measurement of heat capacity of NTD Ge thermistors

12.6.1 Introduction

NTD (neutron transmutation doped) germanium chips are used as resistive thermometers in low-temperature detectors/calorimeters.

In the framework of CUORICINO [41] and CUORE [42] experiments (see Section 16.5), Ge crystal wafers of natural isotopic composition have been doped by neutron irradiation, and the heavy doping led to materials close to the metal insulator transition. Several series of NTD wafers with different doping have been produced [43]. After an implantation and metallization process on both sides of the wafers, thermistors of different sizes can be obtained by cutting the wafers and providing electrical contacts.

When a model for a CUORICINO detector (see Section 15.3.2) was formulated and the pulses simulated by the model were compared with those detected by the front-end electronics, it was evident that a large difference of about a factor 3 in the pulse rise time existed. This discrepancy was mainly attributed to the uncertainty in the values of carrier–phonon decoupling parameter. For the thermistor heat capacity, a linear dependence on temperature was assumed down to the lowest temperatures. As we shall see, this assumption was wrong.

Here we will report measurements on the heat capacity of two NTD Ge 34B wafers, one non-metallized (only doped and annealed) and the other metallized (by B^+ implantation and Au deposition). The comparison of data obtained from the non-metallized NTD Ge wafer and from the wafer with electrical contacts revealed an excess heat capacity, which can be attributed to the implantation process with B ions [44].

12.6.2 NTD process and realization of thermistors

NTD wafers were produced by irradiating natural ultra pure Ge crystals by means of a flux of thermal neutrons (see Section 15.2.2). To realize the electrical contacts, both sides of the wafers (disks, ~3 cm in diameter, 3 mm thick) were doped by implantation with B ions to a depth of ~200 nm. The implanted layers are doped to such a high concentration that the semiconductor becomes metallic. Then a layer of Pd (about 20 nm) and Au (about 400 nm) was sputtered onto the both sides of the wafers. Finally, the wafers were annealed at 200°C for 1 h. The wafers are cut to produce thermistors of length 3 mm between the metallized ends ($3 \times 3 \times 1\ \mathrm{mm}^3$ typical size); the electrical contacts are made by ball bonding with Au wires.

The process discussed in this section was applied both to the metallized wafer to be measured and to the preparation of the thermistor used in the discussed experiments.

12.6.3 Experimental technique

We measured the heat capacity of two NTD Ge wafer samples, which were 3 mm thick and had a diameter of about 3 cm, using the relaxation method (Section 12.2.3).

In the realization of the experiment, we tried to obtain an 'addendum-free' configuration, such as that adopted in the measurement described in Section 12.4. The experimental set up is shown in Fig. 12.15.

The Ge wafer sample were glued with small spots of GE-Varnish onto a copper holder in good thermal contact with the mixing chamber of a dilution refrigerator. Three Kapton foils ($2 \times 2 \times 0.01\,\text{mm}^3$) electrically isolated the Ge wafer samples from the copper holder and realized the thermal conductance G(T) between the samples and the heat sink.

For all the runs, a calibrated NTD Ge 34B thermistor ($3 \times 3 \times 1\,\text{mm}^3$) and a Si heater have been used. Electrical connections were made by means of superconducting NbTi wires 25 μm in diameter. The connection between the gold wires of both thermistor and heater and the NbTi leads was done by crimping the wires in a tiny Al tube (0.1 mg). At the ends of the NbTi wires, a four lead connection was adopted. An AVS47 AC resistance bridge has been used for the thermometry and a four-wire *I–V* source meter (Keithley 236) for the Si heater was used.

The addendum (represented by heater, glue spots and Al tubes) gave a negligible contribution to the total heat capacity (see Table 12.3). The thermistor heat capacity was instead considered as part of the sample. The whole experiment was surrounded

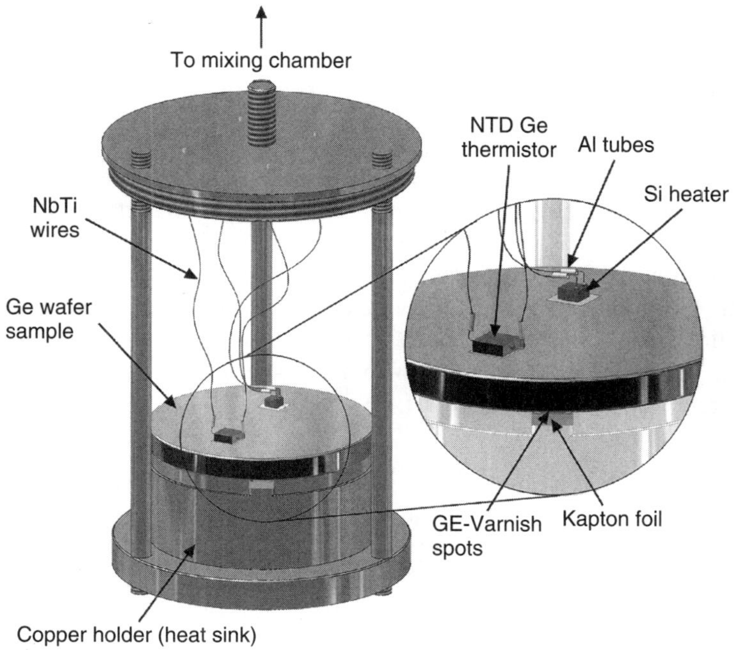

Fig. 12.15. Scheme of experimental set up.

Table 12.3
Estimated heat capacity contributions

Material	Volume [mm^3]	C (50 mK) [J/K]	C (40 mK) [J/K]	C (30 mK) [J/K]
NTD Ge (electrons)	1950	10^{-7}	8×10^{-8}	6×10^{-8}
NTD Ge (phonons)	1950	6.8×10^{-10}	3.6×10^{-10}	1.5×10^{-10}
GE-Varnish	0.52	1.8×10^{-10}	1.45×10^{-10}	1.1×10^{-10}
Au metallization	0.5	1.55×10^{-9}	1.25×10^{-9}	9×10^{-10}
Al tubes	0.157	2.65×10^{-13}	1.36×10^{-13}	6×10^{-14}
NbTi wires	0.12	10^{-12}	5.4×10^{-13}	2.3×10^{-13}

Specific heat data references are in [46].

by a copper shield at the mixing chamber temperature, measured by a calibrated RuO_2 thermometer.

Three measurements of heat capacity have been carried out: the first one on a non-metallized NTD Ge 34B wafer of 12.043 g over a temperature range of $0.024 < T < 0.040$ K, the second one on the same wafer over a temperature range of $0.040 < T < 0.080$ K and the third on a metallized wafer of NTD Ge 34B of 10.39 g over a temperature range of $0.024 < T < 0.040$ K.

For all three runs, the thermal conductance between the Ge wafers and the heat sink was measured by a standard integral method (see Section 11.2).

The fits of the values obtained for the three runs were:

$$\begin{cases} G_1(T) = 1.22\times10^{-4}\cdot T^{2.52}\ [\mathrm{W/K}] \\ G_2(T) = 3.05\times10^{-5}\cdot T^{2.45}\ [\mathrm{W/K}] \\ G_3(T) = 1.84\times10^{-4}\cdot T^{2.68}\ [\mathrm{W/K}] \end{cases} \tag{12.10}$$

It can be observed that these thermal conductances $G(T)$ are typical of phonon conduction between two solids at very low temperature, as already reported [45]. The value of the heat capacity was calculated from equation $C = \tau\cdot G$, where the thermal time constant τ is obtained from the fit to the exponential relaxation of the wafer temperature.

Using the known thermal conductivity data [46] of the wafers, their internal thermal relaxation time was estimated to be less than 1 ms, i.e. much shorter than C/G. Such estimate was confirmed by the fact that, within the experimental errors, a single discharge time constant τ was always observed.

12.6.4 Results

Since in our temperature range, the Debye temperature of Ge is ~370 K [47], the phonon contribution to the heat capacity can be neglected. Hence, the heat capacity of our samples is expected to follow the equation:

$$C = \Gamma\cdot T + C_{\mathrm{M}} \tag{12.11}$$

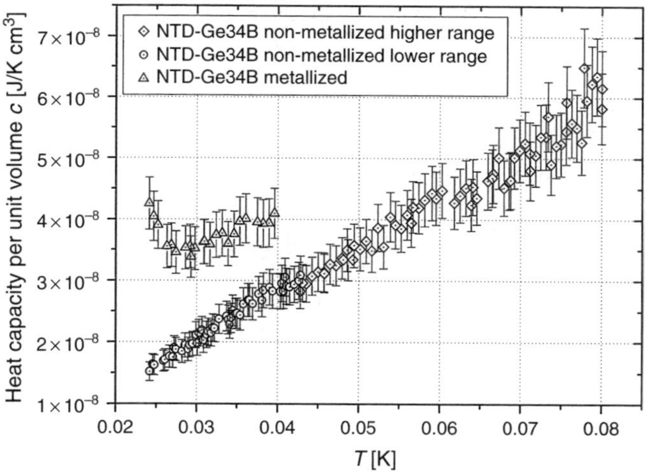

Fig. 12.16. Heat capacity per unit volume of the two neutron transmutation doped (NTD) Ge wafers.

where Γ is the contribution of itinerant charge carriers and C_M is the magnetic contribution due to localized magnetic moments [48]. Heat capacity data per unit volume for the two wafers are reported in Fig. 12.16.

The non-metallized wafer data clearly agree with eq. (12.11) with $C_M \approx 0$.

Instead, an anomalous trend and an excess heat capacity appear at low temperature in the case of the metallized wafer.

It is worth noting that we had several constraints in the measurements. First, the minimum temperature of our refrigerator was about 18 mK. Also, the sampling rate of the AC bridge was limited to ~2 sample/s. Finally, it was impossible to reduce the spurious power below $\sim 3 \times 10^{-11}$ W.

At fixed conductance G between the sample and the heat sink, and for fixed sample masses, the first and third constraints limit the lowest obtainable temperature, whereas second constraint fixes the highest temperature limit.

In fact, since the relaxation constant is $\tau = C/G$ and C is linear in the higher temperature range, while $G \propto T^{2.5}$, τ scales as $T^{-1.5}$. Hence, at higher temperatures, the number of useful data points, obtained in a single thermal discharge measurement, vanishes and the error in the determination of τ increases. For example, at the highest temperatures, τ was as low as $2 \sim 3$ seconds.

12.6.5 Discussion

Let us first consider data from the first and second measurements on the non-metallized sample (see Fig. 12.16). Data can be well represented by a linear fit which crosses the origin within the experimental errors. The heat capacity per unit volume of the wafer sample is expressed by the following formula in the measured temperature range of 24–80 mK:

$$c(T) = \gamma \cdot T = (7.52 \pm 0.08) \times 10^{-7} T \ \left[\text{J/K cm}^3\right] \tag{12.12}$$

where $\gamma = \Gamma$/sample volume. It is likely that the contribution C_M (see eq. (12.11)) of the localized magnetic moment cannot be evaluated [48,49], because of the limited accuracy of measurements. The value of γ is close to most of the Sommerfeld constant values reported in the literature for the NTD Ge of similar doping [50–53]. Note that the theoretical dependence of γ on the compensated dopant concentration is $p_c^{1/3}$ [54].

In Fig. 12.16, the metallized wafer, in contrast to the non-metallized one, shows a large excess contribution in the heat capacity. The excess heat capacity cannot be attributed to the term C_M of eq. (12.11), since the material of the two wafers is the same, and they were produced in the same run of neutron irradiation. Instead, this excess is to be attributed to the metallization process.

Let us examine the data of the third measurement on the metallized wafer. There are two contributions to the heat capacity, a linear contribution and a spurious one. The spurious contribution may be interpreted as the high temperature side of a Schottky anomaly. In this hypothesis, the heat capacity per unit volume of the metallized wafer may be represented as:

$$c(T) = \gamma \cdot T + \frac{A}{T^2} \tag{12.13}$$

where A/T^2 is the high temperature contribution of the Schottky anomaly.

Plotting data of metallized wafer as $c(T) \cdot T^2$ versus T^3, we obtain the graph of Fig. 12.17.

In Fig. 12.17, the slope of the dashed line γ is given by eq. (12.12), whereas the contribution A of eq. (12.13) is given by the intercept of the dashed line with the y-axis.

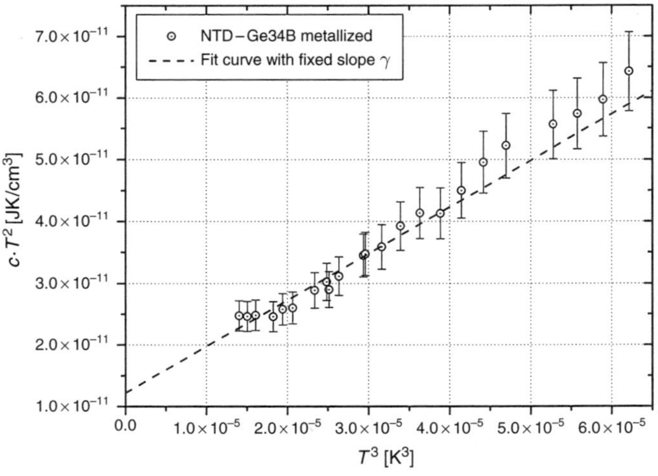

Fig. 12.17. Heat capacity per unit volume of the metallized wafer multiplied by T^2, plotted as a function of T^3.

This spurious contribution could be attributed to the presence of boron nuclei in the metallized wafer. Boron is present in nature with two stable isotopes (^{10}B, ^{11}B), one having nuclear spin 3/2 and abundance of 80.3%, and the other nuclear spin 3 and abundance of 19.7%. The nuclei have non-zero electric quadrupole moments.

As such, nuclear contributions to the heat capacity due to the interaction between germanium crystalline electric field gradients and the quadrupole moments of boron nuclei could account for the observed onset of the Schottky anomaly.

From the value of $A = 1.2 \pm 0.1 \times 10^{-11}$ [J/K cm^3], obtained from the fit, and the estimated number of boron nuclei of about 10^{17} in the sample, we found an order of magnitude for the average energy splitting $\Delta E \sim 10^{-25}$ J.

In the literature [55], typical energies involved in the nuclear quadrupole moments – crystalline electric field gradient interactions range up to $\Delta E \sim 2 \times 10^{-25}$ J. The measured ΔE seems to confirm the hypothesis that the excess specific heat of the metallized wafer is due to boron doping of the Ge lattice.

Low-temperature thermometers are obtained by cutting a metallized wafer of NTD Ge into chips of small size (typically few mm^3) and bonding the electrical contacts onto the metallized sides of the chip. These chips are seldom used as calibrated thermometers for temperatures below 30 mK, but find precious application as sensors for low-temperature bolometers [42,56]. When the chips are used as thermometers, i.e. in quasi-steady applications, their heat capacity does not represent a problem. In dynamic applications and at very low temperatures $T < 30$ mK, the chip heat capacity, together with the carrier-to-phonon thermal conductance $g_{c-\phi}$ (Section 15.2.1.3), determines the rise time of the pulses of the bolometer.

Up to now, in the formulation of a bolometer model, only the heat capacity of itinerant carriers was considered [57]. However, our measurements show that, even at 24 mK, the presence of a spurious heat capacity in the thermometer increases the expected value of the pulse rise time. We expect that the spurious contribution in Fig. 12.17 increases down to the temperature of the Schottky peak at $T = \Delta E/k_B T$ about 10 mK. Since $g_{c-\phi}$ decreases at low temperatures, the total effect on pulse rise time and pulse amplitude can be dramatic at lowest temperatures. In reality, the measured rise time of CUORICINO pulses is about three times longer than that obtained from a model which neglects the spurious heat capacity of the thermistor. For the same reason, also the pulse amplitude is by a factor two smaller than the expected value (see Section 15.3.2).

References

[1] G.R. Stewart: Rev. Sci. Instrum. **54**, 1 (1983)
[2] G.J. Seller, A.C. Anderson: Rev. Sci. Instrum. **45**, 1256 (1974)
[3] K. Albertet al.: Cryogenics **22**, 417 (1982)
[4] R.L. Filler, P. Lindenfeld: Rev. Sci. Instrum. **46**, 439 (1975)
[5] J.P. Harrison: Rev. Sci. Instrum. **39**, 145 (1968)
[6] F.J. Morin, J.P. Maita: Phys. Rev. **129**, 1115 (1963)
[7] K.M. Al-Shibani, O.A. Sacli: Phys. Stat. Sol. (b) **163**, 99 (1991)
[8] D.W. Osborne, H.E. Flotow, F. Schreiner: Rev. Sci. Instrum. **38**, 159 (1967)
[9] O. Hiroo, E. Toshiaki, W. Nobuhiko: J. Phys. Soc. Jpn. **59**, 1695 (1990)

[10] P.F. Sullivan, G. Seidel: Phys. Rev. **173**, 679 (1968)
[11] K. Tashiro et al.: J. Phys. Soc. Jpn. **59**, 4022 (1990)
[12] R.W. Willekers et al.: Cryogenics **31**, 168 (1991)
[13] P. Gutsmiedel, C. Probst, K. Andres: Cryogenics **31**, 54 (1991)
[14] R.E. Schwall, R.E. Howard, G.R. Stewart: Rev. Sci. Instrum. **46**, 1054 (1975)
[15] R. Bachmann et al.: Rev. Sci. Instrum. **43**, 205 (1972)
[16] R.J. Schutz: Rev. Sci. Instrum. **45**, 548 (1974)
[17] E.P. Roth, A.C. Anderson: J. Appl. Phys. **47**, 3644 (1976)
[18] S. Riegel, G. Weber: J. Phys. E Sci. Instrum. **19**, 790 (1986)
[19] J.E. Graebner: Rev. Sci. Instrum. **60**, 1123 (1989)
[20] M. Barucci et al.: J. Low Temp. Phys. **123**, 303 (2001)
[21] E.E. Haller: Infrared Phys. Technol. **35**, 127 (1994)
[22] E. Aubourg et al.: J. Low Temp. Phys.**93**, 289 (1993)
[23] S.J. Laredo: Proc. Roy. Soc. (London) **A 229**, 473 (1954)
[24] J.R. Olson: Cryogenics **33**, 729 (1993)
[25] J.P. Shepherd: Rev. Sci. Instrum. **56**, 273 (1984)
[26] K. White, S.J. Collocott, J.G. Collins: J. Phys. Condens. Matter **2**, 7715 (1990)
[27] Y. Ohmachi, N. Uchida: J. Appl. Phys. **41**, 2307 (1970)
[28] H. Schweppe: Ultrasonics **8**, 84 (1970)
[29] D.D. Betts, A.B. Bhatia, G.K. Horton: Phys. Rev. **104**, 43 (1956)
[30] A.Y. Wu, R.J. Sladek: Phys. Rev. B **2**, 5230 (1981)
[31] M. Barucci et al.: Cryogenics **46**, 767 (2006)
[32] R.C. Richardson, E.N. Smith eds.: *Experimental Techniques in Condensed Matter Physics at Low Temperatures*, Redwood City, CA: Addison-Wesley, Advanced Book Classics (1988)
[33] F. Pobell: *Matter and Methods at Low Temperatures*. 2nd ed., Springer-Verlag, New York (1995)
[34] G. Ventura et al.: Cryogenics **39**, 481 (1999)
[35] G. Ventura et al.: Cryogenics **38**, 453 (1998)
[36] R.B. Stephens: Phys. Rev. B **8**, 2896 (1973)
[37] S. Riegel, G. Weber: J. Phys. E: Sci. Instrum. **19**, 790 (1986)
[38] S. Hunklinger, K. Raychaudhuri: *Progress in Low Temperature Physics*, Vol. 9, ed. by D.F. Brewer, North-Holland Physics Publishing, Amsterdam (1986)
[39] W.A. Phillips: J. Low. Temp. Phys.**7**, 351 (1972)
[40] G. Hartwig: *Polymer Properties at Room and Cryogenic Temperatures*, Plenum Press, New York (1994)
[41] C. Brofferio et al.: Nucl. Phys. B **145**, 268 (2005)
[42] C. Arnaboldi et al.: Nucl. Inst. Meth. A **518**, 775 (2004)
[43] E.E. Haller: Inf. Phys. Tech. **35**, 127 (1994)
[44] E. Olivieri et al.: J. Low Temp. Phys. **143**, 153 (2006)
[45] R.E. Peterson, A.C. Anderson: J. Low Temp. Phys. **11**, 639 (1973)
[46] O.V. Lounasmaa: *Experimental Principles and Methods Below 1 K*, Academic, London (1974)
[47] P.H. Keesom, G. Seidel: Phys. Rev. **113**, 33 (1959)
[48] R.N. Bhatt, P.A. Lee: Phys. Rev. Lett. **48**, 344 (1982)
[49] S. Marnieros et al.: Physica B **259–261**, 862 (1999)
[50] N. Wang et al.: Phys. Rev. B **41**, 3761 (1990)
[51] E. Aubourg et al.: J. Low Temp. Phys. **93**, 289 (1993)
[52] A. Alessandrello et al.: J. Low Temp. Phys. **93**, 207 (1993)
[53] P. Stefanyi et al.: Physica B **161**, 194 (1994)
[54] B.I. Shklovskii, A.L. Efros: *Electronic Properties of Doped Semiconductors*, Springer, Berlin (1984)
[55] Kaye and Laby: tables of physical and chemical constants: http://www.kayelaby.npl.co.uk/chemistry/3_8/3_8_4.html
[56] S. Marnieros et al.: Nucl. Inst. Meth. A **520**, 101 (2004)
[57] M. Pedretti et al.: Physica B **329–333**, 1614 (2003)

13 Measurements of Thermal Expansion

Contents

13.1 Introduction

The study of the thermal expansion coefficient of solid materials over a wide temperature range is a great interest both from a theoretical point of view and for practical applications. The temperature dependence of the thermal expansion coefficients gives precious information on the intrinsic properties of materials (binding forces, lattice dynamics, band and crystal structure, degree of crystallinity, specific heat and phase transitions). On the other hand, the knowledge of the thermal expansion coefficients is indispensable in the project of mechanical structures, in particular in the field of cryogenic engineering, where apparatuses are made up of parts with very different thermal and mechanical properties.

Several methods of measurement of the thermal expansion have been developed as a function of the material, dimension and shape of the sample, temperature range and requested accuracy. The measurement of the linear expansion coefficient $\alpha = 1/L \cdot (\Delta L/\Delta T)$ of a sample is done by recording the length change ΔL (in a definite direction) due to a temperature variation ΔT.

The main limit in this type of measurements is the minimum measurable value of ΔL. In fact, α decreases with temperature and can reach, for many materials, values of the order of 10^{-8}/K at temperatures below 20 K. This means that if we want to measure the linear expansion coefficient with a 1% accuracy, for a temperature change of 1 K, we must be able to detect $\Delta L/L$ of the order of 10^{-10}. It is illusory for this goal, the use of samples with large L, since non-homogeneity of temperature along the sample can take place, which introduces uncontrolled effect on the measurement. Hence, the only way of measuring with great accuracy the linear expansion coefficient of materials for $0 < T < 300$ K is to adopt high-intrinsic sensitivity methods.

The measure of the thermal expansion coefficient below room temperature is particularly difficult for low-expansion materials (see Section 3.9). Remember also how newly produced composite materials show extremely low-expansion coefficient of both sign.

Several methods have been used to measure the linear expansion coefficient for $T < 300\,\mathrm{K}$ [1,2]. These different techniques are:

- Capacitance dilatometer [3–18].
- Interferometric dilatometer [19–30].
- Laser speckle photography [31,32].
- Electronic speckle pattern interferometry [33].
- Optical lever dilatometer [34,35].
- Optical dilatometer [36–38].
- Variable transformer technique [39–41].
- Inductive dilatometer [29,42].
- Strain gauge technique [33–43,44].
- X-ray method [45–48].
- Photometer method [49].
- Microwave cavity method [50,51].
- Comparator method [38].
- Compression dilatometer [52].

Among these techniques, the capacitance dilatometer method may be very sensitive. The change in the sample length produces a capacitance change between the two electrodes of a capacitor: one plate of the capacitor is kept in a fixed position while the other is fixed onto one end of the sample. At the maximum elongation of the sample, the two capacitor plates are practically in contact. When the sample contracts the capacitance varies as $1/\Delta L$. One of the main difficulties in this measurement is the realization of a dilatation-free support.

In the interferometric dilatometer, the change in length of the sample causes the movement of interference fringes. Knowing the laser wavelength and counting the moved fringes, it is possible to deduce the dilatation of the sample. Hereafter, we shall briefly describe a very simple interferometric dilatometer used for the measurement of the linear contraction coefficient of Torlon. For a more detailed description of this dilatometer, see ref. [53].

13.2 A simple interferometric dilatometer

This interferometric dilatometer consists of a rather simple and small Michelson interferometer, in which the two arms are parallel, and of a ^{4}He cryostat, in which the sample to be measured is hold. The sample is cooled to 4 K, and data are taken during the warm up of the cryostat. The optical path difference between the two arms depends on the sample length; hence a variation of the sample length determines an interference signal. The Michelson interferometer consists of a He-Ne stabilized laser ($\lambda = 0.6328\,\mu\mathrm{m}$), two cube corner prisms, a beam splitter, three mirrors and a silicon photodiode detector placed in the focal plane of a 25 mm focal length biconvex lens (see Fig. 13.1).

The laser beam is directed by a mirror towards the beam splitter and splitted into two optical paths: one is transmitted by the beam splitter and directed towards the sample by means of two mirrors; the other is reflected and directed towards the sample holder.

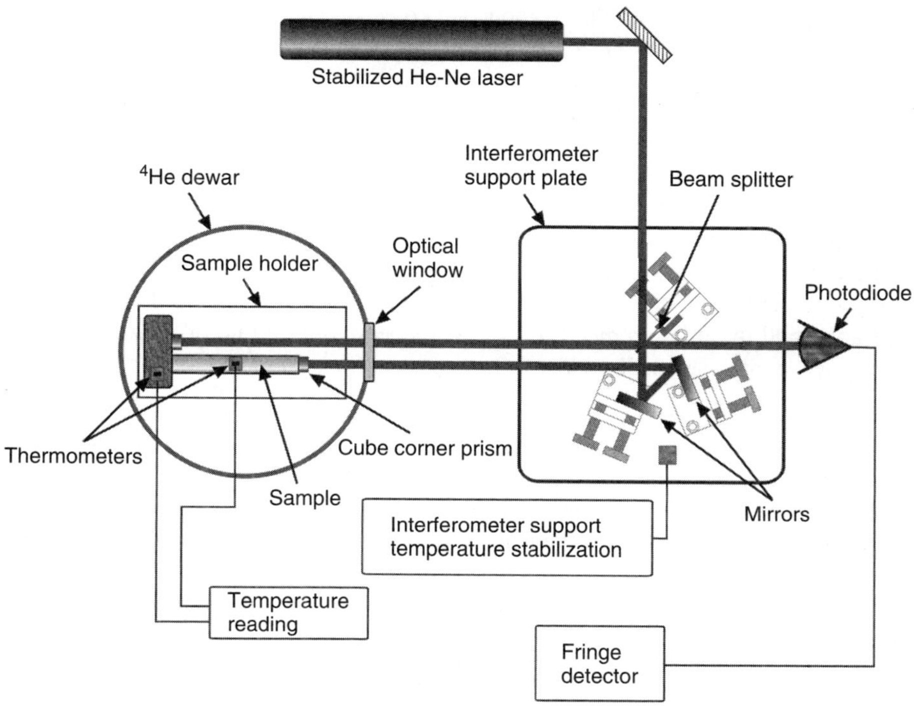

Fig. 13.1. Interferometric dilatometer set up.

The beams are backreflected by the cube corner prisms which are fixed, respectively, on the sample and on the sample holder. Since the cube corner prisms are able to make reflected beam exactly parallel to incident beam, this interferometer is tilt independent. The reflected beams get back to the beam splitter through the same path, but shifted by about 2 mm in the vertical direction. The beam splitter lets a part of the two beams go towards the photodiode sensor and lets the other part of beams reach the laser source (off axis, therefore giving no feedback effect).

The total optical path difference between the two arms of the interferometer, for a sample length of about 50 mm, is of the order of 10 mm or less, minimizing the systematic error due to laser frequency fluctuations. To reduce the thermal effects on the interferometer assembly, the interferometer support plate is stabilized to a temperature slightly higher than room temperature and insulated from air currents by a polystyrene foam shield. The temperature variation of the interferometer support is kept below 0.1 K.

The sample is put in a ^{4}He cryostat and enclosed by a shield thermally connected to the liquid helium reservoir. A second thermal shield connected to a liquid nitrogen reservoir encloses all the liquid helium system, allowing for a slow warming-up cycle in order to ensure thermal homogeneity of sample and holder. A window in the dewar enables the laser beam to enter the chamber and to reach the sample through small bores in both thermal shields. The sample is fixed onto a copper support that is in good thermal contact with the liquid helium reservoir, without any type of glue, to eliminate any spurious contribution to the thermal expansion. The distance between the optical axes of the two arms of the interferometer is made as small as possible (11 mm). Minimizing this distance is also useful to reduce systematic effects due to misalignment. Two calibrated carbon thermometers, one glued on the sample and the other on the sample holder, are used. Since a warm up from 4 K to room temperature takes place in about 8 h, the temperature variation is slow enough to consider the sample and sample holder isothermal.

The fringe interference signal is detected by a photodiode. The readings of photodiode signal are made by means of a simple current to voltage amplifier. A computer system, connected to instrumentation by a GPIB bus, records the interferometric fringe signal, the temperature and the interferometer assembly temperature at a rate of 1 Hz. The analysis of the fringe signal was made by the following steps:

- First the signal was smoothed through a Savitzky–Golay algorithm, whose parameters were optimized considering the mean fringe number for a fixed temperature variation.
- Another smoothing process was performed through a running average, on a window large enough to enclose about five fringes, to obtain the average zero level of the fringe signal.
- This latter level was then subtracted from the smoothed signal obtained in the previous step.
- The number of zero crossings in the obtained data file was then counted.

Since the distance between the zeroes, in terms of sample length variation, is $\lambda/4$, the total contraction of the sample is $\Delta L = N \cdot \lambda/4$, where N is the total number of zero crossing. The resolution of the system is basically $\lambda/4$, but it could be improved with an evaluation of fractions of a fringe.

13.3 Thermal expansion of Torlon between 4.2 and 295 K

The thermal expansion of Torlon between 4.2 and 295 K was measured by means of the dilatometer described in Section 13.2. The sample was a cylinder 8.2 mm in diameter and 3 cm long. The optical set up for the interferometric measurement of the thermal expansion is that shown on Fig. 13.1.

The sample was mounted on a sample holder in good thermal contact with the cold plate of a ^{4}He dewar (see Fig. 13.2). A copper thermal shield was placed around the sample in order to ensure uniformity in the temperature of the cold parts of the experiment.

Measurements were performed by cooling the sample down to 4.2 K and then recording the interference signal versus temperature during the sample warm up.

The uncertainty of the expansion data was evaluated to be less than 5%. The relative thermal expansion $\Delta L/L$ versus T is shown in Fig. 13.3. Curves for other materials are also reported for the sake of comparison. Torlon shows a linear thermal expansion lower than most polymers. As Stycast 2850FT, Torlon thermal contraction closely matches that of some common metals (i.e. aluminium [53] and brass [54]). Smoothed values for the thermal expansion of Torlon relative to 4.2 K are reported in Table 13.1. The data

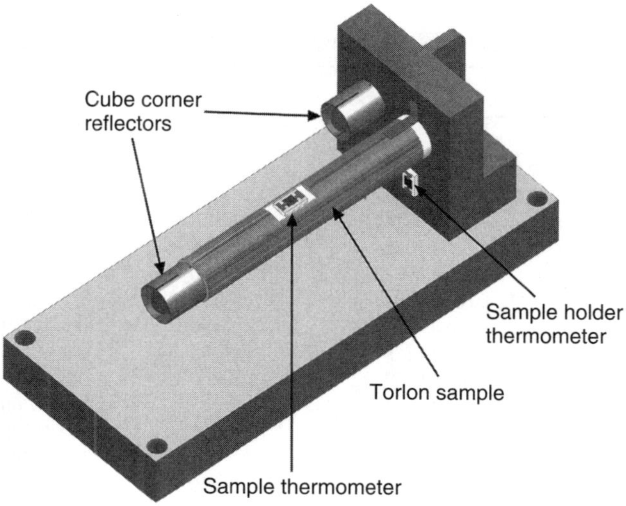

Fig. 13.2. Scheme of the sample holder inside the ^{4}He dewar.

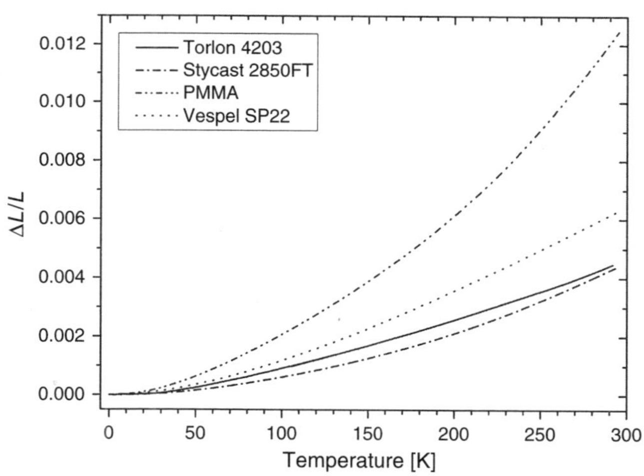

Fig. 13.3. Relative thermal expansion of Torlon compared with those of PMMA [55], Vespel SP22 (polyimide) [54] and Stycast 2850FT [56].

are consistent with the thermal expansivity value (3.06×10^{-5}/K in the 250 K–400 K temperature range) supplied by the Torlon manufacturer.

Data of thermal expansion do not always show a 'regular' positive slope increasing with temperature, as those shown in Fig. 13.3 [57]. Look, for example, in Fig. 13.4, at the data for a cold sintered SiC, a material specially suitable for the realization of high-quality mirrors for space astronomy [58].

Table 13.1
Smoothed values for the thermal expansion of Torlon relative to 4.2 K

T [K]	$\Delta L/L\%$	T [K]	$\Delta L/L\%$
4.2	0	155	0.177
10	0.000260	160	0.186
15	0.000573	165	0.195
20	0.00146	170	0.202
25	0.00284	175	0.212
30	0.00525	180	0.221
35	0.00913	185	0.229
40	0.0138	190	0.239
45	0.0189	195	0.248
50	0.0237	200	0.257
55	0.0295	205	0.266
60	0.0357	210	0.276
65	0.0421	215	0.285
70	0.0485	220	0.296
75	0.0547	225	0.305
80	0.0609	230	0.316
85	0.0682	235	0.325
90	0.0751	240	0.334
95	0.0825	245	0.344
100	0.0900	250	0.353
105	0.0971	255	0.363
110	0.104	260	0.373
115	0.113	265	0.384
120	0.120	270	0.394
125	0.127	275	0.406
130	0.135	280	0.417
135	0.144	285	0.429
140	0.152	290	0.442
145	0.160	295	0.454
150	0.169		

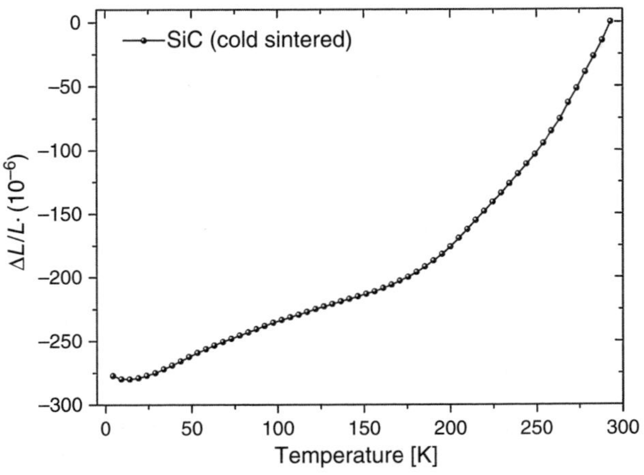

Fig. 13.4. Relative thermal expansion of cold sintered SiC [58].

References

[1] R.F. Cooper: The thermal expansion of solids. Phys. Educ., p. 284 (1976)
[2] S. Kanagaraj, S.Pattanayak: Cryogenics **43**, 339 (2003)
[3] D. Bijl, H. Pullan: Physica **21**, 285 (1955)
[4] G.K. White: Cryogenics **1**, 151 (1961)
[5] G.K. White, J.G. Collins: J. Low Temp. Phys. **7**, 43 (1972)
[6] D.E. Schafer et al.: Phys. Rev. B **12**, 5532 (1975)
[7] H.N. Subrahmanyam, S.V. Subramanyam: Pramana, J. Phys. **27,** 647 (1986)
[8] R. Villar et al.: Rev. Sci. Instrum. **51**, 1 (1980)
[9] S. Kanagaraj, S. Pattanayak: Thermal Conductivity **26**, 540 (2005)
[10] R.D. McCammon, R.N. Work: Rev. Sci. Instrum. **36**, 1169 (1965)
[11] D.C. Heberlein, E.D. Adams, T.A. Scott: J. Low Temp. Phys. **2**, 449 (1970)
[12] S. Kanagaraj, S. Pattanayak: Cryogenics **43**, 451 (2003)
[13] G. Brandli, R. Griessen: Cryogenics **13**, 299 (1973)
[14] C.R. Tilford, C.A. Swenson: Phys. Rev. B **5**, 719 (1972)
[15] F.R. Kroegner, C.A. Swenson: J. Appl. Phys. **48**, 853 (1977)
[16] M. Rotter et al.: Rev. Sci. Instrum. **69**, 2742 (1998)
[17] R. Pott, R. Schefzyk: J. Phys. E: Sci. Instrum. **16**, 444 (1983)
[18] S. Pattanayak, S. Kanagaraj: Thermal contraction of FRPs and its measurements at cryogenic temperature. Proceedings of Beijing International Cryogenic Conference, p. 185 (2000)
[19] B.W. James, B. Yates: Cryogenics **5**, 68 (1965)
[20] S.S. Tompkins, D.E. Bowles, W.R. Kennedy: Exp. Mech. **26**, 1 (1986)
[21] N. Waterhouse and B. Yates: Cryogenics **8**, 267 (1968)
[22] S.J. Feltham, B. Yates, R.J. Matin: J. Mater. Sci. **17**, 2309 (1982)
[23] O. Pirgon, G.H. Wostenholm, B. Yates: J. Phys. D Appl. Phys. **6**, 309 (1973)
[24] A.H. Thomas: J. Appl. Phys. **41**, 5096 (1970)
[25] E.G. Wolff, S.A. Eselun: Rev. Sci. Instrum. **50**, 502 (1979)
[26] E.G. Wolff, R.C. Savedra: Rev. Sci. Instrum. **56**, 1313 (1985)
[27] M. Okaji et al.: Cryogenics **35**, 887 (1995)
[28] M. Okaji et al.: Cryogenics **37**, 251 (1997)
[29] G. Baschek, G. Hartwig: Cryogenics **38**, 99 (1998)
[30] C.G. Tseng, Y.S. Jiang: Meas. Sci. Technol. **16**, 2114 (2005)
[31] S. Nakahara et al.: Adv. Cryog. Eng. **32**, 209 (1986)
[32] S. Nakahara et al.: Adv. Cryog. Eng. (Materials) **38**, 85 (1992)
[33] S. Nakahara et al.: Proc. 16th ICEC/ICMC, p. 665 (1997)
[34] J.M. Shapiro, D.R. Taylor, G.M. Graham: Can. J. Phys. **42**, 835 (1964)
[35] F.N.D.D. Pereika, C.H. Barnes, G.M. Graham: J. Appl. Phys. **41**, 5050 (1970)
[36] W.O. Hamilton, D.B. Greene, D.E. Davidson: Rev. Sci. Instrum. **39**, 645 (1968)
[37] V. Arp et al.: Cryogenics **2**, 230 (1962)
[38] M.D. Campbell: Adv. Cryog. Eng. **10**, 154 (1965)
[39] K.O. McLean, C.A. Swenson, C.R. Case: J. Low Temp. Phys. **7**, 77 (1972)
[40] R.H. Carr, C.A. Swenson: Cryogenics **4**, 76 (1964)
[41] D. Evans, J.T. Morgan: Cryogenics **31**, 220 (1991)
[42] G. Schwarz: Cryogenics **28**, 248 (1988)
[43] R. Grossinger, H. Muller: Rev. Sci. Instrum. **52**, 1528 (1981)
[44] P. Walsh: Proc. 16th ICEC/ICMC, p. 661 (1997)
[45] B.F. Figgins, G.O. Jones, D.P. Riley: Philos. Mag. **1**, 747 (1956)
[46] D.N. Batchelder, R.O. Simmons: J. Chem. Phys. **41**, 2324 (1964)
[47] R.O. Simmons, R.W. Balluffi: Phys. Rev. **108**, 278 (1957)
[48] D.N. Batchelder, R.O. Simmons: J. Appl. Phys. **36,** 2864 (1965)
[49] K. Andres: Cryogenics **1**, 93 (1961)
[50] J.R.G. Keyston, J.D. Macpherson, E.W. Guptill: Rev. Sci. Instrum. **30**, 246 (1959)
[51] V.M. Pudalov, M.S. Khaikin: Cryogenics **9**, 128 (1969)
[52] R.F. Robbins, D.H. Weitzel, R.N. Herring: Adv. Cryog. Eng. **7**, 343 (1962)

[53] G. Bianchini et al.: Meas. Sci. Technol. **17**, 689 (2006)
[54] F. Pobell: Matter and methods at low temperatures, p. 52, Springer, Berlin (1991)
[55] K.G. Lyon, G.L. Salinger, C.A. Swenson: Phys. Rev. B **19**, 4231 (1979)
[56] C.A. Swenson: Rev. Sci. Instrum. **68**, 1312 (1997)
[57] G. Ventura et al.: Cryogenics **39**, 481 (1999)
[58] E. Olivieri et al.: Proc. 8th Int. Conf. Adv. Technol. Particle Phys., p. 593, ed. by M. Barone et al., Como 2003, World Scientific (2003)

PART VI

14 Practical, Industrial and Space Applications of Cryogenics

Contents

14.1 Introduction

Cryogenics finds a variety of applications in physics, chemistry, biology, medicine, engineering and industry. Only few of them, such as the cooling of superconducting magnets or NMR equipments, are usually known.

We want to point out that many analytical apparatus originally designed for room temperature have been modified to allow operation at low temperature with much better performances. An example is the STM (scanning tunnelling microscope) [1,2] which allows to image the morphology of a surface. Spectacular demonstrations have been provided of what could be done with a 4 K STM [2,3]. Today several low-temperature STM of different design are in operation [4–12].

In this chapter, we shall report some of the applications of cryogenics:

- in industry;
- in medicine, biology and chemistry;
- in space technology;
- in cold electronics.

Applications to nuclear physics and astrophysics will be dealt in Chapter 15 and Chapter 16. For the applications in vacuum technology see Section 1.6.4.

14.2 Industrial and technical use of cryoliquids

Nitrogen, oxygen and argon are produced by fractional distillation of air. Liquid nitrogen (LN) is a quite cheap and safe source of cold and finds applications such as:

1. freezing of food (e.g. hamburgers);
2. pressurization of plastic bottles and aluminium cans containing drinks: a small quantity of LN is injected into the liquid just before sealing the bottle. The LN vaporizes

and produces a pressure slightly above the atmospheric pressure, which makes the container very strong and capable of standing piling up ($10\,mm^3$ of LN in a bottle with $10\,cm^3$ over the liquid produces an overpressure of about 0.5 atm);
3. fixing of pipelines by freezing the liquid on either side of the leak, without emptying the whole system;
4. ground freezing, to allow excavation and tunnelling operation in wet unstable soils;
5. deflashing of moulded polymer products: deflashing can be obtained by a tumbling process rather than by treating each piece individually;
6. heat treatment of metals, e.g. steel tools to improve the wear resistance, temper of musical instruments, etc.;
7. working of solid explosive and bomb disposal: freezing of explosives makes them temporarily harmless;
8. cooling of cold traps in vacuum systems;
9. precooling of ^{4}He dewars and, in astronautics, of fuel tanks before filling them with liquid oxygen or liquid hydrogen;
10. Cryo-cleaning, which is a blasting process that uses dry ice. Similar in size to rice corns, the granules are created by expanded liquid carbon dioxide. By impact, they clean and sublime directly into the gas phase;
11. a magnetic energy storage system to provide power to industrial electric loads subjected to short-term voltage disturbances.

Liquid hydrogen (and deuterium) is used in high-energy physics experiments as a target for the particles produced by accelerators and in bubble chambers.

14.3 Biological and medical applications

In a 1930 science fiction novel by Neil R. Jones [13] a living man is freezed and then, after a 'suspension of life', successfully resuscitates. In reality, low temperatures are used to enable biological materials to be frozen [14–25] and stored, particularly semen [26–30], thin tissues and blood [30], but the preservation of big specimens is very difficult, since the cells suffer damage during the cooling and thawing processes [20,29,30]. On the other hand, this damage is put to good use in the elimination of tumours by freezing [31]. Successful results have been obtained, and it is likely that cryosurgery will be more widely used in the future [32–34].

Thanks to the improvement in the construction of superconducting magnets, ESR (electron spin resonance) and NMR (nuclear magnetic resonance) have been successfully applied to chemistry and biology. MRI (magnetic resonance imaging) has come into routine clinical use. In MRI, an RF field is applied to stimulate the resonance of protons (hydrogen nuclei). The patient is placed into a uniform magnetic field of about 1.5 T, created by superconducting coils. Small gradients in three orthogonal directions are superimposed. By changing the frequency of the electric field and correlating the resonant frequency with the local magnetic field, three-dimensional information on morphology can be achieved. The main information is obtained from the resonance intensity, which depends on the local proton density. Since the water and lipid content is different in the various tissues of the body, a three-dimensional image can be obtained, and abnormalities like tumours may be detected. Further information can be obtained from the measurement of the spin–lattice and spin–spin relaxation time.

14.4 Space cryogenics

Liquid hydrogen is used together with liquid oxygen as fuel for space vehicles. It has a high propulsive energy per unit mass, but it needs very large volumes (in comparison with kerosene): rockets fuelled with hydrogen are much larger than those fuelled with kerosene and have more problems of stability during flight.

In addition, liquid or cold supercritical oxygen is carried for life support, and helium may be carried for pressurizing fuel tanks. The technology is similar to that used on earth, except that weight is at a premium and, once in the space environment, only minimal thermal insulation may be needed. However, the absence of gravity poses serious problems, since liquid no longer separates from vapour and convection currents are non-existent. Special devices have to be used to overcome these problems. In the case of rocket motors, the vehicle may be given a small acceleration by an auxiliary rocket to drive the liquid towards the fuel tank outlet so that the engines may be started reliably.

The small-scale applications are mainly concerned with scientific measurements, including astronomy covering the whole range of electromagnetic wavelengths, recording of magnetic fields, observations of the surface of the earth and microgravity experiments. The instruments often include a cooled detector or a superconducting device.

The provision of a small refrigerator is attractive, but the device must be of long life (several years), very reliable and low in power consumption, weight and vibration. The alternative is to provide a store of cryogenic liquid, but the experiment then has a comparatively short lifetime. Both methods are, in fact, used.

On the last three decades, several space experiments with parts at very low temperatures have been flown. Among these, we mention IRAS (Infrared Astronomical Satellite) launched in 1983 (see Fig. 14.1), COBE (Cosmic Background Explorer) launched in 1989, ISO (Infrared Space Observatory) launched in 1995 and Astro-E (X-ray Observatory), launched in 2000 with instrumentation at 65 mK [35]. Some 'cryogenic' space missions are in the preparation or in final phase in Europe, USA and Japan. For example, ESA is going to fly Planck (for the mapping of the cosmic background radiation) and Herschel (called before FIRST 'Far Infrared and Submillimetre Telescope') [36]. These missions will carry experiments at 0.1 and 0.3 K respectively.

Other 'cryogenic' space experiments, such as XEUS (0.1 K) [37] and DARWIN [38], are programmed for the next future. A review of space missions and technologies is reported in ref. [39].

For the space cryostats containing liquid helium, the following precautions are used:

1. The mechanical supports of the tank containing helium, are made of low-conductivity materials such as stainless steal or fibreglass. Sometimes a double system of supports is used: strong supports (fairly conductive) which stand the high launch accelerations (up to 30 g) are added which are eliminated when the satellite is in orbit in the absence of gravity.
2. Thermal shields connected to heat exchangers cooled by the evaporating gas are used to drastically reduce the radiative input.

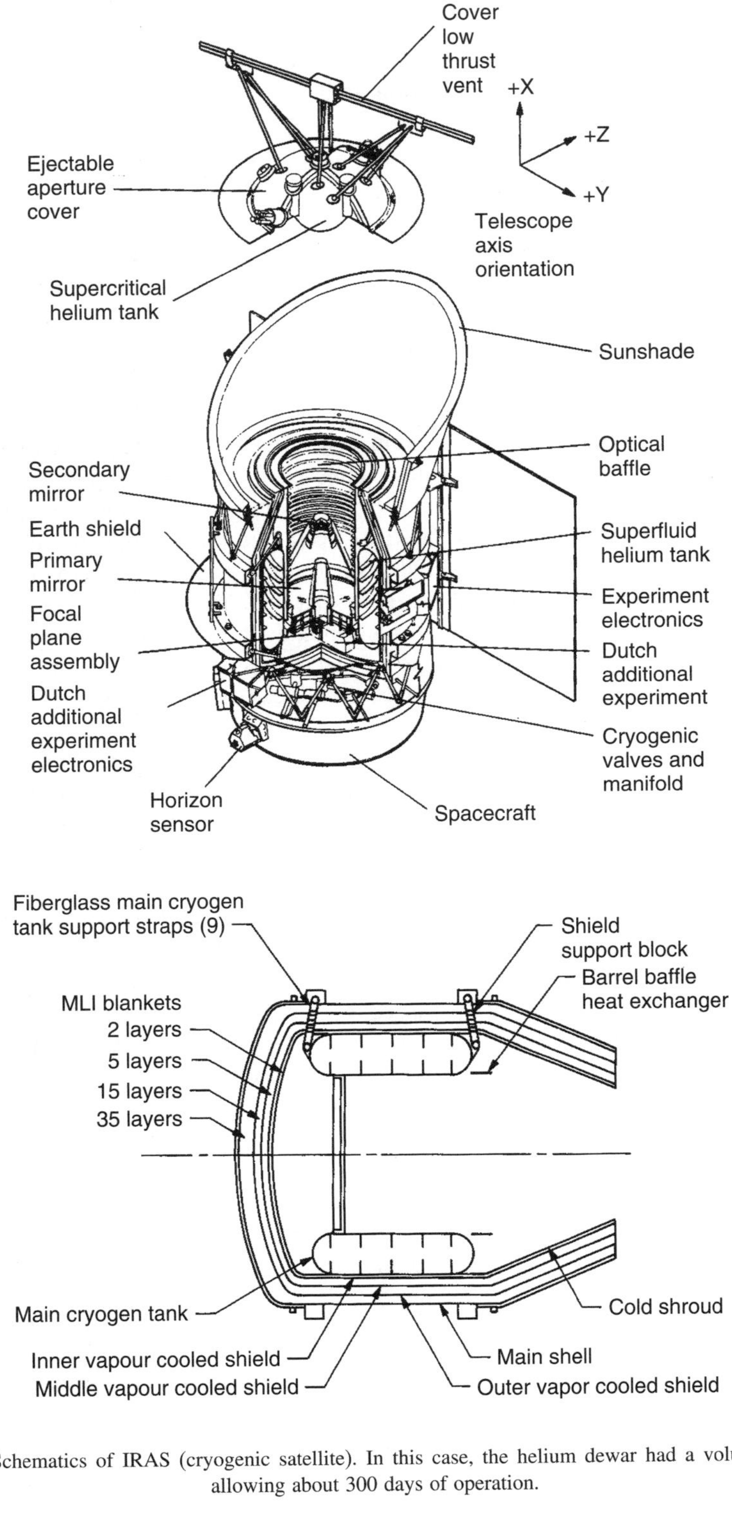

Fig. 14.1. Schematics of IRAS (cryogenic satellite). In this case, the helium dewar had a volume of 600 l, allowing about 300 days of operation.

There are two characteristics that make helium attractive for space applications: the first is weight (about 0.125 kg/l); the second is its superfluidity. Helium becomes superfluid at $T < 2.17\,\text{K}$ ($p < 37.8\,\text{torr}$). Thanks to superfluiduty, helium forms a film that completely covers the walls of the container and guarantees a homogeneous cooling even if most of the liquid does not have a fixed position inside the container (no gravity).

The superfluidity of liquid helium solves also the problem of phase separation in the absence of gravity. The gravity keeps the liquid on the bottom of the container, and the evaporation takes place at the liquid/vapour interface. Helium vapours flow towards the surroundings or are pumped if temperatures lower than 4.2 K are to be obtained. Space is a very efficient pump, but a simple connection to the space vacuum would immediately empty the container. To overcome this serious problem, a 'porous plug' is used. It consists of a slab of porous material which separates the container from the outer space. The diameter of pores is quite critical (usually a few microns). Only the superfluid helium is able to pass through the pores. If the temperature of the external part of the plug is lower than the internal temperature, a thermomechanical force (as in the 'fountain-effect') prevents the liquid from escaping. It must be noted that the external surface of the porous plug is colder because of the liquid evaporation and the porous plug is therefore a passive component.

A review and a performance comparison of liquid helium cryostats flown or planned for space flights is reported in ref. [40]. A survey of low-power cryocooler suitable for space applications is done in ref. [41,42].

Temperatures below 1 K are difficult to obtain in space. ^{3}He refrigerators (0.3 K) suffer from the absence of gravity. The confinement of liquid can be obtained by capillarity, filling the evaporator with sponge materials as cotton or sinterized metals. Adiabatic demagnetization coolers (down to 10 mK) need a sophisticated control and a quite strong magnetic field. Dilution refrigerators are difficult to control on board. The dilution refrigerator in the form described in Chapter 6 does not work in the absence of gravity due to the lack of phase separation. A possible alternative studied for the PLANK/HFI experiment is described in Section 6.8 and in ref. [43].

For the application of cryogenics in space see also ref. [44].

14.5 Cold electronics

We have seen that many electronic components, even not specifically produced for cryogenic applications, can be usefully operated at low temperature: some of them retain their room temperature characteristics like NiCr resistors which do not appreciably change their resistance (less than 10% upon cooling to 4 K) and show a lower noise at low temperature. Other resistors (as RuO_2) and most capacitors change their characteristics with temperature. Mica and polyester film capacitors show a good temperature stability. If capacitors insensitive to temperature are needed, crystalline dielectric or vacuum capacitors must be used.

Usually, inductors and transformer used at low temperatures have nonmagnetic cores, since materials with a high permeability at room temperature usually show a strong temperature dependence and eddy current losses. Superconductors are often used in transformers for low temperature.

Many active electronic devices can be operated at cryogenic temperatures [45]. They are generally of the field-effect transistor (FET) type and are based on silicon (working down to ~100 K) or gallium arsenide (working even below 4 K).

Far instrumentation purposes, there are clear advantages in placing at least some of the electronic circuitry close to the sensor (see Section 15.1) in order to reduce pick-up noise. However, there may also be inherent advantages in operating transistors at low temperatures, such as increased switching speed or lower noise. A serious problem is the effect on device reliability of the stresses induced by thermal cycling.

Superconducting electronic devices are in a different class. They rely on two phenomena: the Josephson effect and the quantisation of magnetic flux, which are described in a simple way in ref. [46], while a more complete account is given in ref. [47].

The quantum of magnetic flux is only 2.07×10^{-15} Wb, which is approximately equal to the amount of the earth's magnetic field enclosed by a ring of 10 μm in diameter. The Josephson effect is observed when two superconductors are separated by a very thin insulating layer (about 20 nm). Single electrons and Cooper pairs can tunnel through such a layer. The characteristics of the Josephson junction are now used to define the volt and have enabled the uncertainty in the maintained standard to be reduced to ~0.1 μV.

The superconducting quantum interference device (SQUID) is formed from a superconducting loop containing at least one Josephson junction. Basically, a SQUID amplifier converts an input current to an output voltage with a transresistance of the order of 10^7 V/A. The input noise is of the order of 10^{-11} A/(Hz)$^{1/2}$. The bandwidth of the SQUID amplifier can be up to 80 kHz. The dynamic range in 1 Hz bandwidth can be 150 dB.

The RF SQUID is formed from a single junction in a superconducting loop, which is inductively coupled to a resonant circuit. This is arranged to drive a current round the loop, so that the voltage across the circuit is a measure of the magnetic flux being measured.

In the DC SQUID, two Josephson junctions are made in the loop, and a dc current is passed through the parallel circuit so formed. The voltage required to produce the current is then a function of the magnetic field trapped inside the loop.

SQUID may be fabricated by technologies similar to those used for integrated circuits [48], although other methods may also be used. The RF device is the easier to make and use; the DC SQUID is the more sensitive, and the quantum limit of sensitivity can be approached. By forming the superconducting loop from two rings with a known distance between their planes, small magnetic field gradients may be measured in the presence of a large uniform field.

Among a variety of applications, SQUID have been used to map local anomalies in the earth's magnetic field (of interest to geologists and archaeologists, for example). They are used as mixers and amplifiers for receiving weak signals from satellites. In medical investigations, SQUID are used to record the varying magnetic fields associated with body activity (fluctuations of the order of 10^{-11} T from the heart down to 10^{-15} T from the brain) [49]. It must be noted that whereas MRI gives information about the structure of tissue, these magnetic field measurements give information about the functional behaviour of the tissue.

Josephson junctions may be arranged in a variety of ways for other purposes. Perhaps, the best-known application is to computers: combinations of Josephson junctions can be designed to act as a very fast switch with low power dissipation or as a memory element. The theoretical switching time is about 10 psec and the power dissipation about 1 nW, giving a product of switching time per power consumption several orders of magnitude better than that of transistors.

The fabrication of logic elements using such devices allows in principle the construction of a large capacity, compact, high-speed computer [50]. Major problems with the technology are that large fan-out ratios are difficult to achieve and that superconducting circuits have a very low inherent impedance and so are difficult to couple with conventional elements at room temperature.

There have been a number of proposals for superconducting devices with transistor-like properties. These include the Gray effect transistor [51], the quiteron [52], the quasiparticle multiplier [53,54] and the quasiparticle trapping transistor (Quatratran) [55]. A basic text for cold electronics is ref. [56]. Advances in low-temperature electronics can be found in WOLTE 1-7 meeting proceedings [57] and in SEMI-THERM 1-22 Symposium proceedings [58].

References

[1] G. Binnig, H. Fuchs, E. Stoll: Surf. Sci. Lett. **169**, 1295 (1986)
[2] D.M. Eigler, E.K. Schweizer: Nature (London) **344**, 524 (1990)
[3] H.C. Manoharan, C.P. Lutz, D.M. Eigler: Nature (London) **403**, 512 (2000)
[4] J.H. Ferris et al.: Rev. Sci. Instrum. **69**, 2691 (1998)
[5] G. Meyer: Rev. Sci. Instrum. **67**, 2960 (1996)
[6] B.C. Stipe, M.A. Rezaèi, W. Ho: Rev. Sci. Instrum. **70**, 131 (1999)
[7] L. Petersen et al.: Rev. Sci. Instrum. **72**, 1438 (2001)
[8] V. Madhavan et al.: Science **280**, 567 (1998)
[9] J.T. Li et al.: Phys. Rev. Lett. **81**, 4464 (1998)
[10] S. Behler et al.: Rev. Sci. Instrum. **68**, 2479 (1997)
[11] S. Heinze et al.: Science **288**, 1805 (2000)
[12] E.T. Foley, A.F. Kam, J.W. Lyding: Rev. Sci. Instrum. **71**, 3428 (2000)
[13] N.R. Jones: *The planet of the double sun*, Ace pub., New York (1967)
[14] *Extreme Cryo 2006 - The 8th Annual Meeting of the Canadian Society for Biopreservation and Cryobiology* – January 27–28, University of Alberta Edmonton, Alberta, Canada (2006), Cell Preservation Technology **4**, 133 (2006)
[15] I. Katkov: Int. J. Refrig. **29**, 341 (2006)
[16] M. Rypka et al.: Cryobiology **52**, 193 (2006)
[17] Abstracts of papers and posters presented at the *Forty-Second Annual Meeting of the Society for Cryobiology*, Cryobiology **51**, 348 (2005)
[18] Abstracts of Papers and Posters Presented at the *Forty-First Annual Meeting of the Society for Cryobiology in Association with the Japanese Societies & Associations for Cryobiology*, Cryopreservation and Cryomedicine, and the Chinese Cryobiological Society, Cryobiology **49**, 291 (2004)
[19] Abstracts of papers presented at the *Fortieth Annual Meeting of the Society for Cryobiology, held in association with the Society for Low Temperature Biology and the Meeting of* CRYOBIOMOL 2003, Cryobiology **47**, 247 (2003)
[20] B. Rubinsky: Heart Fail. Rev. **8**, 277 (2003)
[21] B. Fuller: Cryoletters **24**, 133 (2003)
[22] B. Fuller: Cryoletters **24**, 65 (2003)

[23] Abstracts of Papers and Posters Presented at the *Thirty-Ninth Annual Meeting of the Society for Cryobiology*, Cryobiology **45**, 213 (2002)
[24] Abstracts of Papers Presented at the *Thirty-Eighth Annual Meeting of the Society for Cryobiology*, Edinburgh, United Kingdom – Awards for 2001, Cryobiology **43**, 316 (2001)
[25] J. Wolfe, G. Bryant: Int. J. Refrigeration **24**, 438 (2001)
[26] B. Fuller, S. Paynter: Reprod. Biomed. Online **9**, 680 (2004)
[27] E.J. Woods et al.: Cryobiology **48**, 146 (2004)
[28] L.M. Thurston, P.F. Watson, W.V. Holt: Cryoletters **23**, 255 (2002)
[29] Y. Agca, J.K. Critser: Semin. Reprod. Med. **20**, 15 (2002)
[30] A.W. Rowe: Infusion Ther. Transfus. Med. **29**, 25 (2002)
[31] M.O. Maiwand, J.M. Evans, J.E. Beeson: Cryobiology **48**, 55 (2004)
[32] J.H. Kaouk et al.: Urology **68,** 38 (2006)
[33] R.B. Coleman, R.N. Richardson: Int. J. Refrigeration **28**, 412 (2005)
[34] V.N. Pavlov: Cryogenics **40**, 361 (2000)
[35] Y. Ogawara: *International Astronomical Union, Symposium N° 188*, p. 75, Kyoto, Japan, Kluver Academic Publishers (1998)
[36] B. Collaudin, T. Passvogel: Proc. SPIE, 3358 (1998)
[37] M. Bavdaz et al.: *Proceedings of the Second European Symposium on the Utilization of the International Space Station*, ESA **SP 433**, 621 (1999)
[38] H. Tananbaum et al.: Proc. SPIE, 3765 (1999)
[39] B. Collaudin, N. Rando: Cryogenics **40**, 797 (2000)
[40] W. Holmes et al.: Cryogenics, **41**, 865 (2001)
[41] H.J.M. ter Brake, G.F.M. Wiegerinck: Cryogenics **42**, 705 (2002)
[42] M.R. Emes et al.: Cryogenics **41,** 771 (2002)
[43] A. Sirbi et al.: Cryogenics **39**, 665 (1999)
[44] P. Shirron, K. Shirey: *2005 Space Cryogenics Workshop,* Cryogenics **46**, 73 (2006)
[45] R.K. Kirshman: Cryogenics **25**, 115 (1985)
[46] H.M. Rosemberg: *Solid State*, 2nd edn., Clarendon Press, Oxford (1978)
[47] T.van Duzer, C.W. Turner: *Principles of Superconducting Devices and Circuits*, Edwards Arnold (1981)
[48] M.B. Ketchen: IEEE Trans. Magn. **MAG-17**, 387 (1981)
[49] S.J. Williamson, L. Kaufman: Biomagnetism, J. Mag. Mag. Mater. **22**, 1129 (1981)
[50] J. Matisoo: The Superconducting Computer, Scientific American **282**, 38 (1980)
[51] K.E. Gray: Appl. Phys. Lett. **32**, 392 (1978)
[52] W.J. Gallager: IEEE Trans. Magn. **21**, 709 (1985)
[53] N.E. Booth: Appl. Phys. Lett. **50**, 293 (1987)
[54] D.J. Goldie et al.: Semicond. Sci. Technol. **6**, 203 (1993)
[55] N.E. Booth et al.: Semicond. Sci. Technol. **12**, 538 (1999)
[56] D.E. Gutierrez, J. Deen, C. Claeys eds.: *Low Temperature Electronics: Physics, Devices, Circuits, and Applications*, Elsevier (2000)
[57] Proc. of WOLTE (Workshop on Low-Temperature Electronics):
WOLTE 1: First European Workshop on Low Temperature Electronics, Grenoble, France 29 June–1 July 1994, Cryogenics **35**, 72 (1995)
WOLTE 2: Second European Workshop on Low Temperature Electronics, Leuven, Belgium, June 26–28 1996, ed. by C. Claeys and E. Simoen
WOLTE 3: Third European Workshop on Low Temperature Electronics, San Miniato, Tuscany, Italy, June 24–26 1998, J. de Physique IV 8 (P3): III–IV June (1998) http://www.bo.infn.it/sminiato/sminiato98.html
WOLTE 4: Fourth European Workshop on Low Temperature Electronics, Noordwijk, The Netherlands, Proceedings ESA Proceedings WPP-171, Noordwijk (2000)
WOLTE 5: Fifth European Workshop on Low Temperature Electronics, Grenoble, France, June 19–21 2002, J. de Physique IV 12 (PR3): III–III May (2002)
WOLTE 6: Sixth European Workshop on Low Temperature Electronics, Noordwijk, The Netherlands, 23–25 June 2004 (2006)
WOLTE 7: Seventh European Workshop on Low Temperature Electronics, 21–23 June 2006 at ESTEC, Noordwijk, The Netherlands in print
[58] SEMI-THERM 1–22 Symposium Proceedings: http://www.semi-therm.org

15 Low-Temperature Detectors

Contents

15.1 Introduction

The evolution of low-temperature detectors (LTD) is an attempt to detect excitations of increasingly lower energy. The first particle detectors, Geiger counters, proportional counters and scintillators, were based on ionization and excitation processes requiring typically 20 eV per quantum. In semiconductors, typically 3 eV is necessary to produce one electron-hole pair. The hope in developing the LTD is to effectively use the small binding energy of Cooper pairs in superconductors (typically in the order of meV) or the low energy of phonons at very low temperature (1 peV for thermal phonons at 10 mK) to increase the sensitivity and resolution of detectors.

Obviously, in order not to thermally generate such excitations, these detectors have to be operated at very low temperatures.

Pioneering efforts started as early as the 1960s [1], but is it only in 1980s that we have witnessed a considerable development in LTD as documented by the proceedings of the various conferences on LTD [2].

Three scientific goals have historically driven this technology:

1. The measurement of the cosmic microwave background. Far infrared astronomers were the first to develop thermal detectors. Some of the resulting technologies, such as neutron transmutation doping (NTD) [3], have been transferred to particle detection sensors and have allowed many groups (e.g., ref. [4–11] to make rapid progress).
2. High-resolution spectroscopy: the dramatic improvement in energy resolution obtained by cryogenic detectors could impact a large variety of scientific fields. High-purity germanium- and lithium-doped silicon detectors are limited by Fano fluctuations (roughly 50 eV FWHM at 1 keV). In contrast, cryogenic microcalorimeters already 20 years ago reached 7 eV FWHM [4] coming close to the resolution of crystal (Bragg) spectrometers ($\sim$1 eV) but with much higher quantum efficiency (100% instead of 10^{-3}).
3. The search for weakly interacting particle, which may constitute the dark matter in the universe. The elucidation of the nature of this dark matter is one of the most pressing problems in astrophysics and in cosmology.

About the present challenges with cryogenics detectors we can mention:

- Energy-resolving single-photon counting and imaging for astronomy and other fields of science and technology, from X-rays to far infrared, using microcalorimeters, transition edge sensors (TES) or STJ detector arrays.
- Neutrino mass experiments using large microcalorimeter arrays.
- Neutrinoless double-beta decay and other rare decays studied using massive calorimeters.
- Dark matter searches with detector arrays up to several tens of kilograms.

In the following sections, we will describe cryogenic sensors and cryogenic detectors, while we will describe the cryogenic experiments in Chapter 16.

A cryogenic sensor is, for example, a thermometer used around a fixed temperature where it shows a high sensitivity. It does not usually need a calibration. The realization technology of sensors often differs from that of the corresponding thermometers.

Most of cryogenic sensors have been developed in close connection with detectors and do not find different applications. This is the case of SSG (superheated superconducting granules); others have also independent applications, e.g. Ge thermometers.

A cryogenic detector is a more complex device able to detect some event as the release of energy by a particle or by radiation. In some cases, a formal separation between sensor and detector is difficult.

In a cryogenic experiment, one or several detectors are used for a definite goal for which they have been optimized. For example, in CUORE experiment described in Section 16.5, the sensors are the Ge thermistors, i.e. thermometers used in a small temperature range (around 10 mK). One detector is a bolometer made up of an absorber and a Ge sensor. The experiment is the array of 1000 bolometers arranged in anticoincidence circuits for the detection of the neutrinoless double-beta decay. Note that the sensors, if calibrated, could be used, as well, as very low-temperature thermometers. Also the array of bolometers can be considered a single large detector and used for different purposes as the detection of solar axions or dark matter.

In most cases, a cryogenic detector is developed in view of a definite experiment.

15.2 Cryogenic sensors

Several types of cryogenic sensors have been proposed and used. A detailed discussion of all these devices is beyond the practical size of this book. Most of them are described in the proceedings of LTD [2]. We only mention for instance:

1. Thermoelectric detectors (QVD)
2. Magnetic sensors
3. NIS junctions
4. Tunnel junctions
5. Kinetic inductance sensors
6. Metastable superconducting sensors
7. Schottky sensors
8. Resistive sensors (NTD Ge and TES)

Among the cryogenic sensor, we will describe only the example of resistive sensors (Section 15.2.1) and their use in detectors for nuclear physics (Section 15.3) and astronomy (Section 15.4).

15.2.1 Resistance sensors

Because of their high heat capacity, only few of the thermometers described in Chapter 9 can be used as sensors for detectors. Resistance (carbon) sensors were used for the first time in a cryogenic detector by Boyle and Rogers [12] in 1959. The carbon bolometer had a lot of advantages over the existing infrared detectors [13]. It was easy to build, inexpensive and of moderate heat capacity due to the low operating temperature.

However the carbon specific heat was not as low as the crystalline materials employed later. Also important, the resistor material exhibits an excess low-frequency noise. Nowadays, two types of resistance sensors are used to realize LTD: NTD Ge sensors and TES.

15.2.1.1 NTD Ge sensors

The low-temperature thermometers based on heavily doped compensated germanium (see Section 9.6.2.1) show high stability, good reproducibility, low noise and low specific heat. Ge used for cryogenic sensors is heavily doped ($10^{16} - 10^{19}$ atoms/cm^3), with T_0 of Mott's law ranging between 2 and 70 K (see formula 9.6).

NTD germanium is produced by irradiating ultrapure Ge crystals (usually a disk of about 3 cm in diameter and 3 mm thick) by means of a flux of thermal neutrons.

Natural Ge contains five stable isotopes: ^{70}Ge, ^{72}Ge, ^{73}Ge, ^{74}Ge and ^{76}Ge. When a nucleus captures a neutron, it may become a stable isotope, as in the case of ^{72}Ge and ^{73}Ge, transmutating into ^{73}Ge and ^{74}Ge respectively, which do not contribute to the doping process. Otherwise, the nucleus may undergo a beta decay or a K capture. Hence the doping process produces both acceptors and donors as shown in Table 15.1.

Table 15.1
Induced nuclear decays for neutral absorption in ^{70}Ge, ^{74}Ge and ^{76}Ge isotopes

Isotopical fraction	Reaction	σ [barn]	$\tau_1/2$	Type
20.5%	$^{70}_{32}\text{Ge}\,(n,\gamma)^{71}_{32}\text{Ge} \rightarrow {}^{71}_{31}\text{Ge} + \text{K}$	3.25	11.2 day	p
36.5%	$^{74}_{32}\text{Ge}\,(n,\gamma)^{75}_{32}\text{Ge} \rightarrow {}^{75}_{33}\text{As} + \beta^-$	0.52	82.8 min	n
7.8%	$^{76}_{32}\text{Ge}\,(n,\gamma)^{77}_{32}\text{Ge} \rightarrow {}^{77}_{31}\text{As} + \beta^- \rightarrow {}^{77}_{34}\text{Se} + \beta^-$	0.16	11.3h	n

The dopant concentration depends both on the isotopic abundance and on the neutron absorption cross-section. For natural Ge, a unit neutron flux produces a Ga and As concentration of [14,15]:

- $[\text{Ga}] = 2.94 \times 10^{-2}/\text{cm}^3$ per neutron/cm^2/s
- $[\text{As}] = 8.37 \times 10^{-3}/\text{cm}^3$ per neutron/cm^2/s

The concentration of Se (double donor) is negligible.

Due to the natural isotopic composition of Ge and to the neutron capture cross-section, the resulting doping is of p-type. NTD process has principally two advantages [3]:

- The doping is homogeneous over the crystal, whereas melt doped Ge sensors show dispersive properties even when they are cut from nearby volumes of the same chip;
- The doping process may be controlled by changing the integrated flux of neutrons, making possible the approximate prediction of the resistivity of thermistors as a function of temperature.

The doping homogeneity is due to two reasons:

1. the neutron absorption cross-section is quite low; hence the neutron flux intensity is scarcely attenuated in passing through the crystal.
2. The neutron beam is larger than the dimensions of the sample to be doped; hence the neutron flux impinging onto the crystal is uniform.

On the other hand, the disadvantage of NTD technique, besides the need of a nuclear reactor, consists in the damages induced in the lattice by fast neutron produced by the reactor. To fix such damages, after nuclear irradiation, Ge crystal samples are annealed a few hours at about 450°C.

Another disadvantage of NTD technique is the simultaneous creation of both donors and acceptors giving place to a 'compensation' in the doping. The 'compensation' is the donor/acceptor ratio (in p-type) which depends mainly on the isotopic composition and on cross-sections for neutron absorption.

15.2.1.2 Electrical contacts

The electrical contact with the 'bulk' of the doped crystal is made through a very heavily doped layer, to reduce the height of the Schottky barrier between the bulk and the metal of the external contact (Au). The charge carriers cross this layer by tunnel effect.

To get a very heavily doped layer, two techniques can be used: diffusion or ion implantation.

The diffusion process needs high temperatures and much time. Moreover, the doping takes place along the whole crystal, exponentially decreasing with the distance from the surface.

In the ionic implantation, a beam of trivalent ions (e.g. B^+ ions), is used to produce a p^+ layer. Pentavalent ions (e.g. P^- ions), instead, create an n^+ layer. The main advantage of this technique rely in the fact that the ions penetrate the crystal only for a short distance from the surface (~200 nm). The penetration depth depends on the beam energy ($25 \div 100$ keV). Typical doping level is $10^{18} \div 10^{19}$ ions/cm^3. With such a high dopant concentration, the layer becomes practically a metal.

As we have seen in Section 12.6.4, the B^+ implantation produces an excess heat capacity below ~30 mK. After the implantation process, the crystal undergoes an annealing of about 1 h at 250°C to fix damages to the crystalline structure. Then, a layer of Pd (about 20 nm) and Au (about 400 nm) are sputtered onto the both sides of the wafer. Finally the wafer is annealed at 200°C for 1 h. The metallization process produces low-noise ohmic contacts. The wafers are cut to produce thermistors of length 3 mm between the metallized ends ($3 \times 3 \times 1$ mm^3 typical size), and the electrical contacts are made by ball bonding with Au wires. By the NTD process, doped Ge can be obtained with resistivity ranging over eight orders of magnitude. Figure 15.1 shows a plot of the electrical resistivity for some materials produced by the NTD process. The electrical resistivity is well represented by the Mott's law $\rho = \rho_0 \exp(T_0/T)^p$. In Fig. 15.1, a commonly used p value of 0.5 was adopted, even though $p = 0.6$ has been recently shown to give a better fit of data for $T_0 < 20$ K [16,17].

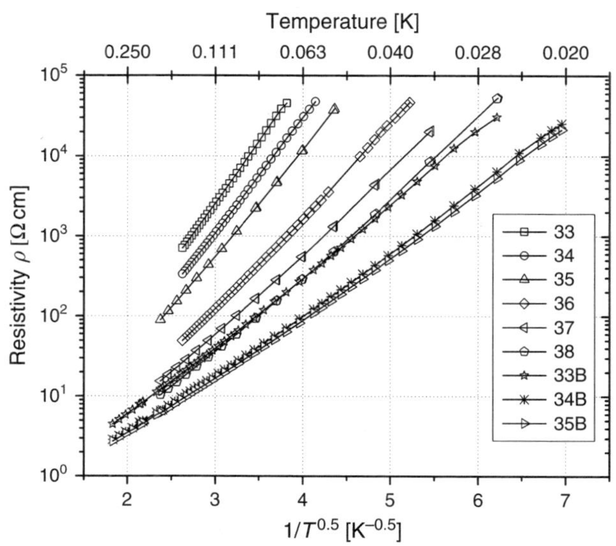

Fig. 15.1. Electrical resistivity for $T < 300$ mK of neutron transmutation doping Ge 33–38 series.

15.2.1.3 Carrier (electron–phonon) decoupling

The interaction between charge carriers (electrons or holes) and phonons (lattice) may give origin to several phenomena in a crystal, e.g. the production of Cooper pairs.

The carrier–phonon interaction decreases with the lowering of temperature, since the emission and absorption of phonons by carriers is proportional to the number of final states available to carriers and phonons. At sufficiently low temperatures, the interaction between the two subsystems can be so weak that there is no thermal equilibrium between them, and the energy is distributed among electrons more rapidly than it is distributed to the lattice, resulting in a different temperature for electron and phonon subsystems, giving rise to the so-called 'electron–phonon decoupling'.

For example in copper, a metal used in nuclear adiabatic demagnetization (see Section 7.4), electron and phonon systems are decoupled: the power transferred between the two systems is [18]:

$$P_{e-\phi} \approx 25 \times \left(T_\phi^4 - T_e^4\right) \,[\mathrm{W/cm^3}] \quad \text{for } T < 50\,\mathrm{mK} \tag{15.1}$$

where T_ϕ and T_e are the phonon and the electron temperatures respectively.

In doped semiconductors, the band structure is not so well known as in metals, and a prevision about carrier–phonon decoupling is difficult.

In V–I DC measurements at constant temperature, non-linear effects appear even at powers as low as 10^{-14} W; they become more evident when temperature is lowered (see Section 9.6.3 and Fig. 9.8) [19]. Such non-linearities can be interpreted by the so-called 'hot-electron model' (HEM).

The 'HEM' is a thermal model which represents a doped semiconductor thermistor (e.g. Ge NTD) as made up of two subsystems: carriers (electrons or holes) and phonons. Each subsystem has its own heat capacity and is thermally linked to the other one through a thermal conductance $G_{e-\phi}$, which takes into account for the electron–phonon decoupling (see Fig. 15.2).

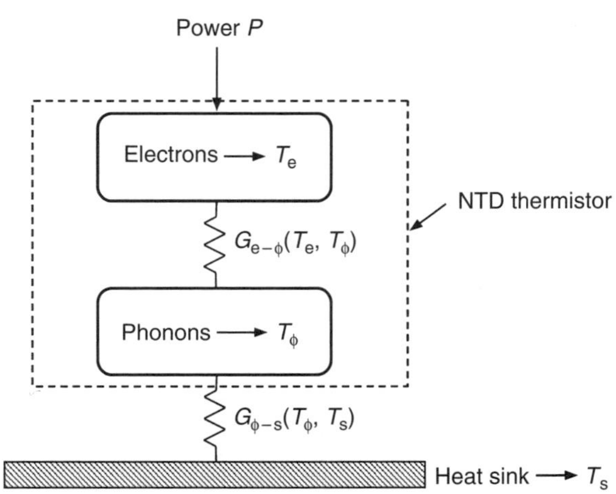

Fig. 15.2. 'Hot electron' model for a Ge thermistor.

The phonon system is linked to the heat sink by a thermal conductance $G_{\phi-s}$.

The 'nature' of $G_{\phi-s}$ is different from that of $G_{e-\phi}$. The latter represents the phenomenon of electron–phonon decoupling, the former represents the contact resistance to the heat sink.

The thermal conductance $G_{e-\phi}$ depends on the temperatures T_e and T_ϕ of the electron system and lattice respectively through a 'conductivity' $g_{e-\phi}(T)$:

$$G_{e-\phi}\left(T_e, T_\phi\right) = \frac{1}{T_e - T_\phi} \int_{T_\phi}^{T_e} g_{e-\phi}(T)\, \mathrm{d}T\, [\mathrm{W/K}] \tag{15.2}$$

The influence of $G_{e-\phi}$ is relevant only below ~30 mK. The dependence on T of $g_{e-\phi}(T)$ is supposed in the form:

$$g_{e-\phi}(T) = (\beta + 1)\, g_{e\phi} \cdot T^{\beta}\, [\mathrm{W/K}] \tag{15.3}$$

where $g_{e\phi}[\mathrm{W/K}^{(\beta+1)}]$ and β are constant.

Analogously, $g_{\phi-s}(T)$ can be expressed as:

$$g_{\phi-s}(T) = (\alpha + 1)\, g_{\phi s} \cdot T^{\alpha}\, [\mathrm{W/K}] \tag{15.4}$$

where $g_{\phi s}[\mathrm{W/K}^{(\alpha+1)}]$ and α are constant.

As we saw in Section 4.4, $g_{\phi-s}(T)$ is expected to depend on temperature approximately as T^3 (there is always an electrical insulating layer in the contact between the thermistor and the metallic support at the temperature of the heat sink) (Fig. 15.3). For a review of the few existing measurements of electron–phonon decoupling, see ref. [20].

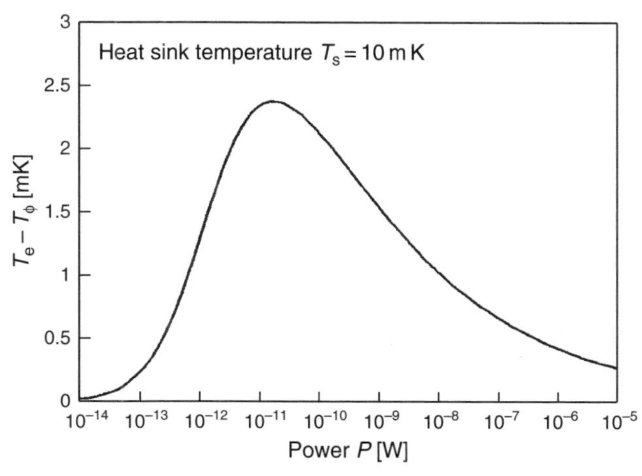

Fig. 15.3. Calculated temperature difference between electrons and phonons as a function of the power P supplied to electrons [15].

15.2.2 TES

TES are based on the steep temperature dependence of the resistance of superconducting metallic films. The useful temperature range is very narrow. These 'thermometers' which may have a very low intrinsic noise, are fabricated by a vacuum deposition process at very low pressure and are patterned either by photolithography technique (see e.g. ref. [21]) or by micromechanical machining (see e.g. ref. [22]). The dimensionless parameter $\alpha = T/R \cdot \mathrm{d}R/\mathrm{d}T$ defines the DC quality of a sensor. TES with α as high as 1000 have been built [23].

For thermal stability reasons, a TES cannot be current biased as was the case of semiconductor sensors. A voltage bias is necessary and the output is a current signal.

In fact, with few exceptions [24], the resistance of TES is very low and the matching to a conventional FET amplifier is impossible. A SQUID amplifier (see Section 14.5) coupled to the TES by a superconducting transformer is the natural solution as schematically shown in Fig. 15.4.

In the case of TES, the joule heating of the superconducting film produces a negative thermal feedback which increases the thermal stability. The thermal equilibrium takes place when joule heating is balanced by the thermal leak to the substrate. If for some reason in a TES, biased by a voltage V at the centre of the transition, the temperature decreases, an increase of the TES electrical resistance R takes place. Consequently, the bias power V^2/R increases, bringing back the TES at the centre of the transition.

TES suffer from some limitations such as the small useful temperature range and the non-linearity of the transition curve. The latter drawback is especially evident in roughly patterned TES, as in the case shown in Fig. 15.5 [25]. Feedback techniques, similar to those used in electronic amplifiers, minimize these drawback, reducing also the TES time response [26]. The superconducting transition temperature (sometimes quite different from those of the bulk metal) of a TES made with one metal layer (single layer) depends on the metal used and on the film thickness.

If transition temperatures other than those allowed by superconducting pure metals are required, two metallic layers can be deposited to form a 'bi-layer' TES. In most cases, only one of the two metals is a superconductor. In this case, the Cooper pairs diffuse from

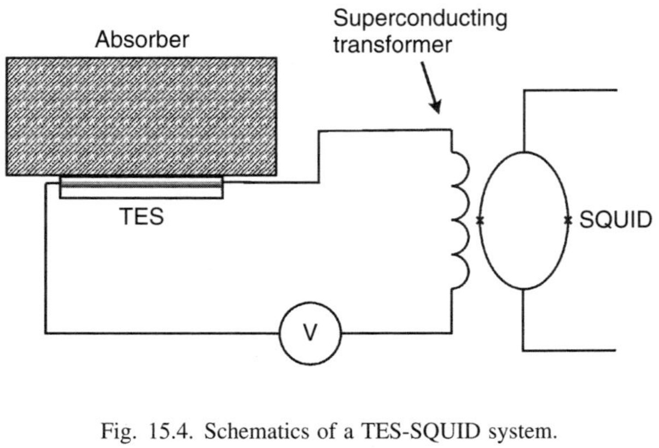

Fig. 15.4. Schematics of a TES-SQUID system.

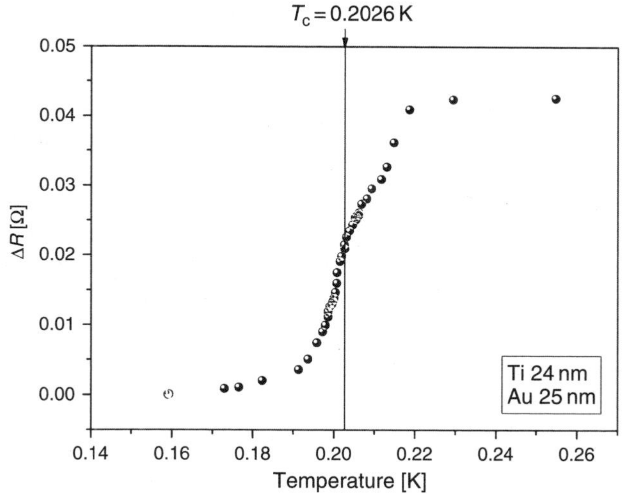

Fig. 15.5. Transitions of a Ti/Au bi-layer ($T_C = 0.2026\,K$).

the superconducting material into the normal metal and electrons diffuse in the opposite sense, giving rise to the 'proximity effect', see ref. [27,28]. It is thus obtained a 'metal' with different properties and a transition temperature which can be somewhat modulated by changing the thickness of the two layers [29].

For example, a bi-layer of Al/Ag gave a transition temperature between 75 and 120 mK [30]. Another very interesting bi-layer is the TES with Mo/Cu [31]. The modulation of the transition temperature may be also obtained by changing the thickness of a single layer [32], but to get high α, very thin layers are to be used and the evaporation system must be extremely clean. There is a huge bibliography about TES and their applications which is mainly reported in LTD proceedings. Some basic references other than those in LTD proceedings are ref. [33–51].

15.3 Examples of cryogenic detectors

A sensor is only one component of a cryogenic detector. In the simplest case, a detector consists of an 'absorber' (for example absorber of energy) and a 'sensor' (for example a thermometer like a TES). Nevertheless, other physical parameters than energy and temperature may be involved in a cryogenic detector. For example, in a cryogenic gravitational antenna (see Section 16.2) the 'absorber' is the cooled bar, whereas the sensors is SQUID-capacitor system.

Hereafter, we will describe in detail two examples of cryogenic detectors:

1. A calorimeter for the CUORICINO experiment (see Section 16.6)
2. A bolometer for millimetric astronomy.

15.3.1 Calorimeters

Cryogenic detectors (calorimeters) were proposed in the 1980s by Fiorini and Niinikoski for searching rare events like neutrinoless double-beta decay (ββ-0v) [52].

The idea of using cryogenic detectors both for measuring the neutrino mass from ^{187}Re end point [53] and for detecting WIMP recoils [54] dates back to the same years. Nowadays, cryogenic detectors are used in the most sensitive ββ-0v and non-baryonic dark matter searches, being considered the 'natural' detectors for these searches. The huge literature about the subject is mainly reported in the LTD proceedings [2].

The physical process occurring in a low-temperature calorimeter as particle detectors are fully described in ref. [55,56]. In the next sections, we will focus on the cryogenic aspect of a detector, examining the type of detector used in CUORICINO experiment (see Section 16.6).

15.3.2 CUORICINO detector model

An extremely simplified scheme of a calorimeter (composite thermal detector) is shown in Fig. 15.6. The temperature of an absorber **A** (T_A) is measured by a thermometer **T**. A thermal conductance G forms a thermal link with the heat sink **B** at the temperature T_s. In the ideal adiabatic situation ($G = 0$), an absorption of an energy ΔE produces an absorber temperature increase:

$$\Delta T = \frac{\Delta E}{C} \tag{15.5}$$

where C is the total (absorber + thermometer) heat capacity.

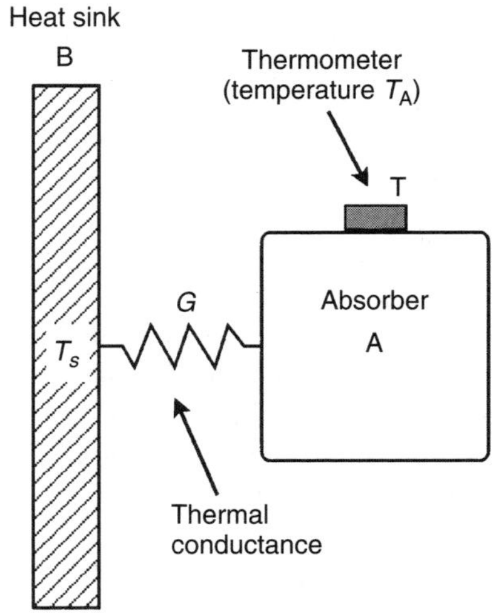

Fig. 15.6. Simplified scheme of a calorimeter.

Since C is usually proportional to temperature, it is evident from formula (15.5) the advantage of working at temperatures as low as possible, to increase the sensitivity of the detector.

A decrease in working temperature also improves the energy resolution:

$$\Delta E_{\mathrm{rms}} = \xi \cdot \left(C \cdot k_{\mathrm{B}} \cdot T^2\right)^{1/2} \tag{15.6}$$

where C is the heat capacity of the detector, k_{B} is the Boltzmann constant and T is the operating temperature. The value of ξ parameter is 1 in the ideal case but, in the real cases, ~1.5 [56].

From eqs (15.5) and (15.6), the energy resolution theoretically scales as $\sim T^{5/2}$.

In the real case, however, several phenomena, not included in the simple scheme of Fig. 15.6, limit the minimal useful temperature to about 10 mK, even if, as we saw in Chapter 6 and Chapter 7, refrigerators can reach much lower temperatures.

The main phenomena which become effective below ~30 mK are:

1. For $G \neq 0$, a noise contribution proportional to G.
2. The unavoidable spurious power P_{S} impinging onto the absorber rises T_{A} over T_{s}. The difference $T_{\mathrm{A}} - T_{\mathrm{s}}$ increases as T_{A} or G decreases.
3. As discussed in Section 15.2.1.3, the thermal coupling between **A** and **T** takes place through a contact thermal resistance R_{C}, which scales as $\sim 1/T^3$ (see Section 4.4). The increase of R_{C} produces several drawbacks, as we shall see in this section, first of all an increase of the fall time of the detector pulses.
4. The carrier–phonon decoupling and the contact resistance between **T** and **A** increase the rise time of the pulse when the temperature is lowered.
5. The excess heat capacity (see Section 12.6.4) also spoils the performance of the detector below about 30 mK.

In order to get an overall idea of the effects of the above-mentioned phenomena on the response of a non-ideal calorimeter, let us now considerer a simplified model of one of the calorimeter of COURICINO (Fig. 15.7) experiment (see Section 16.6).

Figure 15.8 shows the thermal scheme of one detector: there are six lumped elements with three thermal nodes at T_1, T_2, T_3, i.e. the temperatures of the electrons of Ge sensor, TeO_2 absorber and PTFE crystal supports respectively. C_1, C_2 and C_3 are the heat capacity of absorber, PTFE and NTD Ge sensor respectively. The resistors R_1 and R_2 take into account the contact resistances at the surfaces of PTFE supports and R_3 represents the series contribution of contact and the electron–phonon decoupling resistances in the Ge thermistor (see Section 15.2.1.3).

In this simplified model, the thermal elements due to the electrical wiring from the thermistor to the copper frame (heat sink) have been omitted for the sake of simplicity.

Using the model of Fig. 15.8, we have simulated an event leading to an energy adsorption ΔE. To evaluate the corresponding temperature increase ΔT, at different heat sink operating temperatures, a T^3 dependence of the absorber heat capacity was supposed. To obtain the calorimeter response (temperature change on the T_1 thermal node) for a simulated event, a SPICE program was used.

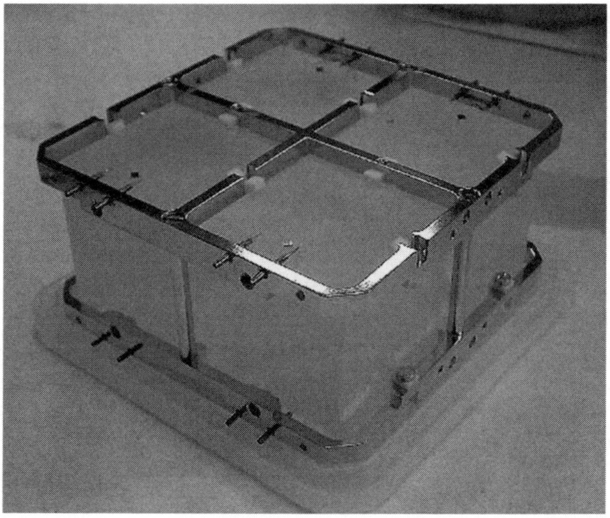

Fig. 15.7. Photograph of four CUORICINO calorimeters in their copper frame [57].

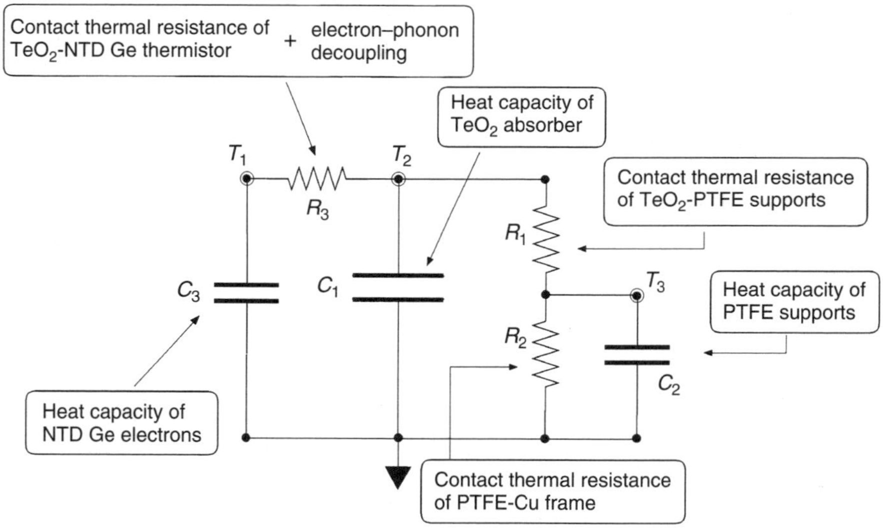

Fig. 15.8. Simplified thermal model of one CUORICINO detector.

Such a program should be used in 'small signal' circuits where the values of the components are constant. For this reason, we supplied a very small 'test' energy ΔE in order to produce very small change ΔT and considered the thermal resistance and heat capacities constant in the calculations for each heat sink temperature.

Seven simulations were carried out at temperatures ranging between 7 and 16 mK. Data used at the various temperatures for the lumped elements are 'plausible' values estimated from very low-temperature measurements. They are reported in Table 15.2.

In particular C_3 has been evaluated (see Section 12.6.4) by the formula:

$$C_3 = 10^{-8} \cdot T + 10^{-13} \cdot T^{-2} \quad [\mathrm{J/K}] \tag{15.7}$$

For R_3, a $1/T^4$ dependence on temperature was chosen as a representative average between the $1/T^3$ (contact resistance) and $1/T^{4.5}$ (electron–phonon decoupling) scaling.

For the other R components, a typical $1/T^3$ dependence was adopted; we put $C_1 \propto T^3$ and $C_2 \propto T$.

The output from the thermal model due to a thermal input ΔE on C_1 is a ΔT on C_3 shown in Fig. 15.9 for seven heat sink temperatures T_s.

From Fig. 15.9, we see that the response has a maximum for $T_s = 10$ mK. Note, however that even at such 'optimal' temperatures, the amplitude of the thermal signal (T_1) on the

Table 15.2
Values of parameters used at seven temperatures to calculated the response from model of Fig. 15.8

T_s [mK]	C_1	C_2	C_3	R_1	R_2	R_3
7	3.77×10^{-10}	3.50×10^{-9}	2.11×10^{-9}	1.46×10^{9}	2.01×10^{8}	2.08×10^{9}
8	5.63×10^{-10}	4.00×10^{-9}	1.64×10^{-9}	9.77×10^{8}	1.35×10^{8}	1.22×10^{9}
9	8.02×10^{-10}	4.50×10^{-9}	1.32×10^{-9}	6.86×10^{8}	9.47×10^{7}	7.62×10^{8}
10	1.10×10^{-9}	5.00×10^{-9}	1.10×10^{-9}	5.00×10^{8}	6.90×10^{7}	5.00×10^{8}
12	1.90×10^{-9}	6.00×10^{-9}	8.14×10^{-10}	2.89×10^{8}	3.99×10^{7}	2.41×10^{8}
14	3.02×10^{-9}	7.00×10^{-9}	6.50×10^{-10}	1.82×10^{8}	2.51×10^{7}	1.30×10^{8}
16	4.51×10^{-9}	8.00×10^{-9}	5.51×10^{-10}	1.22×10^{8}	1.68×10^{7}	7.63×10^{7}

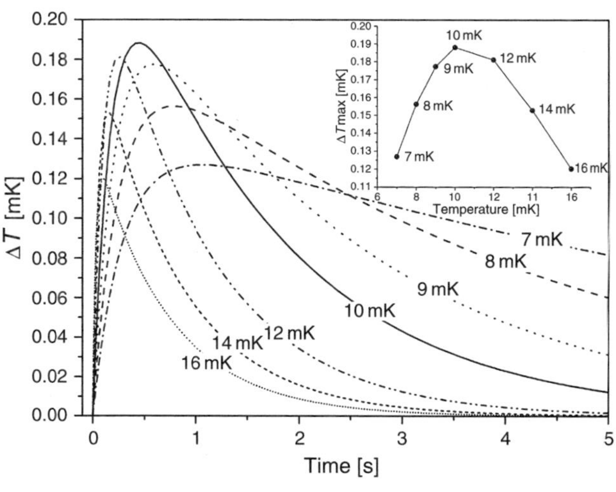

Fig. 15.9. Simulated heat pulses for a CUORICINO detector.

thermistor is only 24% of the amplitude (T_2) on the absorber. Below 10 mK, the pulse amplitude becomes smaller, in contrast with the expectations of formula (15.5), and an increase of rise and fall times is also observed. At 7 mK, the time of recovery of 10% of the base line is about 150 s and T_1 is only 7% of T_2. It is evident that a composite detector of this type cannot be used at temperature below ~10 mK.

The electronics of CUORICINO (see Section 16.6) consists of a complex anticoincidence system. One would hope that signals from all detectors had the same amplitude and shape for the same input. For this reason, all the components for the 56 detectors were carefully selected to be equal. Also the mounting of the components was made with the most reproducible technique. Nevertheless, there are some parameters such as the contact resistances between the thermistors and the absorbers and especially those between the absorbers and the frame which show a wide spread in their values.

To overcome such a problem, a silicon heater of negligible heat capacity was added to each detector to trim its sensitivity by a slight change of the (detector) temperature around the working temperature (see Section 16.6). Due to the steep dependence on T of the R and C parameters, changes in detector temperatures of the order of 1 mK are needed for the equalization of the detector response.

15.4 Infrared bolometers

The astronomical calorimeters for the detection of the infrared radiation (usually called bolometers) do not conceptually differ from the cryogenic detectors used in nuclear physics as those just described for CUORICINO.

The main parameter which characterizes a calorimeter is the resolution, whereas in astronomical bolometers the qualifying parameter is the NEP (noise equivalent power, see later). Also the size is very different: bolometers are usually very smaller. Their more sophisticated technology allows the realization of large arrays of detectors. In operation, bolometers usually handle a chopped signal (tens of Hertz) of much smaller energy than calorimeters.

Infrared radiation was discovered by Herschel [58] in 1800, using a mercury thermometer to detect sunlight dispersed by a prism. However, the Latin poet Lucretius in his 'De rerum Natura' (On the Nature of Things, about 50 BC) clearly showed a clear feeling of the 'infrared' radiation.* Of course Lucretius's terminology was far from the modern one, and he had no thermometer at his disposal!

*[Lucretius: De rerum natura,liber V].

forsitan et rosea sol alte lampade lucens
possideat multum caecis fervoribus ignem
circum se, nullo qui sit fulgore notatus,
aestifer ut tantum radiorum exaugeat ictum.

Perhaps the sun, which shines on high with rosy torch,
possesses about him plenty of invisible fire,
which makes no light,
but, full of heat, increases the power of its rays.

As in the case of calorimeters, a bolometer consists of an absorbing element with heat capacity C, which converts the impinging electromagnetic radiation to heat, and which is linked to a heat sink at temperature T_s via a thermal conductance G. The temperature T_A of the absorber is measured by a thermometer in thermal contact with the absorber.

When an incident radiation power P is absorbed, the temperature T_A initially increases with time at a rate $dT_A/dt = P/C$ (as in formula (15.5)) and approaches the limiting value $T_A = T_s + P/G$, with a time constant $\tau = C/G$.

When the power P is turned off, T_A relaxes back to T_s with the same τ.

Bolometric signals, as we said, are modulated at a frequency ω. Very rarely bolometers are used to detect pulsed signals or steady-state radiation levels.

Early bolometers used, as thermometers, thermopiles, based on the thermoelectric effect (see Section 9.4) or Golay cells in which the heat absorbed in a thin metal film is transferred to a small volume of gas; the resulting pressure increase moves a mirror in an optical amplifier. A historical review of the development of radiation detectors until 1994 can be found in ref. [59,60]. The 'modern' history of infrared bolometers starts with the introduction of the carbon resistor, as both bolometer sensor and absorber, by Boyle and Rogers [12]. The device had a number of advantages over the Golay cell such as low cost, simplicity and relatively low heat capacity at low temperatures.

Nevertheless the heat capacity of a carbon resistor was not so low as that of crystalline materials used later. More important, carbon resistors had an excess noise which limited the bolometer performance. In 1961, Low [61] proposed a bolometer which used a heavily doped Ge thermometer with much improved characteristics. This type of bolometer was rapidly applied to infrared astronomy as well also to laboratory spectroscopy. A further step in the development of bolometers came with improvements in the absorber. In the early superconducting bolometer built by Andrews et al. (1942) [62], the absorber was a blackened metal foil glued to the T_A thermometer. Low's original bolometer [61] was coated with black paint and Coron et al. [63] used a metal foil as substrate for the black-painted absorber. A definite improvement is due to J. Clarke, G. I. Hoffer, P. L. Richards [64] who used a thin low heat capacity dielectric substrate for the metal foil and used a bismuth film absorber instead of the black paint.

The detailed investigation of superconducting (Al) bolometers is due to Clarke and Richards [64,65]. This bolometer had a very low heat capacity and reached the thermal fluctuation noise limits. This bolometer needed a much more complex electronics than those using Ge thermistors.

The next step was the introduction of ion implantation to dope Si for thermometers. Downey et al. [66] used micromachining to realize a Si bolometer with an implanted thermometer. This bolometer had very little low-frequency noise. The use of thermometers doped by neutron transmutation instead of melt doping is described by Lange et al. [67]. The evolution of bolometers sees the replacement of the nylon wires to make the conductance to the bath, used by Lange et al. with a micromachined silicon nitride membrane with a definite reduction in the heat capacity associated to the conductance G [68].

In a further improvement, the NTD Ge 'spider-web' bolometer developed by the Caltech/JPL group [69], and shown in Fig. 15.10, the absorber is a metallized silicon nitride 'spider-web' structure, and the thermometer is a very small NTD germanium crystal [14]. The spacing of the web pattern is much smaller than the wavelength, so it acts

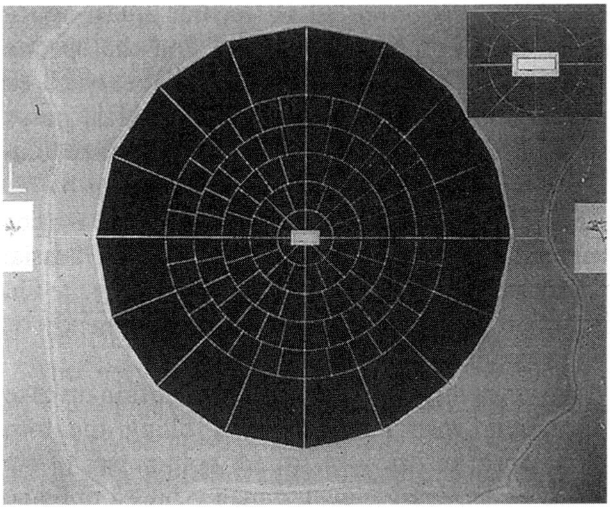

Fig. 15.10. Photograph of a JPL/Caltech spider-web bolometer.

as a plane surface to the incident radiation. Besides reducing the overall heat capacity, the web also minimizes the geometrical cross-section for ionizing radiation. NTD bolometers used by the MPIR group are essentially the same except that the silicon nitride absorber is not etched into a web pattern.

NTD bolometers operating at temperatures above 100 mK can approach quite closely the performance of an ideal detector [70]. Spider-web devices have achieved a NEP of about $1.5 \times 10^{-17}\,\mathrm{W/(Hz)^{1/2}}$, with time constant $\tau = 100\,\mathrm{ms}$ at 300 mK, and a NEP $= 1.5 \times 10^{-18}\,\mathrm{W/(Hz)^{1/2}}$ with $\tau = 65\,\mathrm{ms}$ at 100 mK [71].

Remember that the NEP is defined as the incident signal power required to obtain a signal equal to the noise in 1 Hz bandwidth. Note that NEP is a measure of the S/N ratio, not just noise. The NEP and τ both depend on the thermal conductance G, and sensitivity and speed of response can be traded off by varying G, with the constant NEP, the value of $(\tau)^{1/2}$ being a useful figure of merit.

A wide range of NTD materials is available to tailor the impedance of the device to the desired range (see ref. [14] and Section 15.2.1.1). With a typical bolometer impedance of a few MΩ, low-noise J-FETs preamplifiers (see section 10.6) are used with these detectors. The use of an AC bias at 100 Hz allows the signal information to be extracted without degradation by preamplifier $1/f$ noise [72].

The inherent $1/f$ noise of NTD Ge is very low, and noise spectra have been achieved with no excess noise down to 20 mHz in working systems, still limited by the readout electronics rather than by the detector itself.

The theory of bolometer operation was developed by Mather [73–75] and parameterized analyses of an ideal bolometer have been developed [76,77] and extended to include the effects of electrical non-linearities [78]. For further developments of infrared bolometers into arrays see LTD proceedings [2]. For the coupling of bolometers to a telescope see, e.g., ref. [79].

15.4.1 Simplified calculation of bolometer responsivity

A bolometer absorbs an incident radiative power P_r:

$$P_r = P_1 + P_2 \exp(j\omega_s t) \tag{15.8}$$

The temperature of the bolometer consequently varies as:

$$T_{bol.} = T_1 + T_2 \exp(j\omega_s t) \tag{15.9}$$

The thermistor is biased with a DC current I.

The produced time varying heat is at the first order:

$$I^2 \cdot R(T) = I^2 \cdot \left[R(T_1) + \frac{dR(T)}{dT} \cdot T_2 \exp(j\omega_s t) \right] [\mathrm{W}] \tag{15.10}$$

The bolometer loses a power $\dot{G}(T_{bol.} - T_s)$ to the heat sink at temperature T_s, through the thermal conductance G. Hence, the thermal budget is given by:

$$P_r + I^2 \cdot R(T) = \dot{G}(T_1 - T_s) + G \cdot T_2 \exp(j\omega_s t) + j\omega_s \cdot C \cdot T_2 \exp(j\omega_s t) \tag{15.11}$$

where C is the bolometer heat capacity and G is the dynamic thermal conductance at the temperature T_s. Equating the time-independent terms of eq. (15.11), we get the average working temperature:

$$P_1 + I^2 \cdot R(T_1) = \dot{G}(T_1 - T_s) \tag{15.12}$$

Equating the time varying terms, we obtain:

$$\frac{P_2}{T_2} = G + j\omega_s \cdot C - I^2 \cdot \frac{dR(T)}{dT} \tag{15.13}$$

We define the voltage responsivity of a bolometer as the change in voltage drop per watt of adsorbed signal power:

$$S_A = \frac{V_2}{P_2} = I \cdot \frac{dR(T)}{dT} \cdot \frac{T_2}{P_2} \tag{15.14}$$

From eq. (15.13):

$$S_A = I \cdot \frac{dR(T)}{dT} \times \frac{1}{G + j\omega_s \cdot C - I^2 \cdot \frac{dR(T)}{dT}} \tag{15.15}$$

The responsivity of a bolometer is influenced by the electrothermal feedback [76] (see Section 15.2.2)

The temperature rise $T_2 = P_2/\dot{G}$ expected from an increase of the incident power P_2 is modified by the fact that $R(T)$ changes, and, as a consequence, the bias heating varies. This effect can be expressed as an effective thermal conductance:

$$G_e = G - I^2 \cdot \left.\frac{dR(T)}{dT}\right|_{T_1} \tag{15.16}$$

Introducing the parameter $\alpha = R/T \cdot dR(T)/dT$, calculated for $T = T_1$, we can write:

$$G_e = G - I^2 \cdot R(T_1) \cdot \alpha \tag{15.17}$$

For semiconductor bolometers which have negative $\alpha \rightarrow G_e > G$. For superconductor bolometers which have positive $\alpha \rightarrow G_e < G$.

The electrothermal feedback also influences the time response of the bolometer. The thermal time constant is $\tau_e = C/G_e$. Using these definitions, the responsivity can be written:

$$S_A = \frac{I \cdot R \cdot \alpha}{G_e \cdot (1 + j\omega_s \tau_e)} \tag{15.18}$$

For the measurement of the responsivity of a bolometer, see e.g. ref. [60].

The optimization of a bolometer for a specific experiment is sometimes complicated because of its dependence on the radiative background.

15.4.1.1 Example: design and realization of an infrared bolometer operating at 0.3 K

We shall now describe an infrared bolometer operating at 0.3 K, built by Lange et al. in 1992 [80]. We chose this example because such composite bolometer can be easily described from a cryogenic point of view, whereas the analysis of more recent devices would be quite complicated. Nevertheless the bolometer performances are very good, the NEP being less than $10^{-16}\,\mathrm{W/(Hz)}^{1/2}$ and the time constant $\tau = 11$ ms. It was used for the measurement of the cosmic microwave background (CMB) in an experiment on board of a stratospheric balloon [81].

Let us remember that the bolometer noise consists of two contributions [82]: the phonon noise caused by fluctuations in the transfer of thermal energy between the bolometer and the heat sink and the Johnson noise due to the thermistor resistance.

For an optimally biased bolometer [74] ($T_{bol.} = 1.1 \times T_s$):

$$\mathrm{NEP_{bolo}} = \xi \cdot \left(4k_B T_{bol.}^2 \cdot G\right)^{1/2} \tag{15.19}$$

where $T_{bol.}$ is the operating temperature, G the thermal conductance to the heat sink and $\xi \approx 1$ (see eq. (15.6)).

Usually the infrared signal is modulated at a frequency f, and the responsivity of the bolometer decreases when $G < 2\pi f \times C$, where C is the heat capacity of the bolometer. The optimum NEP occurs when $G \approx 2\pi f \times C$ [74].

If we insert this G in the eq. (15.19):

$$\mathrm{NEP_{bolo}} \approx \xi \cdot \left(8\pi f \cdot C \cdot k_B T_{bol.}^2\right)^{1/2} \tag{15.20}$$

Since $C \propto T^3$ for dielectrics and $C \propto T$ for metals, the NEP of a bolometer scales as $\sim T^{2\pm0.5}$ (compare this result with eq. (15.6) for calorimeter where the $C \propto T$ contribution is usually negligible). Again we can deduce that the performance of the bolometer becomes better when the temperature is decreased. Nevertheless, the atmospheric noise for ground-based observations and photon noise for measurement from satellites put limits to the minimum 'useful' temperature of a bolometer. This limiting temperature is $\sim$0.1 K for ground observations with large telescopes and a few tens of millikelvin for space telescopes. These types of limitations occur at higher temperatures than those affecting calorimeters.

15.4.1.2 Bolometer components

The thermistor was a cubic chip (length 0.2 mm) of Ge NTD#17 (see Fig. 15.11), with $T_0 = 49$ K and $R_0 = 8$ ohm (see eq. (9.15)). It follows that the resistance of a chip at 300 mK is about 3 MΩ. The electrical contact (see Fig. 15.11) between the gold-plated faces of the thermistor and the graphite fibres were made with silver (20%) epoxy ($\approx$40 micron drops). The thermistor was connected to a low heat capacity absorber (diamond, 2 mm × 2 mm × 35 μm, covered by a 0.6 μm Bi film).

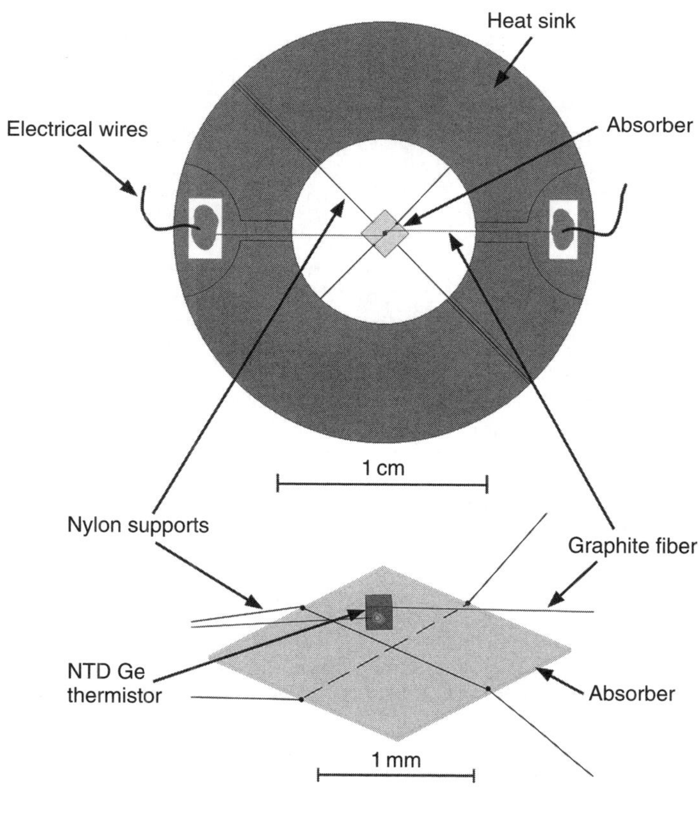

Fig. 15.11. Scheme of an infrared bolometer [80].

The depth of the Bi film and diamond was chosen to give a frequency-independent absorption [65]. The thermistor was glued to the diamond with a 40 μm drop of non-conductive epoxy.

The estimated heat capacities of bolometer components are shown in Table 15.3 (specific heat data references are in [80]).

The heat capacities are evaluated at the biased operating temperature of the bolometer (0.33 K). The heat capacity of the components that thermally connect the bolometer to the heat sink is supposed to contribute by 1/3. Hence, the total heat capacity at 330 mK should be 3.1×10^{-12} J/K.

For operation at 5 Hz, the optimum $G \approx 2\pi f \cdot C$, is approximately 2×10^{-10} W/K.

It was not simple (in 1992) to find out a low heat capacity (mechanically adequate) link giving such optimum G. Graphite fibres (1 cm long, 8 μm in diameters) glued with Ag epoxy to the NTD Ge thermistor (see Fig. 15.11) give a $G = 5 \times 10^{-12}$ W/K at 0.3 K. There is, however, another contribution to G due to the nylon wires which support the absorber. The characteristics of the bolometer are summarized in Table 15.4.

Table 15.3
Estimated heat capacities of bolometer components

Material	$C_{el.}$ [J/cm^3 K^2]	$C_{ph.}$ [J/cm^3 K^4]	Volume [cm^3]	$C_{tot.}$ at 330 mK [J/K]
Ag	6.3×10^{-5}	1.7×10^{-5}	6.7×10^{-9}	1.4×10^{-13}
Au	6.8×10^{-5}	4.5×10^{-5}	3.2×10^{-8}	7.7×10^{-13}
Pd	1.1×10^{-3}	1.1×10^{-5}	1.6×10^{-9}	5.6×10^{-13}
Ge	2.0×10^{-9}	3.0×10^{-6}	8.0×10^{-6}	8.7×10^{-13}
Graphite	6.1×10^{-6}	1.9×10^{-5}	1.3×10^{-7}	3.6×10^{-13}
Diamond	—	5.1×10^{-8}	1.4×10^{-4}	2.6×10^{-13}
Bi	3.8×10^{-7}	5.3×10^{-5}	2.4×10^{-7}	4.9×10^{-13}
Nylon	—	2.6×10^{-5}	1.1×10^{-6}	1.0×10^{-12}
Brass	9.7×10^{-5}	7.9×10^{-6}	1.3×10^{-7}	4.5×10^{-12}

Table 15.4
Bolometer characteristics

Heat sink temperature	296 mK
Responsivity at 0 Hz	8.7×10^7 V/W
Resistance	2.0 MΩ
V_{bolo}	8.2 mV
I_{bolo}	4.2 nA
G	8.8×10^{-10} W/K
C	1.7×10^{-11} J/K
τ	11.4 ms
Bolometer noise calculated at 5 Hz	6.4 nV/$\sqrt{\text{Hz}}$
Noise measured at 5 Hz	8 ± 0.5 nV/$\sqrt{\text{Hz}}$
Amplifier noise at 5 Hz	5 ± 0.5 nV/$\sqrt{\text{Hz}}$
Bolometer noise measured at 5 Hz	6.2 ± 0.8 nV/$\sqrt{\text{Hz}}$
NEP bolometer	7.5×10^{-17} W/$\sqrt{\text{Hz}}$
NEP total	1.0×10^{-16} W/$\sqrt{\text{Hz}}$

15.4.1.3 Note about infrared radiation (IR) filters

In the bolometer just described, the optimum conductance to the heat sink is $G \approx 2 \times 10^{-10}\,\mathrm{W/K}$. This means that an absorbed power of the order of 10^{-10} W saturates the bolometer. Since the bolometer is a broad-band detector, it would receive, e.g., a power of the order of 10^{-7} W from a 30 K black body. Of course, optical filtering is needed to reduce the bandwidth of the impinging radiation. Filtering takes usually place in several steps: a room temperature filter eliminates visible light; an intermediate temperature filter (at about 77 K) rejects the micron wavelengths, whereas the submillimetre or millimetre filter is made up of a low-pass and an interference band-pass filter.

In principle, cryogenics would not be involved in the realization of such IR filters, but the knowledge of thermal parameters of the materials used in the construction of the filters is needed as well as some information about vacuum.

IR filters are described e.g. in ref. [83,84].

References

[1] For a review of early cryogenic detectors see, for instance: B. Sadoulet: IEEE Trans. Nucl. Sci. NS-**35**, 47 (1988)

[2] LTD Proceedings:

LTD-1: *Low Temperature Detectors for Neutrinos and Dark Matter: Proceedings of a Workshop*, Held at Ringberg Castle, Tegernsee, May 12–13, 1987, ed. by K. Pretzl, N. Schmitz, L. Stodolsky, Springer-Verlag, Berlin, Heidelberg (December 1987)

LTD-2: *Low Temperature Detectors for Neutrinos and Dark Matter II; Proceedings of the Second European Workshop on Low Temperature Devices for the Detection of Low Energy Neutrinos and Dark Matter*, Annecy-le-Vieux, France, May 2-6, 1988, ed. by L. Gonzales-Mestres and D. Perret-Gallix, Gif-sur-Yvette, France, Editions Frontieres (1988)

LTD-3: *Proc. 3rd Int. Workshop on Low Temperature Detectors for Neutrinos and Dark Matter*, Gran Sasso, L'Aquila. Italy, Sept. 20 1989, ed. by L. Brogliato, D.V. Camin and E. Fiorini, Editions Frontieres, Gif-sur-Yvette, France (1990)

LTD-4: *Proc. 4th Int. Conf. of Low Temperature Dark Matter and Neutrino Detectors*, Oxford 1991, ed. by N.E. Booth and G.L. Salmon, Editions Frontieres, Gif-sur-Yvette, France (1992)

LTD-5: *Proc. 5th Int. Workshop on Low Temperature Detectors*, Berkeley CA, 29 July–3 August 1993, ed. by N.E. Booth and G.L. Salmon, J. Low Temp. Phys. 93 (1993)

LTD-6: *Proc. 6th Int. Workshop on. Low Temp. Detectors LTD6*, Beatenberg/Interlaken, Switzerland, ed. by H. R. Ott and A. Zehnder, Nucl. Instrum. & Methods A 370 (1996)

LTD-7: *Proc. 7th Int. Workshop on Low Temperature Detectors*, 27 July–2 August 1997, Munich, Germany, ed. by Ed. S. Cooper, F. Probst, W. Seidel, MPI Physik, Munich (1998)

LTD-8: *Proc. 8th Int. Workshop on Low Temperature Detectors*, Dalfsen, The Netherlands, August 15–20, 1999, ed by Piet de Korte and Tone Peacock, North Holland (2000)

LTD-9: *Ninth International Workshop on Low Temperature Detectors*, Madison, Wisconsin, 22–27 July 2001, ed. by F.S. Porter, D. McCammon, M. Galeazzi, C.K. Stahle, AIP Conference Proceedings **605**, American Institute of Physics (2002)

LTD-10: *Proceedings of the 10th International Workshop on Low Temperature Detectors*, Genoa, Italy, July 7–11 2003, Nuclear Instruments and Methods in Physics Research Section A, vol. 520, Issue 1–3, Elsevier (2004)

LTD-11: *11th International Workshop on Low Temperature Detectors*, Tokyo, Japan, 31 July–5 August 2005, Published in Nuclear Instruments and Methods in Physics Research Section A, vol. 559, Elsevier (2006)

[3] E.E. Haller et al.: Infrared Phys. **25**, 257 (1985)

[4] D. McCammon et al.: Jpn. J. Appl. Phys. **26**, 2084 (1987)
[5] S.H. Moseley et al.: IEEE Trans. Nucl. Sci. **NS 35**, 59 (1988)
[6] D. McCammon et al.: NIM A **326**, 157 (1993)
[7] A. Alessandrello et al.: NIM. A **370**, 241 (1996) and references therein
[8] T. Shutt et al.: Phys. Rev. Lett. **29**, 3425 (1992)
[9] T. Shutt et al.: Phys. Rev. Lett. **29**, 353 (1992)
[10] B. Censier: NIM A**559**, 381 (2006)
[11] V. Sanglard et al.: Phys. Rev. D **71**, 122002 (2005)
[12] W.S. Boyle, K.F. Rogers Jr.: J. Opt. Soc. Am. **49**, 66 (1959)
[13] R.A. Smith, F.E. Jones, B. Chasmar: *The Detection and Measurement of Infrared Radiation*, Clarendon, Oxford (1957)
[14] E.E. Haller: Infrared Phys. Technol. **35**, 127 (1994)
[15] E. Pasca et al.: *Proc. of 8th Int. Conf. Adv. Technol. Part. Phys.*, p. 93, ed. by M. Barone et al., Como 2003, World Scientific (2004)
[16] A.L. Woodcraft: J. Low Temp. Phys.**134**, 925 (2004)
[17] M. Barucci et al.: Physica B **368**, 139 (2005)
[18] F. Pobell: *Matter and Methods at Low Temperatures*, Springer, Berlin (1995)
[19] N. Wang et al.: Phys. Rev. B **41**, 3761 (1990)
[20] M. Barucci et al.: ICATPP-7 *Proc. of 7th Int. Conf. Adv. Technol. Part. Phys.*, p. 684, Como 2001, World Scientific (2002)
[21] P. Colling et al.: J. Low Temp. Phys. **93**, 549 (1993)
[22] G.C. Hilton et al.: IEEE Trans. Appl. Supercond. **13**, 664 (2003)
[23] K.D. Irwin, B. Cabrera, B. Tigner, S. Sethuraman: *Low Temperature Detectors for Neutrinos and Dark matter IV*, p. 209, ed. by N. Booth and G. Salmon, Frontieres (1992)
[24] L. Dumoulin et al.: NIM A **370**, 211 (1996)
[25] M. Barucci et al.: *Conference Issue of the IEEE Transactions on Nuclear Science* (TSN), 2000 Nuclear Science Symposium and Medical Imaging Conference, p. 13, Lyon (2000)
[26] K.D. Irwin et al.: J. Appl. Phys. **83**, 3978 (1998)
[27] H. Courtois et al.: J. Low Temp. Phys. **116**, 187 (1999)
[28] V.T. Petrashov et al.: Phys. Rev. B **58**, 15088 (1998)
[29] J.M. Martinis et al.: NIM A **444**, 23 (2000)
[30] F.M. Finkbeiner, et al.: p. 121; and A. von Kienlin, et al.: p. 176 in *Proc. 7th Int. Workshop on Low Temperature Detectors*, 27 July–2 August 1997, Munich, Germany, ed. by Ed. S. Cooper, F. Probst, W. Seidel, MPI Physik, Munich (1998)
[31] K.D. Irwin et al.: Nucl. Instr. Meth. A **444**, 184 (2000)
[32] E. Monticone et al.: in *Proc. of Ninth International Workshop on Low Temperature Detectors*, p.181, Madison, Wisconsin, 22–27 July 2001, ed. by F.S. Porter, D. McCammon, M. Galeazzi, C.K. Stahle, AIP Conference Proceedings **605**, American Institute of Physics (2002)
[33] M. Mohazzab et al.: J. Low Temp. Phys. **121**, 821 (2000)
[34] J.N. Ullom et al.: Appl. Phys. Lett. **84**, 4206 (2004)
[35] G.C. Hilton et al.: IEEE Trans. Appl. Supercond. **11**,739 (2001)
[36] K.D. Irwin: Appl. Phys. Lett. **66**, 1998 (1995)
[37] K.D. Irwin et al.: Appl. Phys. Lett. **69**, 1945 (1996)
[38] F. Pröbst et al.: J. Low Temp. Phys. **100**, 69 (1995)
[39] P.L. Whitehouse, P.J. Shirron, R.L. Kelley: Cryogenics, **44**, 543 (2004)
[40] A.T. Lee et al.: Appl. Phys. Lett. **69**, 1801 (1996)
[41] P. Nagel et al.: J. Low Temp. Phys. **93**, 543 (1993)
[42] C.A. Mears et al.: IEEE Trans. Appl. Supercon. **7**, 3415 (1997)
[43] F.C. Wellstood et al.: Phys. Rev. B **49**, 5952 (1994)
[44] S.W. Nam et al.: IEEE Trans. Appl. Supercon. **9**, 4209 (1999)
[45] A.J. Miller et al.: IEEE Trans. Appl. Supercon. **9**, 4205 (1999)
[46] H. Kraus,et al.: J. Low Temp. Phys. **93**, 533 (1993)
[47] U. Nagel et al.: J. Appl. Phys. **76**, 4262 (1994)
[48] P. Ferger et al.: Phys. Lett. B **323**, 95 (1994)

[49] K.D. Irwin et al.: Rev. Sci. Instrum. **66**, 5322 (1995)

[50] D.A. Wollman et al.: *Proc., Microscopy and Microanalysis 1998 Conf.*, Atlanta, GA: 196-197, Jul (1998)

[51] S.W. Deiker et al.: Appl. Phys. Lett. **85**, 2137 (2004)

[52] E. Fiorini, T.O. Niinikoski: NIM **224**, 83 (1984)

[53] S. Vitale: INFN Internal Report INFN/BE-85/2

[54] A. Drukier, L. Stodolski: Phys. Rev. D**30**(11), 2295 (1984)

[55] D. McCammon: NIM A, 520 (1–3), 11 (2004)

[56] D. Twerenbold: Rep. Progr. in Phys. 59(3), 349 (1996)

[57] C. Arnaboldi et al.: Phys. Lett. B **584**, 260 (2004)

[58] W. Herschel: Philos. Trans. R. Soc. **90**, 284 (1800)

[59] R.A. Smith, F.E. Jones, B. Chasmar: *The Detection and Measurement of Infrared Radiation*, Clarendon, Oxford (1957)

[60] P.L. Richards: J. Appl. Phys. **76**, 1 (1994)

[61] F.J. Low: J. Opt. Soc. Am. **51**, 1300 (1961)

[62] D.H. Andrews et al.: Rev. Sci. Instrum. **13**, 281 (1942)

[63] N. Coron, G. Dambier, J. Leblanc: *Infrared Detector Techniques for Space Research*, p. 121, ed. by V. Manno and J. Ring, Reidel, Dordrecht (1971)

[64] J. Clarke, G.I. Hoffer, P.L. Richards: Rev. Phys. Appl. **9**, 69 (1974)

[65] J. Clarke et al.: J. Appl. Phys. **48**, 4865 (1977)

[66] P.M. Downey et al.: Appl. Opt. **23**, 910 (1984)

[67] A.E. Lange et al.: Int. J. Infrared Millimeter Waves **4**, 689 (1983)

[68] E. Kreysa et al.: Proc. SPIE **3357**, 319 (1998)

[69] P.D. Mauskopf et al.: Appl. Opt. **36**, 765 (1997)

[70] R.V. Sudiwala et al.: NIM A **444**, 408 (2000)

[71] A.E. Lange et al.: ESA **SP-338**, 105 (1996)

[72] M. Piat et al.: *Proc. 7th Int. Workshop on Low Temperature Detectors*, p. 129, 27 July–2 August 1997, Munich, Germany, ed. by Ed. S. Cooper, F. Probst, W. Seidel, MPI Physik, Munich (1998)

[73] J.C. Mather: Appl. Opt. **21**, 1125 (1982)

[74] J.C. Mather: Appl. Opt. **23**, 584 (1984)

[75] J.C. Mather: Appl. Opt. **23**, 3181 (1984)

[76] G. Chanin and J. P. Torre: J. Opt. Soc. Am. **A1**, 412 (1984)

[77] M.J. Griffin and W.S. Holland: Int. J. Infrared Millimeter Waves **9**, 861 (1998)

[78] S.M. Grannan, P.L. Richards and M.K. Hase: Int. J. Infrared Millimeter Waves **18**, 319 (1997)

[79] M.J. Griffin: Proc. of LTD 8, 397 (1999)

[80] D.C. Alsop et al.: Appl. Opt. **31**, 6610 (1992)

[81] M.L. Fisher et al.: Ap. J. **388**, 242 (1992)

[82] F.J. Low: J. Opt. Soc. Am. **51**, 1300 (1961)

[83] C. Lee, P.A. Ade, C.V. Haynes: ESA**SP-388**: *Proceedings of the 30th ESLAB Symposium 'Submillimetre and Far-Infrared Space Instrumentation'* (December 1996)

[84] H.P. Gemünd et al.: ESA **SP-388**: *Proceedings of the 30th ESLAB Symposium 'Submillimetre and Far-Infrared Space Instrumentation'* (December 1996)

16 Large Cryogenic Experiments

Contents

16.1 Introduction

Several large cryogenic experiments have been realized in recent years. Many are under construction, a few very close to the realization, others in the R&D phase.

The description and status of most of these experiments can be found in the proceedings of LTD conferences (see ref. [2] of Chapter 15). Among these experiments we wish to mention:

1. Planck/Herschel [1–5], a space experiment which should be launched in 2007 by an Arianne 5G vector.

   The main goal of the Planck instrument is to improve the accuracy of the measurement of the cosmic microwave background (CMB), in order to extract cosmological parameters that remain poorly constrained after the results of WMAP (Wilkinson microwave anisotropy probe) and of the best ground-based experiments. The basic idea of HFI-Planck is to use all the information contained in the CMB radiation, i.e. to perform a radiometric measurement limited by the quantum fluctuations of the CMB radiation itself. In these conditions, the accuracy is only limited by the number of detectors and by the duration of the observation.

   Herschel, a far infrared and submillimetre instrument (60–670 μm), has the following main goals:

   - Study the formation of galaxies in the early universe and their subsequent evolution.
   - Investigate the creation of stars and their interaction with the interstellar medium.

- Observe the chemical composition of the atmospheres and surfaces of comets, planets and satellites.
- Examine the molecular chemistry of the universe.

Both instrument share the same refrigeration system [6], capable of working in no-gravity, which is described in Section 6.8.

2. *SCUBA-2* [7–15] (submillimetre common-user bolometer array), built by the Royal Observatory Edinburgh for the James Clerk Maxwell Telescope (JCMT), is a new generation, wide-field camera capable of carrying out large-scale surveys of the submillimetre sky, as well as deep, high-fidelity imaging of selected regions. To achieve these fundamental goals, the instrument must deliver the following:

   - Detector sensitivity dominated by the sky background. This largely dictates the cryogenic design of the instrument in that the detectors need to operate in the 100 mK regime.
   - The largest field-of-view achievable with the JCMT. Large re-imaging optics are needed to match the 660 mm diameter telescope field to a suitable array size (~100 mm). This also has consequences for the size of the cryostat.
   - Detector focal planes: the 850 μm focal plane instantaneously fully samples an area of sky, whilst the 450 μm is undersampled by a factor of 2. With over 10,000 pixels in total, the volume of readout electronics is considerably in excess of any previous instrument.
   - Dual wavelength imaging: imaging at two wavelengths simultaneously means that there will be two separate arrays of detectors fed via a dichroic beam splitter.

   The delivery of SCUBA-2 instrument to the JCMT is due for the end 2006.

   SCUBA-2 cryogenics is based on a dry dilution refrigerator (DR) (see ref. [7]): the main cryostat cooled by pulse tube coolers with 60 and 4 K stages, and the 1 K box and detector arrays with DR for 1 K and millikelvin stages. The PT system has a cooling power of 50W at 45K on the first stage and 1.0 W at 4.2 K on the second stage. The coolers were chosen for their inherently low vibration levels.

3. *CRESST* [16–31] (cryogenic rare event search with superconducting thermometers). This experiment is a search for WIMP (weakly interacting massive particles) dark matter via their elastic scattering on nuclei. WIMPs are massive particles which weakly interact with matter. The nuclei are in the absorber of a cryogenic detector, capable of detecting the small energy of the recoiling nucleus. Such a search for very rare interactions with a low energy deposit involves two major aspects:

   - A sensitive detector and a highly efficient suppression of background. Due to the low event rate anticipated for WIMP-nucleus elastic scattering, an extremely low background environment is required. Thus, to shield against cosmic radiation, the set up is installed in a deep underground site: under the Gran Sasso massive in Italy, in average covered by 1400 m of rock;
   - Ambient radioactivity originating from the surroundings is suppressed as much as possible by multiple passive shielding layers.

Cryodetectors are extremely sensitive and can measure the total energy deposited by an interacting particle. To achieve the millikelvin temperatures necessary to detect the low energies involved, the detector is mounted in a DR, which can reach temperatures below 10 mK. This system worked well in the first phase of the CRESST experiment, with sapphire calorimeters as target materials. Since recent work has indicated that the most important remaining background is neutrons, the CRESST II setup has recently installed an additional 50 cm thick polyethylene neutron shield around the cryostat, together with muon veto panels. However, to reach the level of background suppression needed for the next generation of experiments, such measures are not sufficient in themselves, and in CRESST II a new type of detector module, giving a high degree of active background suppression, is being installed. Thus, detectors were developed based on scintillating $CaWO_4$ crystals as absorbers. In these crystals, a particle interaction produces mainly phonons or heat, as for sapphire, but, in addition, a small amount of the deposited energy is emitted as scintillation light. Therefore, when a second smaller calorimeter is added to detect this light, most common backgrounds can be eliminated through their light signal. In tests, this system was found to give a very efficient active background discrimination.

The CRESST II setup will consist of up to 33 modules with both light and heat detection, reaching up to 10 kg of active target mass. The system is read out by a 66-channel SQUID system, two readout channels for each module.

Each module consists of two cryogenic calorimeters:

1. A phonon/heat detector to measure the total energy deposit. This is a cylindrical 300 g $CaWO_4$ crystal of both diameter and height 40 mm.
2. A second, smaller, calorimeter functioning as a light detector for the scintillation light.

The combination of phonon and light signals leads to a very efficient suppression of non-nuclear recoil backgrounds on an event by event basis. Data from early 2004 with two 300 g prototype modules, run for less than 2 months still without neutron shield, gave first significant limits on WIMP dark matter by the phonon-light technique [16]. As a by-product, the alpha decay of the 'stable' ^{180}W isotope with a half-life of $(1.8 \pm 0.2) \times 10^{18}$ years and an energy release of $Q = (2516.4 \pm 1.1(\text{stat.}) \pm 1.2(\text{syst.}))$ keV was unambiguously identified [17].

4. *MARE* [32–38]. Neutrino oscillation experiments have proved that neutrinos are massive particles, but cannot determine their absolute mass scale. Therefore, the neutrino mass is still an open question in elementary particle physics. An international collaboration is growing around the project of microcalorimeter arrays for a rhenium experiment (MARE) for directly measuring the neutrino mass with a sensitivity of about $0.2\,\text{eV}/c^2$.

In 1984, Fiorini and Niinikosky invented a 'true bolometric' low-temperature approach for rare event searches (e.g. $\beta\beta$-0ν see Section 16.4.1). Following this idea, Vitale proposed to adopt the new technique for single-beta decay studies applied to ^{187}Re. ^{187}Re is the lowest-Q known β-transition ($E < 2.5$ keV), and thus the most interesting for neutrino mass measurements [39]. ^{187}Re is the major isotope in natural rhenium (62%). The β transition is forbidden, but possible as a Gamow–Teller process in which the pair electron-antineutrino must carry at least one unit of orbital angular momentum. The resulting half-life time ($t_{1/2} > 40\,\text{Gyr}$) exceeds the age of the universe.

5. *EDELWEISS* II [40–48]. EDELWEISS (Expérience Pour Détecter Les Wimps En Site Souterrain) looks for:

- The dark matter in the form of particles (WIMP).
- The existence of supersymetry.

EDELWEISS experiment is dedicated to the WIMP direct search using heat-and-ionization Ge cryogenic detectors. Detectors are 300 g Ge disks with Ge neutron transmutation doping (NTD) temperature sensors and aluminium electrodes to detect ionization signals. The EDELWEISS experiment is located in the Modane Underground Laboratory (LSM) in the Fréjus tunnel connecting France and Italy under ~1800 m of rock. In the laboratory, the muon flux is $4\,\mu/\text{m}^2/\text{day}$ and the measured fast neutron flux $\sim 1.6 \times 10^{-6}\,\text{cm}^2/\text{s}$. The EDELWEISS II goals are to gain two orders of magnitude in sensitivity in comparison with EDELWEISS I. This gain in sensitivity depends on improvements on the background reduction and rejection and on the increase of the detector mass. The realization of EDELWEISS II is in two steps: a first with 28 bolometers and a second with 120. To reduce the radioactive background in the cryostat, all the materials are tested for radiopurity in a Ge dedicated detector, with very low radon level. A clean room surrounds the experiment.

A new generation of detectors has been developed with NbSi thin-film sensors (instead of the NTD sensors for present detectors). Each detector consists of a Ge crystal with two NbSi sensors acting also as electrodes for charge collection. These thin-film sensors are sensitive to the athermal component of the phonon signal, acting as near-surface interaction tag [49].

In the following sections we will describe in more detail two other large cryogenic experiments: MiniGRAIL (Section 16.3) and CUORE (Section 16.5).

16.2 Gravitational waves

One of the consequence of the Einstein's theory of general relativity (1916) is that gravity effects propagate with a finite velocity equal to the light speed c. In this (geometrical) theory, the three-dimensional space is combined with time into a four-dimensional space–time. The gravitational field is represented by the curvature of space–time. The way a source of space–time curvature moves is perceived at a distance in the form of waves. Because of conservation of momentum of an isolated system, there are no dipolar waves as in electromagnetism. Since the first non-zero time derivative in the multi-pole expansion of the energy–momentum tensor is the quadrupole, the total power I radiated by a source of density ρ and quadrupolar tensor D_{ik} is given by:

$$I = \frac{G}{5c^2} D_{\text{ik}}^2 \tag{16.1}$$

with $G/5c^2 = 5.5 \times 10^{-54}$ [1/W].

This is an extremely small quantity, which combined with the also extremely small interaction of gravitational waves (GWs) with matter makes it impossible to generate and detect GW on earth. Fast conversions of solar-size masses are required to produce signals with amplitudes that could be detectable. Astrophysical sources are for instance supernova explosions or a collision of two neutron stars or black holes.

The amount of mass that may be converted into GWs depends on the ellipticity of the event. A perfectly spherical event will not emit GWs.

A GW gives rise to a quadrupolar deformation normal to the direction of propagation. The deformation can be described by means of a dimensionless strain amplitude $h = \Delta L/L$, where ΔL is the deformation of a region of space–time separated by a distance L. For example, a supernova explosion, with a mass conversion into GWs of 1% of the total mass, at a distance of 10 kpc (roughly in the centre of our galaxy), would cause a strain on earth of $h \sim 3 \times 10^{-18}$ [50].

A GW detector of 1 m length will thus be deformed by $\sim 10^{-18}$ m, while the earth diameter will be deformed by about 10^{-11} m!

Considering that supernova explosions happen in average only once in ~40 years in our galaxy, we need to look for such events in a much larger volume of space, actually as far as the Virgo cluster (~10 Mpc = 3×10^{23} m distant), and has about 2500 galaxies.

Since the strain amplitude is inversely proportional to the distance (and thus the energy flux is inversely proportional to the square of the distance), a strain amplitude sensitivity of $\sim 10^{-21}$ would be needed to see several events per year.

The best antennas have, at present, a burst sensitivity of 3×10^{-19}, which is quite enough for events of that type in our galaxy and in neighbouring ones, but not enough to reach the Virgo cluster. Of course, there should also be much more energetic events like collisions of small black holes and/or neutron stars that could be detectable even at much larger distances. It is worth noting that events capable of causing strain amplitudes of 10^{-18} would cause a very high-power flux on earth. The energy flux can be estimated from [50–51]:

$$F = 30\,\mathrm{W\,m^2}\left[\frac{f}{1\,\mathrm{kHz}}\right]^2 \frac{\langle h^2\rangle}{(10^{-20})^2} \tag{16.2}$$

Thus, an event causing $h \approx 10^{-18}$ at 3 kHz (whatever the source distance) produces a power flux of 2.7 MW/m^2. Even so, the amount of energy deposited in any amount of matter will be extremely small.

A mechanical analogy can be done: we must anticipate that there will be very little coupling between GW and ordinary matter. In fact, we know that sound waves in common materials couple efficiently from one body of density ρ_1 and sound velocity ν_1 to another of density ρ_2 and sound velocity ν_2 if the product $\rho_1\nu_1 = \rho_2\nu_2$. The product $\rho\nu$ is called the sound impedance per unit area and has dimension of kg/m^{2}s. For an antenna made of copper with a surface area of ~1 m^2, the impedance is 3×10^7 kg/s. For space–time this would be:

$$\frac{C^3}{8\pi G} \approx 10^{34}\,\mathrm{kg/s} \tag{16.3}$$

so the mismatch is huge and only a fraction of order 10^{-27} will interact with the solid. This weak interaction makes the detection extremely difficult, but it has the attractive aspect that the waves reach us essentially in their original state, although the sources might be at large distances from us and the waves might have passed billions of stars. A GW contains direct information about its source, for instance the angular momentum and the mass of a black hole or neutron star and possibly information about the equation of state of very dense matter.

Until now, no certain evidence of the existence of GWs has been found. However Hulse and Taylor, studying the PSR 1913+13 binary system of pulsars, found that experimental data were fully coherent with the hypothesis that part of the system energy was lost through gravitational radiation [52–54].

Moreover, coincidence was measured between resonant bar detectors Nautilus and Explorer [55] in 2001.

16.2.1 GW detectors

16.2.1.1 Introduction

During the past four decades, two types of GW detectors have been developed: the resonant bar detectors and the laser interferometers. Joseph Weber started to study the feasibility of detecting GWs in the late 1950s [56]. His first antenna was an aluminium bar, suspended in vacuum at room temperature, with a mass of 1500 kg and a resonant frequency of 1600 Hz. He used piezo-ceramic crystals to monitor the vibrations of the first longitudinal mode of the cylinder, which should be excited by GWs. The strain $h = \Delta L/L$ caused by a GW will give a displacement $1/2\,\Delta L$ of the face-end of the cylinder with length L (thus a total of ΔL for the total bar, considering the centre immobile). The first antenna went in operation in 1966 and within 2 years, he built a second, similar detector at a distance of 1000 km to be able to do coincidence measurements. In 1969, he achieved a strain sensitivity of the order of 10^{-15}. He claimed to have measured coincidences caused by GWs [57], but his results were not verified by other groups. Since then, the sensitivity of the bar detectors has been improved with almost four orders of magnitude, mainly due to cooling the detectors down to 1 K or even 30 mK, using resonant transducers and SQUIDs (superconducting quantum interference devices) for the transducer readout (see Section 14.4).

Five bar detectors Nautilus (Rome, Italy), Explorer (CERN), Auriga (Padova, Italy), Allegro (Louisiana, USA) and Niobe (Perth, Australia) have been operating for several years now, having a strain sensitivity h between 3 and 6×10^{-19}. In order to better compare with the interferometers, which are broad-band detectors, the spectral amplitude $h\ (\mathrm{Hz})^{-1/2}$ is often used. For the resonant antennas, this is about $3 \times 10^{-22}\ (\mathrm{Hz})^{-1/2}$. All resonant detectors are being improved to increase the bandwidth (that at present is only about 1 Hz). The Explorer detector has already achieved a bandwidth of 10 Hz. The strain sensitivity is improved, mainly by using a lower noise SQUID and a better coupling of the transducer to the bar. Multimode capacitive transducers are also being studied. They are expected to reach a strain sensitivity h in the 10^{-20} range.

It was also Weber who proposed to use an interferometer to detect GWs. In 1969, his student Robert Forward was the first to build a prototype interferometric GW detector. In 1972 started the technology development for LIGO. At present, four groups are constructing laser interferometers for GW detection. The LIGO (USA) project consists of two large interferometers having arm lengths of 4 km, based in Livingston and Hanford, USA, and one smaller, 2-km detector, also in Hanford. VIRGO, a French/Italian collaboration, is building a 3-km interferometer in Cascina, near Pisa in Italy. GEO600 (Hannover, Germany) [58] and TAMA (Japan) [59] are two smaller interferometers with an arm length of 600 and 300 m respectively. A prototype interferometer of 80 m length arms AIGO is being built at 85 km north of Perth (Australia) and is being used for development work in view of 1-km-long interferometer. Laser interferometers can in principle achieve very high sensitivities, of the order of 10^{-22}–10^{-24} h $(\mathrm{Hz})^{-1/2}$, with a bandwidth of more than 1 kHz, which makes them ideal instruments to determine the waveforms. Nevertheless, it is important to realize that resonant antennas could also reach the same bandwidth.

Since cryogenics is not involved in interferometer detector (but the proposal of cooling mirrors has been done for VIRGO), we no longer deal with this type of detector.

16.2.1.2 Resonant mass detectors and resonant transducers

The principle of operation of a resonant mass detector is based on the fact that any mode of oscillation that has a quadrupole moment will be excited by a GW. The GW will deposit a fraction of its energy in the detector, and the components of the wave that match the quadrupole resonant frequencies of the antenna will excite those modes. The resonant detector will oscillate, during and after the GW has passed, for a time that is determined by the mechanical quality factor of the antenna. The traditional shape of a resonant mass detector is a cylinder with a typical length of a few meters having a mass of 1 or 2 tons. The fundamental quadrupole mode of the cylinder will interact most strongly with a GW that hits the cylinder perpendicularly to the axis. It is important to note that, being an oscillating mass spring system, the antenna will amplify the incoming signal by a factor that depends on the quality factor of the mass at $\omega = \omega_0$, where ω_0 is the resonance frequency of the antenna. A resonant detector is thus most sensitive at ω_0, but the sensitivity bandwidth is usually much larger (as the ratio of the narrow-band noise to the wide-band noise). The bandwidth can in fact be as large as the resonant frequency itself if the deposited (mechanical) energy could be completely transformed in electromagnetic energy at the motion transducer. Converting the mechanical energy into electromagnetic energy is one of the difficulties of the resonant antennas. The resonant frequency of oscillation of the cylinder is determined by its dimensions and by the sound velocity of the detector material. Since there is a large difference in propagation speed between the sound velocity of a metal and the speed of light, the dimension of a resonant detector will be much smaller than the wavelength of the GW.

With analogy to electric circuits, a *transfer function* of the antenna can be calculated and the response of the antenna to an incoming wave obtained. The output signal is usually expressed as antenna cross-section. It is defined as the ratio between the total energy absorbed by the antenna and the incident spectral density function of the incident wave. In the case of Nautilus antenna (2300 kg, 3×0.6 m) the cross-section is of the order of 10^{-25} $\mathrm{m^2\,Hz}$.

There is a tremendous insertion loss in converting the mechanical signal into an electric signal. Following Paik's suggestion [60], a resonant transducer is usually introduced which amplifies the mechanical signal before it is converted. A resonant transducer consists of a secondary mechanical resonator coupled to the antenna with a spring tuned to the frequency of the mode of interest. The displacement of the transducer is converted into an electronic signal by various techniques.

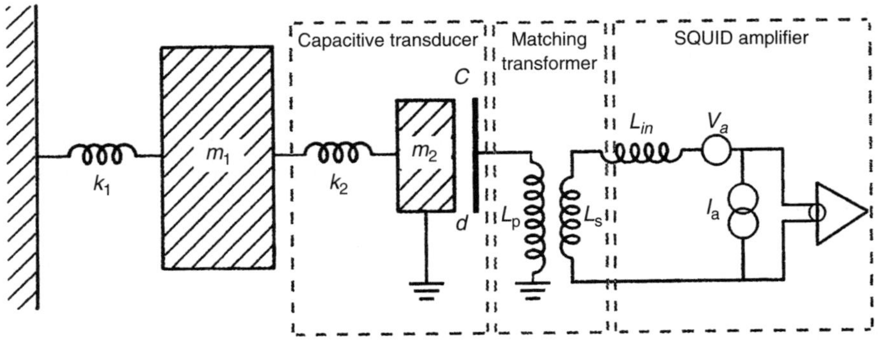

Fig. 16.1. Capacitive transducer for resonant gravitational waves detectors.

An example of capacitive transducer is shown in Fig. 16.1.

The sources of noise which limit the sensitivity of a gravitational antenna are:

- vibrations as seismic noise;
- thermal noise;
- amplifier noise.

Burst signals (supernovae) have been the primary target for the resonant mass detectors. The minimum detectable perturbation of the metric sensor caused by a burst of GW of duration τ is:

$$h_{\min} = \frac{L}{\left[(\nu_s)^2 \tau\right]} \left(\frac{\Delta E_{\min}}{M}\right)^{1/2} \tag{16.4}$$

where $\Delta E_{\min}$ is the minimum energy change detectable in the (bar) antenna, L and M are the length and mass of the antenna respectively and ν_s is the speed of sound in the bar material. $\Delta E_{\min}$ depends on the apparatus characteristics and is approximately:

$$\Delta E_{\min} \approx 4 \left(\frac{T_n T_e}{\beta Q}\right)^{1/2} \tag{16.5}$$

where T_n is the noise temperature of the amplifier, β is the ratio of the electromagnetic energy in the transducer to the total vibration energy in the antenna, Q is the mechanical quality factor of the bar and T_e is the temperature of the detector.

From the last two formulas, we see the need to use a large mass of high Q and high speed of sound material, reduce its temperature as much as possible, use a high β transducer and an amplifier with the lowest possible noise temperature.

A practical limit to the sensitivity is also given by noise source external to the apparatus. Multistage mechanical filters are always used. We wish now to examine the possibility of realizing a gravitational antenna of the spherical shape.

The idea of a spherical detector was suggested in the early 1970s by R. Forward [61]. The response of an elastic sphere was studied by a number of authors [62–70]. A sphere has five degenerate quadrupole modes that interact with a GW, and each mode can act as a separate antenna. Relatively small spherical GW detectors are very promising for frequencies above 1 kHz, where the laser interferometers start to lose sensitivity due to shot noise.

A sphere has a larger mass than a typical bar ($L = 5\text{D}$) resonating at the same frequency, and because it is equally sensitive for all directions and polarizations it has a cross-section (for the same material) that is about 75 times larger. A single sphere is also capable of determining the source direction and polarization. A spherical detector is the only detector for GWs with isotropic sky coverage and the capability of finding the location of the source. Both laser interferometers and bar detectors are unable to do this with just one detector: six bar detectors would be needed to build an omni-directional observatory.

16.3 MiniGRAIL

Figure 16.2 shows a schematic picture of MiniGRAIL (Spherical Gravitational Antenna In Leiden) [68–70] set up. The sphere is made of a Cu alloy (see Table 16.1).

A Q of 2×10^7 was measured.

The sphere is suspended from the centre and hangs from a vibration isolation stack, as depicted in Figs 16.2 and 16.3.

Fig. 16.2. Schematic picture of the MiniGRAIL setup (courtesy of Leiden Cryogenics).

Table 16.1
MiniGRAIL material properties

Material	CuAl6%
Density	$\rho = 8000\,\text{kg/m}^3$
Diameter	$\text{Ø} = 0.65\,\text{m}$
Mass	$M = 1150\,\text{kg}$
Sound velocity	$\nu_s = 4000\,\text{m/s}$
Resonant frequency	$f = 3160\,\text{Hz}$

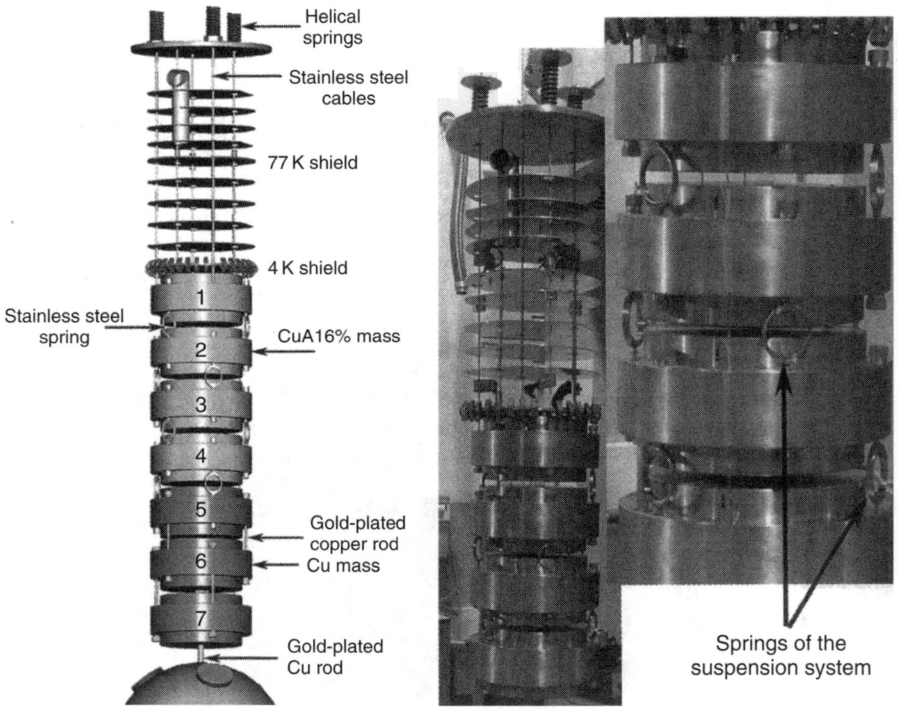

Fig. 16.3. Left: Schematic picture of the damping system of MimiGRAIL. The suspension consists of seven stages, the upper four made of CuAl followed by three copper masses. The upper CuAl mass is suspended from the top flange of the cryostat by stainless steel cables hanging from helical springs. Mass number 5, the first copper mass, will be cooled by the dilution refrigerator. Right: Picture of the four CuAl masses hanging from the top flange (courtesy of Leiden Cryogenics).

16.3.1 Cryogenics for MiniGRAIL

The cryogenic system for MiniGRAIL differs from standard dilution system. The outer diameter of the cryostat is about 1 m and it is 3 m high. The upper part of the cryostat contains the helium and nitrogen reservoirs. The bottom part of the cryostat consists of three radiation shields at 4 (indium seal), 77 and 300 K. The volume of the nitrogen and the helium reservoir is 200 and 340 l respectively. Four clear shot tubes give access to the IVC (inner vacuum can). Two of them, with a diameter of 19 mm, are used for the wires and for the input of a forced helium flow for rapid cooling down to 4.2 K.

The two 60 mm diameter feed-throughs are used as the 1 K pot pumping tube and as a feed-through for the SQUID wiring. The fifth tube is used to guide the ^{3}He pumping tube. The pumping system used for the circulation of ^{3}He consists only of dry pumps to prevent plugging of the condensing lines with oil.

Four Varian 550 turbo pumps are used in parallel. They are backed up by an Edwards mechanical booster (2500 m^3/h) and a Pfeiffer uni-dry (60 m^3/h). A maximum circulation rate could be achieved at about 7 mmole/s. The ^{3}He pumping tube is inserted from the top of the cryostat.

A special care was devoted to the minimization of the Kapitza resistance in the mixing chamber (Fig. 16.4) of the DR. The sinter slabs are arranged in a star shape on the cold plate of the mixing chamber. Everything is machined out of one piece of copper. The sinter powder (650 g of DC10 [71] powder) is pressed from the side on to the copper slabs, giving a homogeneous sinter distribution in the vertical direction. The housing of the mixing chamber was made of stainless steel, welded to two stainless steel rings which were silver soldered to the copper plate before the sintering process.

The diameter of the copper slabs is about a centimetre smaller than the inner diameter of the stainless steel housing, leaving a channel for the helium flow. The top of housing has a conical shape, to reduce the amount of ^{3}He necessary to have the phase separation in the right position. The total Kapitza resistance at 20 mK is about 45K/W (e.g. a heat leak of 50 μW on the cold plate of the mixing chamber would give a temperature difference of about 2 mK between the liquid and the cold plate).

Another innovative aspect of the cryogenics for MiniGRAIL is the cooling down to 77 K with a forced helium flow. Nautilus and Auriga experiments need about 20 days to cool a 2.3 ton Al5056 antenna with a surface area of about 6 m^2 from room temperature

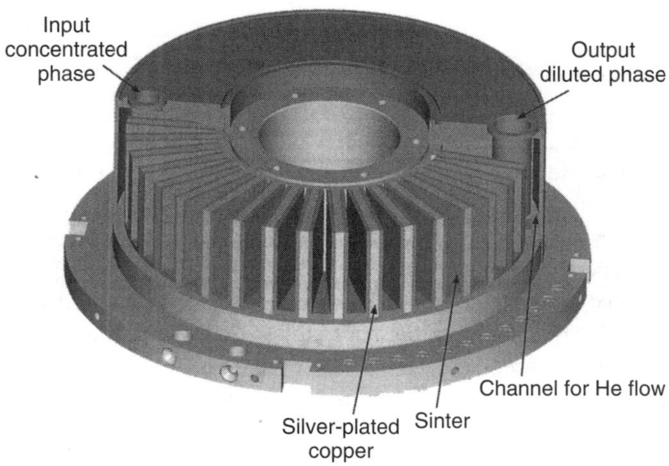

Fig. 16.4. Schematic view of the mixing chamber of MiniGRAIL (courtesy of Leiden Cryogenics).

Table 16.2
The cool down of MimiGRAIL compared to Nautilus

Antenna	$T_{antenna}$ [mK]	T_{mc} [mK]	$\Delta T_{mc\text{-}antenna}$ [mK]	W_{mc} [μW]
MiniGRAIL	81	21	60	45
Nautilus	95	45	50	10

to 77K [72]. The specific heat of Al is twice as high as the specific heat of copper at room temperature. The total heat capacity of MiniGRAIL ($M = 1150\,kg$) is a factor of four lower than the total heat capacity of a traditional bar, but with a contact area of a factor of 6 smaller. Assuming the same exchange coefficient as for the bar antennas, cooling the sphere with exchange gas would take about 1 month. The total enthalpy that has to be removed is in the order of 10^8 J. Using a forced helium flow which is able to remove typically $500\,W/m^2$, the time needed to cool down the sphere to 77 K can be reduced to about 2.5 days.

The enthalpy of copper at nitrogen temperature is $H_{77K} = 6\,J/g$, so the total entropy of the sphere will be about 6×10^6 J. The time needed to cool from 77 K down to 4 K is of the order of 4 h. The total helium consumption from room temperature to 4.2 K would be about 600 l. The temperatures reached in a test run are reported in Table 16.2. The expected final sphere temperature is about 20 mK. A comparison of MiniGRAIL and Nautilus cool down is made in Table 16.2. The high power leak on the sphere has been attributed to a time-dependent heat leak caused by the ortho–para conversion (see Section 2.2) of molecular hydrogen present in the copper of the sphere (see Fig. 16.5) (the Nautilus bar instead is made by Al). A similar problem has been found in the cool down of the CUORICINO Frame (see Section 16.6).

A new generation of cryogenic GWA are under study with the goal of improving sensitivity of two orders of magnitude [73–76].

16.4 Neutrino Physics

In the last decade, neutrino experiments have demonstrated that neutrinos are massive particles which may oscillate among three autostates. Such experiments [77–82] have evidenced the mass difference between the autostates, but not the neutrino mass scale value. The only way to determine the neutrino mass is the knowledge of the shape of the end point of energy spectrum in beta decays. In the hypothesis of the Majorana neutrino (neutrino coincides with antineutrino and its rest mass is different from zero), the measure of the decay half-life in the neutrinoless double-beta decay (DBD) would be necessary. A number of recent theoretical interpretations of neutrino oscillation experiments data imply that the effective Majorana mass of the electron neutrino (as measured in neutrinoless DBD) could be in the range 0.01 eV to the present bounds.

Massive cryogenic calorimeters are considered of great importance in experiments of neutrinoless DBD. In particular, they can investigate about several nuclei candidate for DBD, taking advantage from the fact that the detector may be rich in nuclei under study.

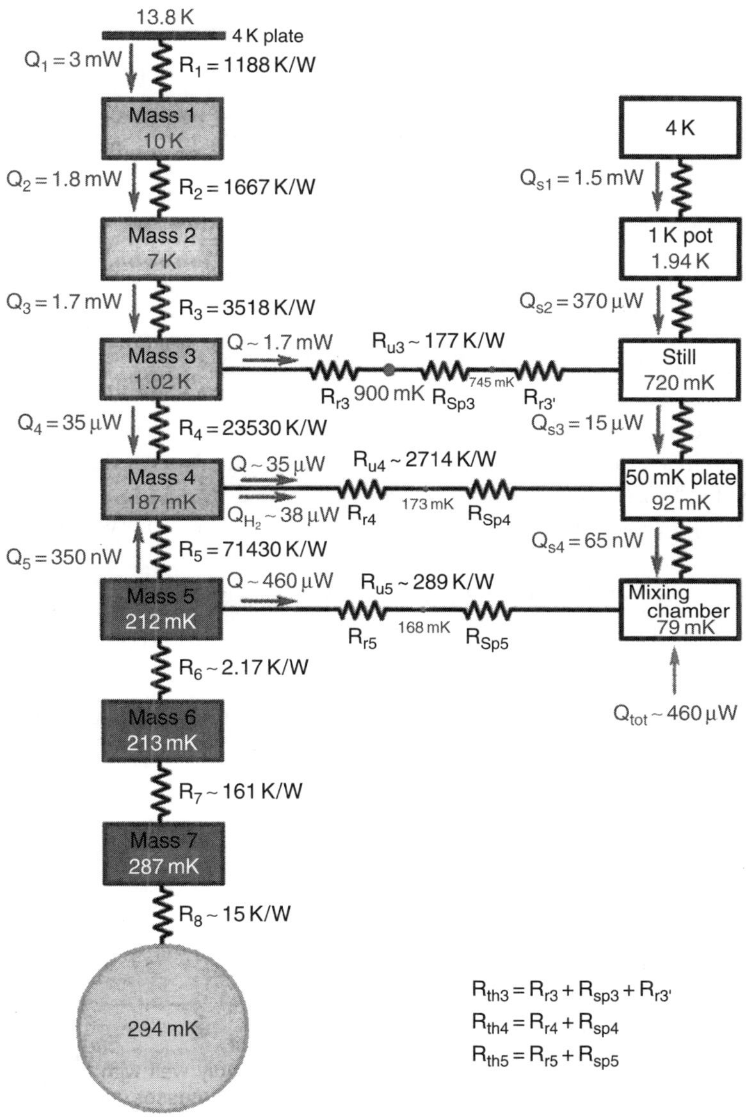

Fig. 16.5. An overview of the minimum temperature of the different elements of the system. An estimate is made on the heat flows Q due to conduction between the different stages that are all connected with stainless steel rods or tubes. The total heat leak on the mixing chamber is estimated to be 45μW. This heat leak decreases in time and comes from the sphere and copper masses. We will see further on that this can be explained by ortho–para conversion of 70 ppm hydrogen impurities in the copper (courtesy of Leiden Cryogenics).

16.4.1 The DBD

The DBD is a rare radioactive process allowed by the standard weak interaction theory [83].

A nucleus (A;Z) decays with the emission of two electrons (e^-) and two antineutrinos ($\bar{\nu}_e$) following the equation:

$$(A;Z) \rightarrow (A;Z+2) + 2e^- + 2\bar{\nu}_e \tag{16.6}$$

This transition has been observed for a few nuclei [84–87].
Instead, the 0ν DBD:

$$(A;Z) \rightarrow (A;Z+2) + 2e^- \tag{16.7}$$

does not respect the law of conservation of the leptonic number. The observation of such decay would violate the Quark Lepton Symmetry of the Standard Model and hence would bring to the birth of a new neutrino physics.

To observe process of eq. (16.7), the energy associated with the two electrons, higher than for process of eq. (16.6), can be measured.

Two types of detectors are possible:

(a) the source of the decay is external to the detector,
(b) the source is part of the detector.

The decay rate for the process involving the exchange of a Majorana neutrino can be expressed as:

$$\left[T_{1/2}^{0\nu}\right]^{-1} = \frac{|\langle m_\nu \rangle|^2}{m_e^2} G^{0\nu} \left|M^{0\nu}\right|^2 \tag{16.8}$$

where $|\langle m_\nu \rangle|$ is the effective Majorana electron neutrino mass, $G^{0\nu}$ the two-body phase-space factor including the coupling constants and $M^{0\nu}$ the nuclear matrix element of the neutrinoless DBD. Some values of $|\langle m_\nu \rangle|$ have been recently evaluated on the base of the latest results of neutrino oscillation experiments [77].

According to most theoretical analyses of the present neutrino experiment results, next-generation DBD experiments with mass sensitivities of the order of 10 meV may find the Majorana neutrino with a non-zero effective electron neutrino mass, if the neutrino is self-conjugate and the neutrino mass spectrum is of the quasi-degenerate type or it has inverted hierarchy [83]. Majorana massive neutrinos are common predictions in most theoretical models, and the value of a few 10^{-2} eV predicted for its effective mass, if reached experimentally, will test its Majorana nature.

DBD experiments with even better sensitivities (of the order of meV) will be essential to fix the absolute neutrino mass scale and possibly to provide information on CP violation. It is therefore evident that next-generation neutrinoless DBD experiments are the next important steps necessary for a more complete understanding of the physics of neutrinos. In the next section, we will describe the CUORE experiment and show how it could reach the required sensitivity.

16.5 CUORE

Rare event physics is playing a significant role in modern physics: the rare event signals, if detected, would be an evidence for the need of a new physics, beyond the standard model of particle Physics, and would have far-reaching consequences in Cosmology.

Due to the low rates of the proposed processes, the essential requirement of these experiments is to achieve an extremely low radioactive background and large mass. In some of these phenomena, a very small amount of energy is deposited, and the sensitivity needed to detect the events must be much large.

The proposed CUORE (cryogenic underground observatory for rare events) [83] experiment is a large arrays of detectors for the research of $\beta\beta(0\nu)$ decay.

CUORE belongs to the type (b) detectors (source is part of the detector). The detection of the DBD is done by means of direct measurement of the energy of the electrons emitted in the process.

The candidate material is ^{130}Te. The ^{130}Te decay:

$$^{130}\text{Te} \rightarrow \ ^{130}\text{Xe} + 2\text{e}^- \tag{16.9}$$

should produce 2528 keV.

CUORE is a proposed tightly packed array of 988 TeO_2 detectors, each being a cube 5 cm^3 on a side with a mass of 750 g. The array consists of 19 vertical towers with 52 crystals each, arranged in a cylindrical structure as shown in the Fig. 16.6.

Each tower will consist 13 frames of 4 crystals (see Fig. 16.6).

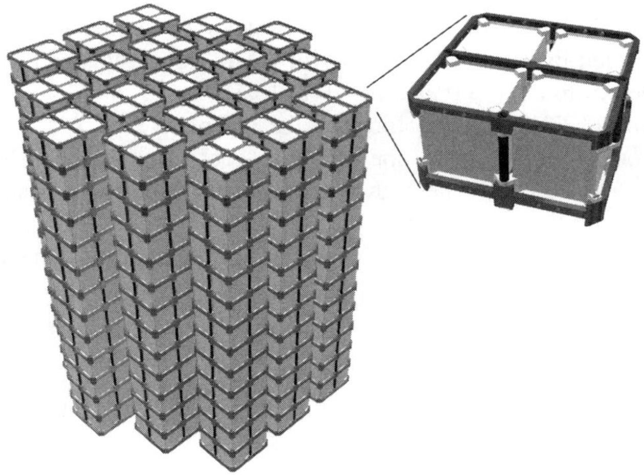

Fig. 16.6. The CUORE detector array and the base module of 4 crystals.

Each detector working at about 10 mK and (see Section 15.3.2 for approximate model) is made of:

- An energy absorber: a TeO_2 crystal ($5 \times 5 \times 5\,cm^3$) containing a high percentage of ^{130}Te.
- An NTD Ge thermometer.
- A Si heater for calibration and equalization purpose.
- A mechanical support (PTFE) and a Cu frame.

A single CUORE detector consists of a single crystal of TeO_2 that acts both as source and absorber. The NTD Ge thermometer and the heater are glued on the TeO_2 absorber by means of araldite glue. An energy release causes an increase of the temperature of the absorber which is measured by the thermometer.

Resistors of 100–200 kΩ, realized with a heavily doped meander implanted on a 1 mm^3 silicon chip, are attached to each absorber in order to calibrate and stabilize the gain of the bolometer over long running periods. The towers will be mechanically connected to an OFHC copper top plate, elastically suspended from the DR mixing chamber.

Taking into account the temperature of operation of the detectors and the total mass to cool, a powerful DR (similar to that of the Nautilus Collaboration that cools a 2-ton gravitational antenna) will be needed.

CUORE detectors will be cooled by a ^3He–^4He refrigerator with a cooling power of 3 mW at 120 mK. Refrigerators with the required characteristics are technically feasible (see Section 6.1). A cryogenic configuration, which allows to keep the 4 K stage temperature constant for the whole duration of the experiment, is foreseen. This solution is based on new commercial pulse tube coolers (see Section 5.8).

CUORE refrigerator will not use cryogenic liquids: five pulse tube refrigerators for the 45 K and the 4 K shields, and a 'dry DR' (see Section 6.7) for the 4 K and lower temperature stages.

In the present state of art, the CUORE cryostat will consist of a room temperature enclosure, radiation shields at 45 K and at 4 K, both cooled by pulse tubes. The 'Still' radiation shield at about 700 mK (including lead shielding), 50 mK and a mixing chamber shield will be cooled by the different stages of a DR. All shields and the crystal + lead will be suspended from the top flange (see Fig. 16.7).

For the cooling of CUORE, it is foreseen the use of five PT 415 (Cryomech) pulse tubes (see Section 5.8). Each pulse tube delivers a cooling power of 40 W at 45 K at the first stage and 1.5 W at 4.2 K. The high power requested by the first stages depends on the thermal load from the shields.

As an example, let us consider the Cu shield at 45 K, with a diameter of 154 cm, a height of 297 cm, and a wall thickness of 0.5 cm. We have:

- Internal surface area = $15.15\,m^2$
- Volume = $0.079\,m^3$
- Mass = 705 kg

And considering the closing plate of the shield:

- Total surface at 45K = $17.16\,m^2$
- Total mass at 45K = 1064 kg

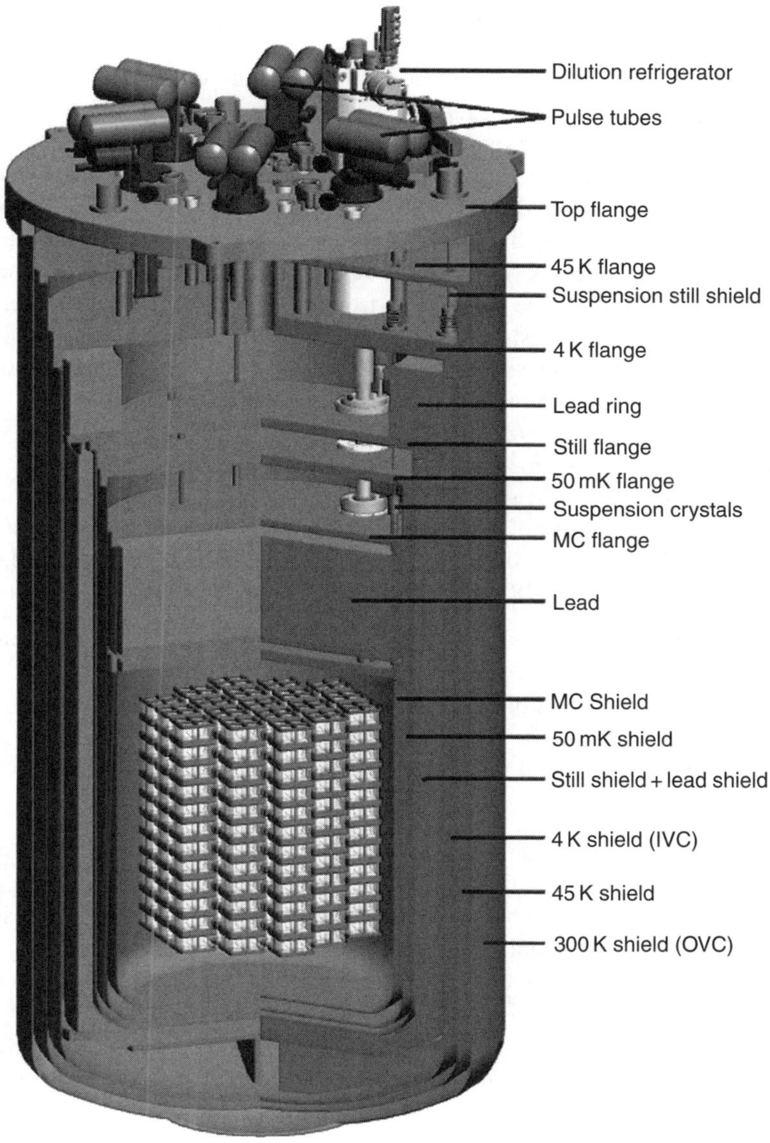

Fig. 16.7. Schematic of CUORE cryostat.

The estimated steady-state heat loads on the shield at 45 K are:

- Conduction = 44.8 W
- Radiation = 98.5 W
- DL mixture = 3 W

For the total cooling power of 146.3 W, at least four pulse tube PT415 are needed.

All the materials used to build the experiment (detectors, cryostat, shielding) are materials with a low radioactive contamination.

Once machined, all the copper and teflon pieces will undergo a surface cleaning procedure that will guarantee the required low level of surface radioactive contamination for those parts that directly face the detectors.

The thousand bolometers will drive an anticoincidence electronic system to eliminate spurious signals such as those due to cosmic radiation. For the same purpose, the bolometer structure will be surrounded by three shields of roman lead (without ^{210}Pb contamination): the first, 10 cm thick, at 10 mK; the second and third at 50 mK and 600 mK respectively. A Faraday cage will protect the experiment from electromagnetic interference. CUORE is now in an initial realization stage.

CUORE will be located in an underground hall of Gran Sasso National Laboratory (L'Aquila – Italy) at a depth of 3400 mwe, where the muon flux is only $\sim 3 \times 10^{-8}\,\mu/\mathrm{cm}^2/\mathrm{s}$ and neutron flux $\sim 10^{-6}\,\mu/\mathrm{cm}^2/\mathrm{s}$.

The goal of CUORE is to achieve a background rate between 0.001 and 0.01 counts/(keV · Kg · y) at the ββ-0ν transition energy of ^{130}Te (2528 keV).

Radioactive contaminations of individual construction materials, as well as the laboratory environment, were measured and the impact on detector performance was determined by Monte Carlo computations [83]. The background sources which were considered are:

1. bulk and surface contamination (^{238}U, ^{232}Th, ^{40}K and ^{210}Pb) of the construction materials;
2. bulk contamination of construction materials due to cosmogenic activation;
3. neutron and muon flux in the Gran Sasso Laboratory;
4. gamma ray flux from natural radioactivity in the Gran Sasso Laboratory;
5. background from the ββ(2ν).

A very conservative evaluation of the background attainable for CUORE has been estimated. This rather pessimistic approach in background evaluation is the only one that presently guarantees a reliable prediction. The expected performance and sensitivity indicate that CUORE will be able to test the 0.02–0.05 eV region for $\langle m_\nu \rangle$.

The expected sensitivity for CUORE is $2.93 \times 10^{26}\,\sqrt{t}$ year, where t is the live running time of the experiment in years [88].

A single CUORE tower was built for a smaller scale experiment called CUORICINO (which in Italian means small CUORE), operating since December 2002 in the Gran Sasso National Laboratory.

16.6 CUORICINO

The main goal of CUORICINO was to demonstrate the feasibility of CUORE, which is to find out and solve the eventual problems, to test the purity of materials and radioactive background, mechanical vibration level, etc.

CUORICINO is an array of 62 TeO_2 detectors designed to search for the neutrinoless DBD of ^{130}Te. With a total mass of 40.7 kg of TeO_2, corresponding to about 11 kg of ^{130}Te, it is one of the more sensitive experiments presently running. In few years, CUORICINO will be able to reach a sensitivity on the Majorana mass of neutrino of the order of 0.1–0.5 eV.

Designed on the basis of the experience gained by with the MiDBD experiment (a 20 bolometers array) [89], CUORICINO is an array of 44+18 TeO_2 crystals arranged in a 13-plane tower structure (see Fig. 16.8) [90–94]. Eleven of these planes are four-crystal

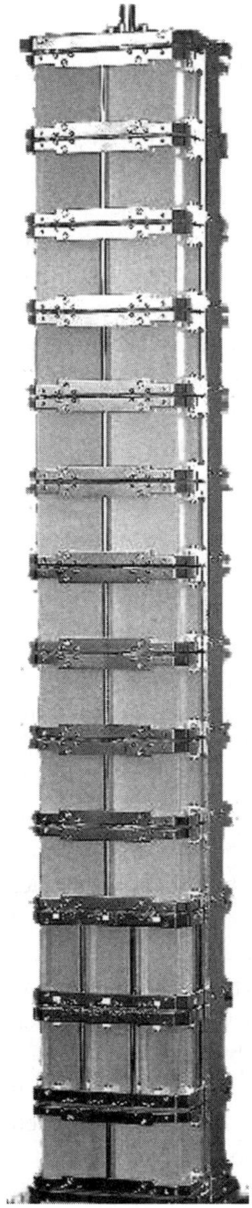

Fig. 16.8. CUORICINO array completely mounted. Two different modules are used: the one with the large size crystals (four-crystal module) and that with the small size ones (nine-crystal module) [90].

modules similar to those foreseen for CUORE. Two additional planes are made of nine-crystal modules; in these two planes 18 of the 20 crystals used for the MiDBD experiment have been housed. Therefore CUORICINO contains 44 TeO_2 crystals of size $5{\times}5{\times}5\,cm^3$ and mass 790 g and 18 TeO_2 crystals of size $3{\times}3{\times}6\,cm^3$ and mass 330 g. All the crystals are made of natural tellurium, except four of the small size ones. These are made with enriched materials: two of them are enriched to 75% in ^{130}Te and two are enriched to 82.3% in ^{128}Te.

To fulfil the background requirements typical of rare event physics, particular care was dedicated to the selection and treatment of the materials used for the construction of CUORICINO: the crystals where grown with low contamination materials in China and sent by ship to Italy where they have been optically polished with specially selected low-contamination powders. The mechanical structure of the array was made exclusively in OFHC copper and PTFE: both these materials have an extremely low radioactive content. All the copper and PTFE parts of the mounting structure underwent a chemical treatment to remove any possible surface contamination. Finally, the array was assembled in an underground clean room in a N_2 atmosphere to avoid Radon contamination.

CUORICINO crystals are grouped in elementary modules of four elements (see Fig. 15.7) held between two copper frames joined by copper columns. PTFE pieces are inserted between the copper and TeO_2, as a heat impedance and to clamp the crystals. There is a 6 mm gap between crystals with no material in between. The four detectors are mechanically coupled; some of the PTFE blocks and springs act simultaneously on two crystals.

The absorbers of CUORICINO detectors are diamagnetic crystals (TeO_2) of $5{\times}5{\times}5\,cm^3$ (44 of them) with a heat capacity $C(T) \propto (T/\theta_D)^3$.

The heat capacity C (at 10 mK) of an absorber is 2.3×10^{-9} J/K (see Section 12.4). An absorption of 10 keV produces a temperature increase of about 1 μK, detected by a very sensitive NTD Ge thermometer (see Section 9.6.2).

^{130}Te is an excellent candidate for the search of DBD because of its high transition energy (2528.8 ± 1.3 keV [95]), its great isotopic abundance in natural tellurium (33.80 ± 0.01% [96]) and its relatively high Debye temperature (~232 K) [97].

The thermometers were developed and produced at the Lawrence Berkeley National Laboratory (LBNL) and UC Berkeley Department of Material Science; they are unique in their uniformity of response. The TeO_2 crystals are produced by the Shanghai Quinhua Material Company (SQM) in Shanghai, China, which will supply also the 750 g TeO_2 crystals for CUORE.

Once closed inside its copper frame, the array was mounted and thermally connected to the mixing chamber of the Oxford DR installed in Hall A of the underground Gran Sasso National Laboratory. The tower was mechanically decoupled from the cryostat in order to avoid vibrations induced by pumps on the detectors and preamplifiers [98].

The damping of such vibrations may also heat the detectors. For this reason, the vibration spectrum of the cryostat has been measured with piezoelectric accelerometers placed on the 1 K pot, the Still, the mixing chamber and the crystal copper frame [98].

These vibration measurements have linked the vibration modes in several working conditions to the bolometer noise.

In classic DR (see Section 6.3), the main source of noise is the continuous refilling of the 1 K pot through a capillary. Noise spectra have been measured in three different situations

(a) capillary valve open (working situation);
(b) valve closed;
(c) inserting a flow impedance in series with the capillary in order to obtain a constant pressure gradient between the helium bath at ~1 bar and the 1 K pot (at ~1 mbar) where ^{4}He is superfluid.

Figures 16.9 and 16.10 show the vibration spectra in the three cases.

To reduce the 'high frequency' noise of Fig. 16.9, and also to minimize 'low frequency' vibrations of the tower, the crystal copper frame has been mechanically decoupled from the cryostat.

To do this, the tower is therefore connected, through a 25 cm copper bar passing through the mixing chamber, to a stainless steel spring fixed at the 50 mK plate. This experimental configuration has appreciably reduced the 'mechanical' noise on the detectors

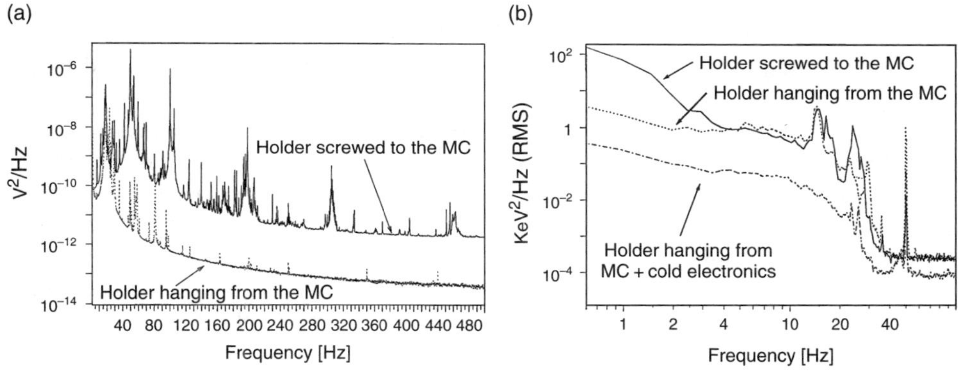

Fig. 16.9. (a) Vibrational noise spectra of the same crystal holder, measured before and after the mechanical suspension. (b) Energy noise spectra obtained with the same detector in different runs. The electronics has a four-pole 12 Hz Bessel filter [98].

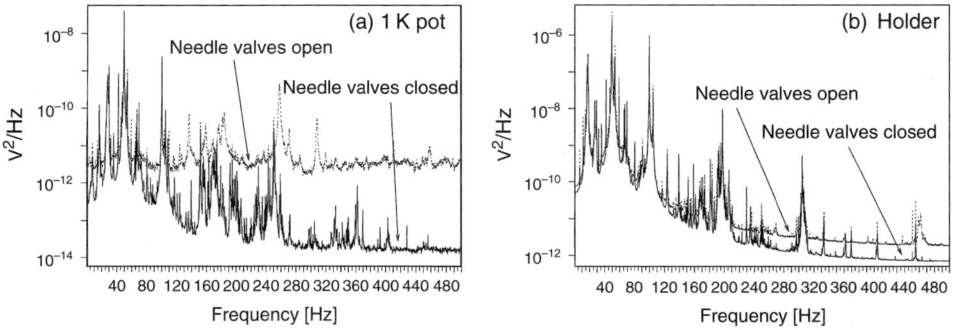

Fig. 16.10. Different vibrational spectra obtained in different cryostat conditions on: (a) 1 K pot. (b) Crystal holder [98].

(see Fig. 16.9). The longitudinal oscillation frequency of the system is ~1.8 Hz. The thermal link is ensured by two $40 \times 10 \times 0.05\,mm^3$ copper (99.995%) foils connecting the MC to the tower. The heat conductance between MC and the tower is $\sim 4.2 \times 10^{-5}$ W/K at 8 mK. The temperature of the tower is stabilized through an especially designed feedback device [99].

This new mechanical (and thermal) configuration has overcome the vibration problems at the expense of a longer cold down time due to smaller thermal conductance of the thermal link. It has also pointed out phenomena not observed in the initial configuration.

One of them is the ortho–para conversion of hydrogen trapped in the frame copper during the production process (see ref. [100–102] and Section 2.2).

The array is surrounded by a ~1 cm thick roman lead cylindrical shield closed with a bottom and a top lead discs of thicknesses of 7.5 cm and 10 cm respectively.

The refrigerator itself is shielded with a 20 cm thick low activity lead and a 10 cm thick borated PET. Nitrogen is fluxed between the external lead shield and the cryostat to avoid any Rn contribution to the detector background. Measurements of residual radioactivity have been carried out on several Pb samples [103]. For the 'roman' lead, a contamination lower than 4 mBq/Kg (in ^{210}Pb) is to be compared with ~250 mBq/Kg of a modern lead (Johnson & Mattey).

CUORICINO has 24 cold electronics acquisition channels and 38 channels with room temperature amplifiers. The formers are realized by JFET pairs working in the cryostat at ~120 K [104]. The differential configuration and their position close to the detectors (between 1.5 K and 600 mK stages) have minimized interferences and microphonic noise from the detector wires.

CUORICINO was cooled at the beginning of year 2003 (Run 1); during the cooling procedure, some of the signal wires disconnected, so that only 32 of the large-size crystals and 16 of the small ones could be read. The problem was solved in 2004 by warming and opening the cryostat to fix the electrical connections. Only for two 790 g crystals, this was not possible. The Run 2 of CUORICINO started in September 2004 with 40 big-size and 18 small-size crystals (two big-size crystals had an excess noise and were no longer used). In both runs, the detectors were at a temperature of about 10 mK. To search for $\beta\beta$-0ν, the detectors are operated in anticoincidence, rejecting a fraction of background events.

The performances of the detectors are quite good: the average FWHM resolution measured during the calibration of the detectors with a ^{232}Th source was ~8 keV for the big-size crystals and of ~10 keV for the small-size crystals. Both these values were measured on the ^{208}Tl gamma line at 2615 keV, just above the position (2528.8 keV) where the $\beta\beta$-0ν peak should appear. Similar resolutions were measured on the sum background spectra of big and small crystals (see Fig. 16.11).

The sum background spectra (in the so-called 'gamma region') of big- and small-size crystals are shown below in Fig. 16.11. The peaks, clearly appearing in the spectra, are due to gamma emission of radioactive contaminants present in the materials: the ^{238}U and ^{232}Th chain gammas as well as the 40K gamma are clearly evident.

In the DBD region, two peaks appear, one below the $\beta\beta$-0ν Q value and one above. They are due respectively to the ^{60}Co sum line (2505.5 keV) and the ^{208}Tl line (2615 keV). The first is ascribed to a ^{60}Co contamination of the copper structures due to cosmogenic

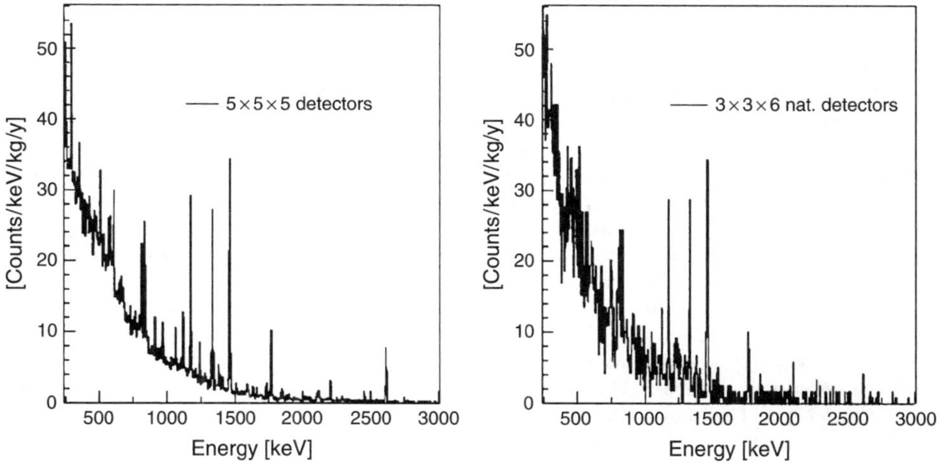

Fig. 16.11. Sum background spectra of big- and small-size crystals [90].

activation, whereas the second is ascribed to ^{232}Th contamination outside the internal roman lead shield. The first line does not give any contribution to the ββ-0ν background; the second proves that a contribution from ^{232}Th through multicompton of the 2615 keV line is present. This background cannot however explain the whole background counting rate in the 0νDBD region that – according to the present interpretation – receives a non-negligible contribution from alpha (from U and Th chains) surface contaminations of the array (the crystals themselves and the inert materials directly facing the crystals). A detailed study of the background sources is on the way.

No evidence for ββ-0ν decay was found up to 2006 and a new lower limit, $T_{1/2}^{0\nu} \geq 1.8 \times 10^{24}$ year (90% C.L.) is set, corresponding to $\langle m_\nu \rangle \leq 0.2 - 1.1$ eV, depending on the theoretical nuclear matrix elements used in the analysis [105].

The limit is evaluated using anticoincidence sum spectra and considering separately the sum spectra of big, small and enriched crystals and distinguishing the two runs.

References

[1] F.R. Bouchet et al.: *SF2A-2005 Semaine de l'Astrophysique Francaise*, p. 675, Strasbourg, France, June 27 – July 1, 2005, ed. by F. Casoli, T. Contini, J.M. Hameury and L. Pagani, EdP-Sciences, Conference Series (2005)

[2] M. Giard, L. Montier: Astrophys. Space Sci. **290**, 159 (2004)

[3] J.M. Lamarre et al.: New Astronomy Reviews **47**, 1017 (2003)

[4] M. Piat et al.: A. & A. **393**, 359 (2002)

[5] J.M. Lamarre et al.: Astrophys. Lett. Commun. **37**, 161 (2000)

[6] S.Triqueneaux et al.: Cryogenics **46**, 288 (2006)

[7] W.S. Holland et al.: in *Proceedings of SPIE Optomechanical Technologies for Astronomy*, vol. **4855**, p. 1, ed. by T.G. Phillips, J. Zmuidzinas (2003)

[8] E. Atad-Ettedgui et al.: "Opto-mechanical design of SCUBA-2", in *Proceedings of SPIE Optomechanical Technologies for Astronomy*, vol. 6273, ed. by E. Atad-Ettedgui, J. Antebi, D. Lemke, 62732H (July 6, 2006)

[9] A.L. Woodcraft et al.: in *Proceedings of SPIE Optomechanical Technologies for Astronomy*, vol. **5498**, p. 446 (2004)

[10] C.D. Reintsema et al.: Rev. Sci. Instrum. **74**, 4500 (2003)

[11] M.D. Audley et al.: *Proceedings of SPIE Optomechanical Technologies for Astronomy*, vol. **5498**, p. 63, ed. by J. Zmuidzinas, W.S. Holland, Stafford Withington (2004)

[12] A.L. Woodcraft et al.: Rev. Sci. Instrum. **78**, 024502 (2007)

[13] R.S. Le Poole, H.W. van Someron Greve: *Proceedings of SPIE Optomechanical Technologies for Astronomy*, vol. **3357**, p. 319, ed. by Tom Phillips (1998)

[14] A.G. Gibb et al.: in *ASP Conf. Ser.*, vol. **347**, p. 585, ADASS XIV, ed. by P.L. Shopbell, M.C. Britton, R. Ebert, San Francisco (2005)

[15] T. Jenness et al.: Mon. Not. R. Astron. Soc. **336**, 14 (2002)

[16] G. Angloher et. al. : Astropart. Phys. **23**, 325 (2005)

[17] C. Cozzini et al.: Phys. Rev. C **70** ,064606 (2004)

[18] H. Wulandari et al.: Neutron Background Studies for the CRESST Dark Matter Experiment, hep-ex/0401032

[19] G. Angloher et al.: Astropart. Phys. **18**, 43 (2002)

[20] F. Pröbst et al.: *Proceedings of TAUP2001*, LNGS, Italy, 8-12 Sept. 2001, Nuclear Physics B (Proc. Suppl.) **110,** 67 (2002)

[21] M. Altmann et al.: *Paper Contributed to the X International Symposium on Lepton and Photon Interactions at High Energies*, 23rd–28th July 2001, Rome, Italy, astro-ph/0106314

[22] *Update of the Proposal to LNGS for a Second Phase of the CRESST Dark Matter Search*, MPI-PhE/2001-02

[23] M. Sisti et al.: NIM A **444**, 312 (2000)

[24] M. Bravin et al.: Astroparticle Physics **12**, 107 (1999)

[25] S. Cooper et al.: *Proceedings of the 2nd International Workshop on the Identification of Dark Matter* (IDM-98), p. 359, Buxton, England, 7–11 Sept. 1998, ed. by N. Spooner and V. Kudryavtsev, World Scientific (1999)

[26] M. Sisti et al.: *Proceedings of 7th Int. Workshop on Low Temperature Detectors*, Munich, 27 July – 2 August 1997, pub. by MPI Physik, ISBN 3-00-002266-X

[27] A. Gabutti et al.: Astropart. Phys. **6**, 1 (1996)

[28] M. Bühler et al.: *Proceedings of the Sixth Int. Workshop on Low Temperature Detectors (LTD-6)*, Beatenberg/Interlaken, Switzerland, 28 August – 1 September 1995, NIM A**370**, 237 (1996)

[29] P.C.F. Di Stefano et al.: Phys. Lett. A **356**, 262 (2006)

[30] G. Angloher et al.:NIM A **520**, 108 (2004)

[31] F. Petricca et al.: NIM A **559**, 375 (2006)

[32] Ch. Kraus et al. : Eur. Phys. J. C **40**, 447 (2005)

[33] V. Lobashev et al.: Nucl. Phys. B Proc. Suppl. **91**, 280 (2001)

[34] M. Sisti et al.: Nucl Instr. Meth. A **520**, 125 (2004)

[35] E. Cosulich et al.: Nucl. Phys. A **592,** 59 (1995)

[36] G. Hilton et al. : IEEE Trans. Appl. Supercond. **11,** 739 (2001)

[37] C. Enss et al. : J. Low Temp. Phys. **121**, 137 (2000)

[38] P.K. Day et al.: Nature **425**, 817 (2003)

[39] C. Arnaboldi et al.: Phys. Rev. Lett. **96**, 042503 (2006)

[40] Broniatowski et al.: NIM A **444**, 327 (2000)

[41] R. Lemrani: Phys. Atom. Nucl.**69**, 1967 (2006)

[42] B. Censier: Nucl. Instrum. Meth. A **564**(1), 614 (2006)

[43] A. Broniatowski et al.: Nucl. Instrum. Meth. A **559**(2), 378 (2006)

[44] B. Censier: Nucl. Instrum. Meth. A **559**, 381 (2006)

[45] X.F. Navick et al.: NIM A **559**, 483 (2006)

[46] F. Schwamm et al.: NIM A **559**, 669 (2006)

[47] V. Sanglard et al. : Phys. Rev. D **71,** 122002 (2005)

[48] K. Eitel: Prog. Part. Nucl. Phys. **57**, 366 (2006)
[49] S. Marnieros et al.: Nucl. Inst. Meth. A**520**, 185 (2004)
[50] C.W. Misner, K.S. Thorne, J.A. Wheeler: *Gravitation*, W.H. Freeman and Co. (1973)
[51] D.G. Blair: *The Detection of Gravitational Waves*, Cambridge University Press, Cambridge (1991)
[52] J.H. Taylor: Rev. Mod. Phys. **66**, 711 (1994)
[53] R.A. Hulse, J.H. Taylor: Astrophys. J. (Lett.) **201**, L 55 (1975)
[54] R.A. Hulse: The discovery of the binary pulsar, *Les Prix Nobel 1993, 58–79*. The Nobel Foundation (1994)
[55] P. Astone et al.: Classical and quantum gravity **19**, 5449 (2002)
[56] J. Weber: Phys. Rev. **117**, 306 (1960)
[57] J.Weber: Phys. Rev. Lett. **22**, 1320 (1969)
[58] F. Ricci: Contemp. Phys. **39**, 107 (1998)
[59] http://tamago.mtk.nao.ac.jp
[60] H.J. Paik: J. Appl. Phys. **47**, 1168 (1976)
[61] R. Forward: Gen. Rel. Grav. **2**, 149 (1971)
[62] G. Frossati, E. Coccia: Suppl. ICEC **34**, 9 (1994)
[63] N. Ashby, J. Dreitiein: Phys. Rev. D **12**, 336 (1975)
[64] C.Z. Zhou, P.F. Michelson: Phys. Rev. D **51**, 2517 (1995)
[65] T.R. Stevenson: Phys. Rev. D **56**, 564 (1997)
[66] S.M. Merkowitz, W.W Johnson: Phys. Rev. D **51**, 2546 (1995)
[67] A. de Ward et al.: Physica B **280**, 535 (2000)
[68] A. de Waard: *MiniGRAIL The First Spherical Gravitational Wave Antenna* Ph.D. Thesis Leiden University, The Netherlands (2003)
[69] A. de Waard, L. Gottardi, G. Fossati: Classical and quantum gravity **19**, 1935 (2002)
[70] A. de Waard et al.: Classical and quantum gravity **20,** S143 (2003)
[71] CLAL 45 rue de Paris 93136 Noisy-le-sec Cedex France
[72] P. Astone et al.: Europhys. Lett. **16**, 231 (1991)
[73] M. Bonaldi et al.: Phys. Rev. D **74,** 022003 (2006)
[74] F. Marin et al.: Phys. Lett. A **309,** 15 (2003)
[75] M. Bonaldi et al.: Phys. Rev. D **68**, 102004 (2003)
[76] T. Briant et al.: Phys. Rev. D **67,** 102005 (2003)
[77] S. Pascoli, S.T. Petcov: Phys. Lett. B **544**, 239 (2002) and Addendum, arXiv:hep-ph/0310003
[78] M. Maltoni, T. Schwetz, M.A. Tortola, J.W.F. Valle: arXiv:hep-ph/0309130
[79] F. Feruglio, A. Strumia, F. Vissani: Nucl. Phys. B **637**, 345 (2002) and Addendum, Nucl. Phys. B **659,** 359 (2003)
[80] S.M. Bilenky, S. Pascoli, S.T. Petcov: Phys. Rev. D **64,** 053010 (2001)
[81] H.V. Klapdor-Kleingrothaus, H. Päs, A.Yu. Smirnov: Phys. Rev. **D 63**, 073005 (2001)
[82] A. Osipowicz et al.: arXiv:hep-ex/0109033
[83] Cuore Proposal: arXiv:hep-ex/0501010, v1 (2005)
[84] A. Morales: Nucl. Phys. B Proc. Suppl.**77**, 335 (1999)
[85] H. Ejiri: Nucl. Phys. B Proc. Suppl. **91**, 255 (2001)
[86] E. Fiorini: Nucl. Phys. B Proc. Suppl. **91**, 262 (2001)
[87] V.I. Tretyak e Yu. Zdesenko: Atomic Data and Nuclear Data Tables, vol. **61**, 43 (1995)
[88] C. Arnaboldi et al.: Phys. Rev. Lett. 95, 142501 (2005)
[89] S. Pirro et al.: NIM A **444**,71 (2000)
[90] C. Arnaboldi et al.: Phys. Lett. B **584**, 260 (2004)
[91] A. Alessandrello et al.: Nucl. Instrum. Methods Phys. Res. Sect. A **142**, 163 (1998)
[92] C. Arnaboldi et al.: IEEE Trans. Nucl. Sci. **50**, 979 (2003)
[93] C. Arnaboldi et al.: IEEE Trans. Nucl. Sci. **52**, 1630 (2005)
[94] C. Arnaboldi et al.: Nucl. Instrum. Methods Phys. Res. Sect. A **520**, 578 (2004)
[95] G.R. Dyck et al.: Phys. Lett. B, vol. **245**, 343 (1990)
[96] R.B. Firestone: *Table of Isotopes*, ed. by V.S. Shirley, John Wiley and Sons, New York (1996)
[97] M. Barucci et al.: J. Low Temp. Phys. **123**, 303 (2001)
[98] S. Pirro et al. : NIM A **444**, 331 (2000)
[99] C. Arnaboldi et al. : IEEE Trans. Nucl. Sci. **52,** 1630 (2005)

[100] L. Katz, M. Guinan, R. Borg: Phys. Rev. B **4**, 330 (1971)
[101] U. Männig et al.: Appl. Surf. Sci. **149**, 217 (1999)
[102] W. Schilling, K. Sonnenberg: J. Phys. F **3**, 322 (1973)
[103] A. Alessandrello et al.: NIM A **409**, 451 (1998)
[104] L. Lanzi, K. Sassoli, G. Ventura: Cryogenics **34**, 959 (1994)
[105] S. Pirro et al.: NIM A **559**, 352 (2006)

Index